MANAGEMENT OF HEALTH RISKS FROM ENVIRONMENT AND FOOD

ALLIANCE FOR GLOBAL SUSTAINABILITY BOOKSERIES
SCIENCE AND TECHNOLOGY: TOOLS FOR SUSTAINABLE DEVELOPMENT

VOLUME 16

Aims and Scope of the Series

The aim of this series is to provide timely accounts by authoritative scholars of the results of cutting edge research into emerging barriers to sustainable development, and methodologies and tools to help governments, industry, and civil society overcome them. The work presented in the series will draw mainly on results of the research being carried out in the Alliance for Global Sustainability (AGS).
The level of presentation is for graduate students in natural, social and engineering sciences as well as policy and decision-makers around the world in government, industry and civil society.

For other titles published in this series, go to
www.springer.com/series/5589

Management of Health Risks from Environment and Food

Policy and Politics of Health Risk Management in Five Countries — Asbestos and BSE

Edited by

Hajime Sato
The University of Tokyo, Japan

Springer

Editor
Hajime Sato
The University of Tokyo
Graduate School of Medicine
Department of Public Health
Tokyo, Japan
hsato@m.u-tokyo.ac.jp; hsato-tky@umin.net

ISBN 978-94-007-3079-3 e-ISBN 978-90-481-3028-3
DOI 10.1007/978-90-481-3028-3
Springer Dordrecht Heidelberg London New York

Softcover reprint of the hardcover 1st edition 2010

Printed on acid-free paper

Springer is part of Springer Science+Business Media (www.springer.com)

THE AGS
The Alliance for Global Sustainability

Aphorisms

Ad primum ergo dicendum quod omnia corporalia obediunt pecuniae, quantum ad multitudinem stultorum, qui sola corporalia bona cognoscunt, quae pecunia acquiri possunt. Iudicium autem de bonis humanis non debet sumi a stultis, sed a sapientibus, sicut et iudicium de saporibus ab his qui habent gustum bene dispositum.

Sancti Thomae de Aquino. (1265–1272). *Corpus Thomisicum. Summa Theologiae, prima pars secundae partis a quaestione I ad quaestionem V.*

Die menschliche Vernunft hat das besondere Schicksal in einer Gattung ihrer Erkenntnisse: daß sie durch Fragen belästigt wird, die sie nicht abweisen kann; denn sie sind ihr durch die Natur der Vernunft selbst aufgegeben, die sie aber auch nicht beantworten kann; denn sie übersteigen alles Vermögen der menschlichen Vernunft.

Immanuel Kant. (1787). *Kritik der reinen Vernunft, Vorrede.*

The central concern of administrative theory is with the boundary between the rational and the nonrational aspects of human social behavior. Administrative theory is peculiarly the theory of intended and bounded rationality – of the behavior of human beings who *satisfice* because they have not the wits to *maximize*.

Harbart A. Simon. (1945). *Administrative Behavior, Commentary on Chapter V.*

Economists tell you what you get for what you give up. Political scientists tell you who gets what and why. Far from being contradictory or incompatible, politics and markets are twin forms of competitive redundancy that complement one another by learning from social interaction.

Aaron Wildavsky. (1979). *Speaking Truth to Power: The Art and Craft of Policy Analysis.*

As all politics, however consequential, is local, so however ambitious, is all understanding. No one knows everything, because there is no everything to know.

Clifford Geertz. (1991). *Local knowledge and its limits: Some obiter dicta.*

Preface

The Alliance for Global Sustainability

The Alliance for Global Sustainability (AGS) is a unique, international partnership between four of the world's leading science and technology universities: Swiss Federal Institute of Technology, Zurich, Massachusetts Institute of Technology, The University of Tokyo, and Chalmers University of Technology.

Formally created in 1997, the AGS today brings together hundreds of university scientists, engineers, and social scientists to address the complex issues that lie at the intersection of environmental, economic, and social goals. Together, we seek to meet these challenges through: Improving scientific understanding of global environmental challenges; Developing technology and policy tools to help societies reconcile ecological and economic concerns; and Educating of a new generation of leaders committed to meeting the challenges of sustainable development.

Since its inception, the AGS has pioneered a new research paradigm that brings together multi-disciplinary research teams from the partner institutions. Strong, local programs engage faculty, students and senior research staff from across their respective institutes. These research teams have developed a significant body of new knowledge on critical issues in sustainability in the areas of energy and climate, mobility, urban systems, water and agriculture, cleaner technologies, and policy and communications.

Since the first set of AGS-sponsored research projects was launched in 1997 with support by the Avina Foundation, the AGS has worked with farsighted leaders from global businesses and industries, governments, and NGOs worldwide to provide innovative and practical solutions to real and urgent environmental problems around the world.

(Adapted from AGS website: http://www.theags.org/about/)

Acknowledgements

This book is based on the research project entitled "Strategic management of health risks" (2005–2009; Principal investigator [PI]: Hajime Sato, project researches: Andrew Webster, Bernard Reber, Pierre-Benoit Joly, Rose Campbell, and Domyung Paek), funded by the Alliance for Global Sustainability (AGS), the University of Tokyo. The research project entitled "Management and Communications of Health Risks from an International Comparative Perspective" (2006–2008; PI: Hajime Sato) was supported by a grant from the Japan Society for the Promotion of Science. The project entitled "A comparative study on health risk management" (2007: PI: Hajime Sato and Andrew Webster), was supported by a Daiwa Small Grant from the Daiwa Anglo-Japanese Foundation. Finally, a research on comparative journalism was funded from Japan Study and Bultler University Holcomb Awards Committee (2006–2007; PI: Rose Campbell).

AGS board members at the University of Tokyo (UT) provided continuous supports for the project. Special thanks should go to Hiroshi Komiyama (President of the UT, April 2005 – March 2009, the AGS Governing Board), Akimasa Sumi (AGS Office, Integrated Research System for Sustainability Science, and Center for Climate System Research, UT), and Yuji Togami (AGS Office, UT). In various phases of the research project, many people offered generous supports, advice and insights. They include Drs./Profs. Philip J. Landrigan (Mount Sinai School of Medicine, New York), Julian Peto (London School of Hygiene and Tropical Medicine), Catherine Labrusse-Riou (Paris University I), Jean-Paul Gaudilliere (INSERM, Villejuif- Île-de-France), Takashi Onodera (Graduate School of Agriculture and Life Sciences, UT), John D. Montgomery (John F. Kennedy School of Government, Harvard University), and several members of the Society for Policy Sciences, to name a few.

Partial results from these projects were presented by Domyung Paek and Hajime Sato on July 9–10, 2007, in Seoul, Korea, at the US-Korea Workshop, entitled "Understanding Bioenvironmental Complexity" sponsored by the School of Public Health at Seoul National University, Rutgers – The State University of New Jersey, the US National Science Foundation, and the Korea Science and Engineering Foundation; and by Rose Campbell and Hajime Sato on November 15–18, 2007, in Chicago, USA, at the 93rd Annual Convention, National Communication Association.

Contents

Part I
Introduction

Chapter 1
The Policies and Politics of Health Risk Management

Hajime Sato

Health Risks and Their Management

Risk is a part of life; it is the potential for harm and the probability of encountering negatively-valued events. The language of risk has become an intrinsic part of the political rhetoric in many industrialized countries. Safety, on the other hand, is the degree to which, in a group of people, one or more of the following three conditions are controlled, avoided, prevented, or made less frequent or probable: (1) temporary ill health or injury, (2) chronic or permanent ill health or injury, and (3) death (Adams 1995; Siddall 1981). As people become wealthier, it becomes easier for them to avoid what previously would have been seen as involuntary hazards. Wealthier societies are thus characterized by a reduction in the fear of numerous health hazards, as our ability to control them correspondingly improves. However, this desire for safety is unsatiable. Whereas scientists and an attentive public are searching for real and potential health issues, politicians might be happy with those emerging issues, as they give them opportunities to take actions that provide them with political rewards.

Risk management has become a dominant concern in public policy. Since the 1970s, environmental hazards have continued to be conspicuous as a social issue (Chaffe 1985; Mintzberg 1988; Pauchant & Mitroff 1992; Porter 1980). Starting in the 1990s, perceived risks associated with genetically modified foods, Bovine Spongiform Encephalopathy (BSE) and variant Creutzfeldt Jakob Disease (vCJD), and emerging pathogens such as *E coli* O157, as well as increasingly complex information about appropriate nutrition, continue to be foci of public fear and cynicism about how food risks are managed (Frewer et al. 1996, 1997, 2001). Study of these issues continues to shed light on the management of technological advances and

H. Sato
Department of Public Health, Graduate School of Medicine, The University of Tokyo, Tokyo, Japan
e-mail: hsato-tky@umin.net; hsato@post.harvard.edu

H. Sato (ed.), *Management of Health Risks from Environment and Food*,
Alliance for Global Sustainability Bookseries 16,
DOI 10.1007/978-90-481-3028-3_1,

industrial development, and they call for the efficient application of scientific knowledge, efforts to address the changing health concerns of the public, and more effective policies. Also, failure in risk management can erode a government's legitimacy. As is expected of a government, the first function of management is "prevoyance", a word that means on the one hand to forecast and foresee, and on the other hand to secure and make reliable (Foyol 1916).

The quest for safety is a balancing act: how can we use risk to get more of the good and less of the bad (Wildavsky 1988)? The controversy over risks and their management inevitably involves a confusing mixture of science and politics, including debates about which substances and technologies present risks, which margins of safety are achievable and prudent, and what costs are necessary and affordable for prevention (Graham et al. 1988). The controversy about known and potential risks arises in part from differences in people's values and interests. In some cases, government actions are based on the law of sacrifice: the safety or macrostability of the whole is dependent on the riskiness (risk taking) or instability of the parts. For example, the mere possibility of an increase in the risk of cancer, no matter how small or speculative, is sufficient to justify severe regulatory restrictions on the use of a suspect substance.

Research Project: Strategic Management and Communication of Health Risks

This research project was intended to examine a set of health-related risk management cases from a comparative perspective. Concrete case histories were constructed with the help of an analytic framework drawn from political science, regulatory science, and sociology.

The regulation of asbestos use and the policy toward BSE (better known as the mad cow disease) were selected for examination. The hazard of asbestos use was scientifically debated as early as the 1960s, but its usage has not been denied or regulated promptly and universally in all countries. It is still an important social issue in many countries, both industrialized and economically developing ones. BSE policy is related to the issues of health, agriculture, and trade. Although scientific information is shared among countries, information about the perception and management of risk is not. Policies are not always concerted, and remain to be internationally disputed, as is exemplified by the import ban on US beef.

To examine the process of policy making on these issues, a comprehensive and exhaustive search was conducted for relevant historical documents. The search included archives of public records, both printed and on-line, as well as newspaper articles and reports of opinion polls. Databases of medical articles were also searched for publications on asbestos and BSE, and other types of documents and books, such as theses, essays and recollections, were assembled. Finally, supplementary interviews were conducted with (ex-)bureaucrats, politicians, researchers, and non-governmental organizations (NGOs).

Concepts and Models as Perceptive Lenses

Studying politics, as Edelman argues, is not simply an effort to learn what is happening. It is also a process of making observations that conform to sets of assumptions called models (Edelman 1971). Models state the relationships that have been observed among the conditions and patterns of political life, and they accumulate credibility insofar as evidence is gathered to support them (Almond & Powell 1966, p. 15). These models differ in terms of the perceived decision makers (actors), as well as their perceived efficacy; these serve as the independent variables of the model. In this study, such concepts and models have been employed to examine the socio-political process of risk management concerning asbestos and BSE.

Policy Process Analysis

In the framework of Policy Process Analysis, policy is considered to be a combination of several processes, which, although interrelated, still can be conceived as distinct elements determining government action (Snare 1995). Although the details of these conceptions vary, they all have several attributes in common. The first stage of the traditional policy process, problem definition, involves the recognition of an emerging problem or crisis. Second, a policy to address the problem is formulated by various governmental and non-governmental actors, such as legislators, executive branch officials, the courts, citizens, and special-interest groups. Specific policy proposals are then adopted in the third stage. The fourth stage is policy implementation, wherein the adopted policies are executed by administrative units. Finally, in the policy evaluation stage, policy makers determine whether the policy has achieved its goals (Altman & Petkus 1994).

This study adopts the conceptualization of these processes proposed by Jones and Kingdon (Jones 1978; Kingdon 1973; Lindblom & Woodhouse 1993). Each process is divided into several stages: Problem Finding and Issue Definition (identification and description of issues and problems), Agenda Setting (deciding what issues are to be officially addressed), Development of Alternatives (formulation of policy proposals), Decision and Legitimization (selection of policy proposals and legitimizing their adoption), and Implementation and Appraisal (initial implementation and continued application of the decision, and evaluation of the government's actions). Each of these component processes has been extensively studied (Anderson 1978; Axelrod 1973; Bauer et al. 1963; Cobb & Elder 1972; Cobb et al. 1976; Connolly 1984; Crenson 1971; Elder & Cobb 1983; Farrar et al. 1980; Freedman 1978; Hall 1973; Jones 1975, 1978; Lasswell 1954; Levine 1982; Light 1982; Lindblom 1959; Lipset 1981; Rochefort & Cobb 1994; Rogers & Bullock 1972; Thompson 1981, 1984; Van Meter & Van Horn 1975; Walker 1981; Weiss 1970; Weiss 1989).

As noted above, the Policy Process Analysis breaks down policy making into functionally distinct processes and provides a useful conceptual analysis of this complexity. The efforts of various coalitions to translate their beliefs into government action, as well as the strategies of policy entrepreneurs, can be explored in the light of these processes (Jenkins-Smith 1991). The processes do not necessarily have causal relationships, but advocacy coalitions have to work on at least most of them to effectively induce policy change. If one deals with each process as a "function" rather than just a step in a sequence, the examination of policy processes becomes a potent tool for analyzing how institutional environments change, how coalitions develop, and how policies change (Lasswell et al. 1977; Lasswell & McDougal 1992). The activities of social groups and coalitions can influence every stage of the policy process. Policies can be adopted and then implemented through the direct participation of, and consultation with, the affected groups, such as the industry. Even if the policy is not changed there may be reasons why certain coalitions remain less powerful, because they lack the kinds of resources that help keep certain other coalitions dominant.

The Garbage Can Model of Policy Making

The Garbage Can Model of policy making was proposed by Cohen et al. (1972), and further developed by Kingdon (1995) as an alternative to linear, comprehensive, and rational models. This model assumes that policy windows open only when the streams of problems, policies, and politics converge, which usually flow independently of one another. In such a convergence, problems are brought to the attention of people in and around the government by applying systemic indicators, by focusing on events such as crises and disasters, and by providing feedback on the operation of current programs. Alternate policy proposals are developed from the many ideas floated by people both in and out of government. The proposals that survive and achieve serious consideration must meet several criteria, including technical feasibility, conformance to dominant values and the current national mood, budgetary feasibility, and political support. The latter can be affected by a variety of influences, including swings in the national mood, administrative or legislative turnover, campaigns by interest groups, and social movements.

In this context, policy windows provide opportunities to push one's preferred policy proposals or conceptions of a problem. They can open either as a result of events in the political stream or the emergence of important problems. In the Garbage Can Model, solutions and problems have equal status as separate streams in the system, and the popularity of a given solution at a given point in time often influences which problems are focused on. In other words, issues and feelings are looking for decision situations in which they might be aired, solutions (collections of choices) are waiting for the issues to which they might be answers, and decision-makers are looking for work. Agendas are determined by the relevant problems or politics, and alternatives are generated within the policy stream. Thus, outcomes are heavily dependent on the coupling of the streams (Hall 1993; Mucciaroni 1992). Sometimes the "policy entrepreneurs" – people who invest their resources in pushing their pet proposals or problems – are responsible not only for prompting important people to pay attention to an issue, but also for coupling solutions to problems and putting both the problems and the solutions in a functional political context (Moon 2005; Wise 1991).

When there are competing social interests in a multi-step policy process, the steps do not always proceed sequentially; rather, they proceed haphazardly and independently. Alternatives are developed well before the official agenda is set, and programs are sometimes developed prior to possible official involvement. Policy change is either implemented or blocked, whether by sequential steps, a top-down process, or in a fragmented, bottom-up fashion. The coalitions can thus exercise their power in all stages of the process, and there thus are many points at which they can influence the government's action (Sabatier 2007). The activities of coalitions might significantly affect the degree of official involvement in the issue, while assuring that the programs are responsive to relevant social interests.

Science and Technology Studies, and Public Policy

Scientific advances are expected to improve policy-making and effective, efficient problem solving. However, the process is not always straightforward. From the perspective of the Garbage Can Model, problems, policies, and politics typically evolve in separate, unconnected streams. When science is considered as a fourth factor, it is seen as influencing all the streams, changing the perception of the problem, helping develop policy, and transforming the political environment. Thus, science can be seen as a potent force approving or disapproving certain (existing) policies and opening policy windows.

However, science does not automatically create these effects by itself, nor does it operate in a political vacuum. Applied research, including public health and epidemiological research, is more likely than basic research to follow an agenda driven by forces other than science. This means that the immediacy of an issue, or its priority on an agenda, can affect the creation and use of scientific knowledge (Kazancigil 1998; Sato 2002; Sato & Frantz 2005). Some initiating event not directly connected with science defines a policy issue that must be decided on. Then a debate ensues on the possible policy options and possible scientific views of the issue, followed by a scientific assessment of the issue leading to the development of a rationale for the policy chosen. Thus, science may legitimize policies developed for nonscientific reasons. If there is a lack of scientific consensus or a consensus does exist but contradicts current policy, its conclusions might be ignored. It should also be noted that scientific disputes are not always resolved during the period that the scientific issues are considered relevant to the policy discussion. Thus, the context in which scientific research is conducted, and the process in which scientific knowledge is incorporated in policy are always important determinants of how risk is managed, both publicly and privately.

Researchers in the sociology of science, as well as in Science and Technology Studies (STS), have examined the nature and context of knowledge production, and they have conceptualized several modes of knowledge production and the changing nature of research. There are two ideal ways in which knowledge can be acquired scientifically: traditional disciplinary research (Mode 1) and transdisciplinary, problem-oriented, context-sensitive activities (Mode 2) (Nowotny et al. 2001). In Mode 1, the goal is the pursuit of the truth about natural phenomena (Newton 1997). In Mode 2, knowledge is sometimes produced through a process of negotiation among different agents with different interests, which might reflect or result in collaborative learning by different actors (Roux et al. 2006). It has also been argued that knowledge is produced and transmitted in the context of application (Gibbons 2000; Graham and Dickinson 2007).

Interactions between science and government have become more prominent as government capacities grew, just as was the case for science and politics (Mannheim 1936; Wittrock 1989). Expertise in production, diffusion, and utilization is sometimes institutionalized. Knowledge is no longer exogenous, but it can still be regarded as a problem-solving tool and a serviceable "truth" that offers a certain degree of scientific acceptability while assuring stakeholders that their interests are not being sacrificed to unacceptable policy goals (Jasanoff 1990). Recently, the triple-helix model has been developed to describe the inter-relationships among universities, industry, and state and local government. These inter-institutional networks have a recursive effect on their sources, as well as on the larger society (Etzkowitz 2002; Fuller 2001). They play an increasingly important role in structuring research and reconfiguring knowledge.

When scientific knowledge is used for policy development, scientists and/or policymakers may choose to err either on the side of public safety or on the side of patient liberty and dignity. Science and scientific arguments are sometimes employed instrumentally (Elder & Cobb 1983; Majone 1992). During policy debates, especially the adversarial process of rule-making, knowledge claims are sometimes deconstructed, exposing areas of weakness or uncertainty. These revealed weaknesses provide justification for political decision-makers to assert their right to re-interpret scientific findings, especially in controversial areas. Again, uncertainty within science itself is also a subject for negotiation, decision-making, and argumentation in policy development, as the quality and extent of knowledge are subject to social negotiation.

This partial transfer of decision-making authority to the legal and political arena may be seen as a way of assuring that the interpretation of (indeterminate) facts reflects the public values embodied in legislation as much as the norms of the scientific community. The connections between scientific data, expert interpretation of these data, and policy are like linked chains of arguments and beliefs. This process might be termed co-evolution or the mutual validation of policy and science. When science and policy are co-constructed through processes that occur in tandem, it becomes difficult to explain one by using the other (Frantz & Sato 2005; Sato & Frantz 2005).

Government and the Public in Risk Management

Public attitudes and societal reactions to risks, as well as government involvement in their management, sometimes differ, depending on the issues. Studies of risk perception and management provide additional insights into their characteristics.

The public and technical experts often come to different conclusions about what is risky and what is not, as well as how identified risks should be managed (Dietz & Rycroft 1987; Slovic 1987). These differences of opinion result in a conflict between the technological values of rationality, efficiency, and expertise on the one hand, and the democratic values of personal experience and individuals' right to voice their views about governance on the other (Fiorino 1989; Kasperson 1983). Technical experts are likely to take the objectivist view that there is only one true risk, which is defined by the probability and severity of hazard. This perspective leads to the conclusion that hazard management is basically a matter of systematically applying the results of analytic risk assessments. Public reactions to hazards are seen as a reflection of scientific illiteracy, and better scientific education is advocated. In contrast, advocates of the constructivist view regard hazards as a threat to people and what they value; thus, the hazards are seen as intrinsically a product of the interaction between physical and psychological characteristics of (risky) events (Kates & Kasperson 1983). Public concern about risks and hazards therefore might include considerations of their broad and high-order effects, which are usually not taken into account in technical risk assessments (McClelland et al. 1990).

In risk management, outrage and a lack of social trust, whether on the side of regulatory and hazard-mitigation agencies or the public or both, can seriously compromise the success of risk management, especially when either party lacks sufficient knowledge about the risks and hazards (Siegrist & Cvetkovich 2000). Public trust in government agencies and policies depends on the public's perception of the knowledge and expertise of the regulators, their openness and honesty, and their concern and caring about the problem and the public (Peters et al. 1997). The erosion of trust in established institutions, either public or private, could result in a loss of legitimacy and paralyze their functioning.

In their examination of risk-related social processes over time, Renn et al. (1992) conceptualize "social amplification." Events pertaining to hazards interact with psychological, social, institutional, and cultural processes, and they can increase or decrease the public's perception of risk and shape their behavior, which, in turn, can have secondary socio-economic and political consequences. When the initial influence of a risk dissipates and these secondary consequences grow, the risk is said to be socially amplified. The opposite process is called "social attenuation" of risk (Pidgeon 1999). These social processes are affected by the nature of the hazard, the actions of the regulatory agencies, and the configuration of various societal actors such as stakeholders and interested members of the public.

Thus, both on the individual and collective level, the relationship between the government and the public regarding risk management (by which I mean the characteristics of risks, the social structure vis-à-vis those risks, and the process by which risks are handled) can be a potent determinant of how issues emerge as social problems, what policies are introduced to solve them, and how these policies work (Etzioni 1976; Hinchliffe 2001).

Case Studies

With these analytical tools in mind, we have carefully examined and will discuss below relevant characteristics of policy-making and associated political processes in different countries during various periods of time. These characteristics include the social status of issues, the setting of the policy agenda and its contents, the development of alternatives (prevention, screening and monitoring, treatment, and compensation) and how these are discussed or rationalized (e.g., scientifically, economically, ideologically, with reference to liability or negligence), political mobilization around the issue, final policy decisions (action or inaction with respect to laws, acts, guidance, guidelines, court orders, etc.), and the efforts to implement policies that are approved. In addition, the process of scientific knowledge production and the utilization of this knowledge in policy making will be examined, when relevant.

Management of Environmental Risk: Cases of Asbestos Regulation

Asbestos is the fibrous form of the mineral silicate. The natural resistance of asbestos to heat and acid, its tensile strength, and its remarkable thermal, electrical and sound insulating properties have led to numerous applications (Tweedale 2000, pp. 3–4). However, exposure to asbestos can lead to asbestosis (pneumoconiosis), lung cancer, and mesothelioma. During the 1950s and 1960s, clinicians and researchers reviewed the medical records of asbestos workers and found positive correlations between exposure to extraordinarily high concentrations of asbestos in the air and specific diseases such as asbestosis, bronchogenic carcinoma, and mesothelioma. Gradually, many countries have introduced regulations for the safe use of asbestos products, eventually banning their production, sale, transfer, and use (Table 1.1).

Asbestos in Japan

In Japan, the industrial use of asbestos started in the 1880s, 20-30 years after European countries did so. During the First and Second World Wars, asbestos became widely used for warships, water pipes, and boilers. Starting in 1958,

Table 1.1 Asbestos types and their characteristics

	Chrysotile	Crocidolite	Amosite
Main sources	Canada, South Africa, Russia	South Africa	South Africa
Fiber color	White	Blue	Gray-yellow
Characteristics	Silky, soft, excellent flexibility and tough	Harsh, reasonably flexible and tough	Coarse, usually brittle
Attributes	Very good spinning potential, excellent resistance to heat and all liquids except strong acids	Resistant to heat and to very strong acids	Heat resistant
Main uses	Asbestos textiles, asbestos cement	Asbestos textiles, spray	Preformed slab-type insulation, spray

Source: Tweedale 2000, p. 3.

asbestos-containing cements were sprayed on ceilings, walls, and other parts of buildings; this usage rapidly spread throughout the country. As a result, asbestos consumption skyrocketed after 1960, reaching a plateau from the late 1970s to the early 1980s.

The first case of asbestosis was reported in 1929, and in the late 1930s outbreaks were reported in a few factories where asbestos was handled. In several areas, especially the Kansai area, health checkups were conducted by health professionals in the workplace. These examinations again disclosed cases of asbestosis. Furthermore, cases of lung cancer, peritoneal mesothelioma, and pleural mesothelioma associated with asbestos exposure were confirmed in 1960, 1973, and 1974, respectively. In the early 1970s, when the International Labor Organization (ILO)/International Agency for Research on Cancer (IARC) published its report on the carcinoginity of asbestos, small-scale screenings for lung cancer were initiated at workplaces where asbestos was manufactured; screening for mesothelioma followed in the 1980s.

In response to the reports of these health hazards, asbestosis became considered as an occupational health issue, and in 1954, for the first time, official compensation for asbestosis was provided for the victim (this was an administrative measure, based on the Labor Standard Law). However, in contrast to the case of silicosis, official policies specifically targeting asbestosis were not introduced until 1960 with the passage of the Pneumoconiosis Law. To improve the workplace environment, local air ventilation was also mandated. In 1972, laws were passed limiting workplace concentration of asbestos fibers in the air. Finally, in 1976, following the ILO report, asbestos spraying was prohibited. At about this time, compensation for lung cancer and mesothelioma, as asbestos-related occupational hazards, started to be provided on a case-by-case basis.

In the late 1970s, possible exposure to asbestos in school buildings, and the resulting health hazard, became a public issue in the US. When news of this reached Japan, it triggered public concern about asbestos hazards in that country. Legislators and bureaucrats introduced programs such as asbestos monitoring in the outside air near factories, construction sites, and industrial waste-disposal facilities (1985-1987). These studies, as well as the World Health Organization (WHO)/Environmental Health Criteria (EHC) report of 1986, concluded that the health effects of asbestos in the general environment were negligible. Nonetheless, the government decided to provide elementary schools, junior high schools, and public hospitals nationwide with subsidiary payments to ablate asbestos-containing construction materials, especially if they were old and friable, from their buildings.

In 1989, the WHO recommended a ban on the use of crocidolite and amosite, and many European countries followed suit in the 1990s. In Japan, manufacturers voluntarily stopped using crocidolite (1989) and amosite (1995), but the government did not officially ban them until 1995. Official ban was then justified by citing the reports on elevated incidences of lung cancer and mesothelioma among workers who had handled asbestos. By 1997 asbestos had come to be regarded as the most frequent cause of work-related cancers. In 2003, a policy was introduced that in principle banned the use of asbestos for new buildings and automobile brake pads. In practice, however, asbestos continued to be heavily used because of the many exceptions allowed by the policy.

The asbestos issue appeared on the public's agenda in 2005 when a construction company disclosed the fact that its (ex-)workers and people living near its factories were suffering from mesothelioma at a disproportionately high rate. Subsequent government surveys revealed many other cases of workers contracting these diseases, especially if they were engaged in construction or shipbuilding. It was argued that it is quite difficult to ascertain the work histories (in particular, asbestos-handling) for many of these workers, because many of the manufacturers had gone out of business. Therefore, the general public could not ascertain the exact degree of exposure, which made it very difficult to determine appropriate compensation for the victims. In 2006, with the aid of the government and the asbestos-related companies still in business, a fund was established to accomplish this objective, while exempting the companies from liability. Soon thereafter, the government overcame industry resistance and introduced a complete ban on the use of asbestos.

Asbestos in the United Kingdom

In the UK, the use of asbestos by small-scale industry started in the late nineteenth century. Shortly thereafter, the possible health hazards resulting from exposure to asbestos dust and fibers became an official concern. In 1989, the Inspector of Factories warned of asbestos danger and by 1910 the danger had been confirmed. However, measures to prevent asbestos hazards were left to the morality of the employer. The first regulation of asbestos, introduced in 1931, was ironed out by the Home Office, the factory inspectorate, industry representatives, and the workers' unions. Dust control was at the core of the regulations, which were intended to prevent asbestosis. Many corporations established funds to compensate workers for the health hazards resulting from asbestos, and some introduced health checkups.

However, the implementation of the regulations was largely left to the industry itself, and probably for that reason it was very ineffective.

The industry established the Asbestos Research Committee in 1942 and the Asbestos Research Council (ARC) in 1952, inviting medical schools and universities, and started research on dust measurements, and on the effect of asbestos in animals. Even though elevated rates of lung and pleural cancers among asbestosis patients had been reported in 1947, and later confirmed by Doll in 1955, this was not a topic that either the Committees or the Council took up. The British Occupational Hygiene Society (BOHS) remained silent until 1968, when, following a request by the government, it published a report on environmental standards for chrysotile.

From 1930 to 1960, a period of industrialization and war, asbestos consumption increased remarkably. It was used widely in the making of electrical equipment and industrial products, shipbuilding, and other such enterprises. By the 1960s, many scientists accepted that there is a causal relationship between asbestos exposure and asbestosis, lung cancer, and mesothelioma. The risks of asbestos were widely reported by the mass media, and after the 1960s there was an increase in lawsuits filed by workers for compensation.

In 1969, the asbestos regulations were revised. Medical checkups and compensation were added to the existing core of measures, namely, the dust control. Environmental standards were established based on threshold limit values (TLVs) for asbestos exposure that had been proposed by the BOHS. However, only the standards for chrysotile were official, because the industry had already ceased to use crocidolite as a result of rigorous standards proposed by the BOHS for its use. In practice, however, the BOHS's TLVs for chrysotile, as specified in the regulations, were interpreted as a goal to be aimed at rather than a standard to be rigidly enforced. In response to the government's action, the ARC in 1971 chose the University of Edinburgh as its partner, and it started providing large amounts of grant money to the university's Institute of Occupational Medicine. It was later reported that the topics of the supported research, and the disclosure of the reports of this research, were decided by the ARC, with the industry looking over its shoulder. Epidemiological studies were never conducted.

Following the recommendations of the Robens Committee in 1972, the Health and Safety at Work Act (HSWA) was adopted in 1974. The Act established the Health and Safety Commission (HSC), which was supposed to propose new laws and standards while inviting opinions from the public at large, and the Health and Safety Executive (HSE), which was responsible for upholding the new laws and standards. In 1976, the WHO and the IARC jointly concluded that any amount of any kind of asbestos can cause lung cancer and mesothelioma (without safety thresholds). In the same year, the parliamentary and health-services ombudsman reported the dreadful conditions in one asbestos factory. Consequently, the Advisory Committee on Asbestos was established to study asbestos-related health hazards and the effectiveness of the 1969 regulations. It concluded that a doctrine of controlled use was feasible for chrysotile, and it recommended a ban on the use of crocidolite and insulation spraying, both of which had already been abandoned by the industry.

Since the 1970s, citizens groups have been forming to advocate for victims of asbestos, calling for a political commitment to the issue. The mass media have also attracted public attention, criticizing the insufficient government efforts to prevent the hazard as well as irresponsible corporate behavior. The regulations on asbestos were tightened: the use of crocidolite and amosite was banned by law in 1984, and the environmental standards for asbestos were made more rigid. After the late 1990s, the UK policies were brought into conformity with EU regulations and recommendations. The HSE served as the main vehicle for the effective implementation of these policies. The Control of Asbestos Regulations passed in 2006 prohibit the importation, supply, and use of all forms of asbestos, although existing materials in good condition can be kept if they are monitored.

Asbestos in France

In France, high mortality rates among asbestos workers were observed in 1906, but the French government did not become involved in efforts to prevent the hazard for a long time. The listing of asbestosis as a compensable occupational hazard in 1945 was the first step of this kind in France. Despite reports in the 1950s-1960s citing asbestos as a cause of lung cancer and mesothelioma, the government was almost completely silent. Meanwhile, the consumption of asbestos increased rapidly, reaching a plateau in 1975.

Then, in the mid-1970s, the ex-workers of a closed asbestos factory staged a political demonstration, demanding re-employment and compensation for their asbestos-related health problems, especially asbestosis. When the presence of asbestos fibers in the air was reported in the buildings of Paris Universities, their workers demanded that the government take prompt action to abolish the asbestos risk. The ex-workers of the factory and the university workers joined forces to demand compensation for their lost salaries, re-employment, and payment for their medical costs resulting from asbestos exposure. Other workers also joined the movement, and thus the issue of asbestos-related health hazards became a broad societal concern. Following scientific and political disputes between groups for and against regulation of asbestos, the first French regulations were adopted in 1977, but they were only loosely enforced.

The Permanent Asbestos Committee was established in 1982. It was composed of representatives from the asbestos industry, the medical community, trade unions, and government agencies. Until the mid-1990s, it took a leadership role in discussions about asbestos-related policy. For the members from trade unions, employee salaries and job security were the top priorities. To protect employments, they tried to dismiss any idea that could threat their industry, repeatedly insisting that there was insufficient data on the health effects of asbestos. The French government finally adopted a policy of controlled use, opposing the ban on asbestos proposed by the European Economic Community (EEC). However, the health risks of asbestos continued to be a public concern. Victims of asbestos-related health hazards started filing law suits, claiming compensation, and a number of citizens groups were formed.

In the mid-1990s, lawsuits filed by the victims of asbestos were widely reported in the media, corporate liability was called for, and it was demanded that the government take the necessary action to preserve public health. In 1996, the government required notification of asbestos use in buildings. Following consultations, a national institute publicized its estimates on future health hazards due to asbestos exposure. Then in 1997, a political decision was made to ban the use of crocidolite and amosite, overriding the long-lasting scientific controversy. The French Institute for Public Health Surveillance was established in 1998, as was an official registry for mesothelioma.

In 2000, the government acknowledged its responsibility for the inadequacy of its past actions, and it established a fund to compensate asbestos victims. In 2002, the French Court of Cassation also reversed its previous rulings concerning employer liability (excusable fault), which opened the way for victims to file compensation claims. The French Agency for Environmental and Occupational Health Safety continues to conduct research on asbestos risks, as requested by the government.

Asbestos in the United States

In the US, asbestos mining started in 1894. More than 60 asbestos mines, mainly on the East Coast, were active until 2002, when the last chrysotile mine was closed. As early as the 1900s there were reports of a high incidence of lung disease among workers who handles asbestos. Around 1918, insurance companies began to set higher premiums for those workers or even deny them coverage. The first clinical case report of asbestosis was published in 1930, and many other case reports and epidemiological studies appeared in the 1930-1940s. During this same period, many vermiculite mines were opened. This substance contains amphibole, which is widely used for house insulation, land improvement, and cement production. Furthermore, the mining of talc, which contains asbestos, dates back to 1878 and was widely used until the 1940s.

Many citizens groups emerged, which, with the support of law firms, filed suits against asbestos mining and manufacturing companies, claiming compensation for illnesses caused by asbestos exposure. Especially after the 1950s, when lung cancer and mesothelioma were found to be associated with asbestos, many class-action lawsuits were filed. Gradually, the discussions in the courts came to focus on possible employer negligence. The plaintiffs contended that the employers, although knowing the hazardous nature of asbestos, did not take appropriate measures to protect employees' health.

Government involvement in the prevention of asbestos started with the Clean Air Act of 1970, in which the Environmental Protection Agency (EPA) designated asbestos as a hazardous pollutant. In 1972, the Occupational Safety and Health Administration (OSHA) published steps that employers must take to protect workers. The Clean Air Act of 1973 led to the creation of the National Emission Standards for Hazardous Air Pollutants. Then in 1976, the Toxic Substance Control Act was passed, authorizing the EPA to regulate asbestos in schools and in public and commercial buildings. In the same year, the Consumer Product Safety Commission banned two consumer products containing inhalable asbestos. In 1978, most spray-applied surfacing asbestos-containing materials (ACMs) were banned.

Despite the lack of firm scientific evidence on its health hazards, of the use of asbestos in school buildings continued to be a public concern. In 1984, Congress passed the Asbestos School Hazard Abatement Act, which authorized the funding of loans and grants to schools to abate asbestos during renovations. Also adopted at this time was the Asbestos Hazard Emergency Response Act, which set guidelines for asbestos inspection and abatement procedures. Based on the argument that asbestos at any level is unsafe, the EPA banned new uses of asbestos through the Toxic Substance Control Act of 1989 (Asbestos Ban and Phase-out Rule), which was appealed and never put into effect. This decision left the management of asbestos safety largely to consumers.

In recent years, the Fairness in Asbestos Injury Resolution Act, which is intended to establish an Office of Asbestos Disease Compensation within the Department of Labor to process all work-related injury claims related to asbestos exposure, has been discussed by Congress. In 2007, the Ban Asbestos in America Act was proposed; this legislation requires the National Institute for Occupational Safety and Health (NIOSH) to conduct scientific studies on asbestos and at the same time empowers the EPA to regulate the import, manufacture, processing, and distribution of asbestos-containing products.

Asbestos in Korea

In Korea, asbestos mining started in the 1920s, and most of the minerals taken from the mines were then exported. Mining for domestic consumption started in 1960. The amount of ore extracted was at first limited, because the mines were small and the quality of the ore was low. Therefore, most of the country's asbestos was imported. Later, in the 1960s-1970s, foreign companies started manufacturing construction, friction, and fiber materials in Korea (the companies moved to Korea soon after occupational safety and health regulations were introduced in their homeland). Nationwide, many houses were built using asbestos-containing slates for their walls and roofs. Consequently, as the consumption of asbestos rapidly increased, so did both domestic and foreign production (for import). Domestic production reached its plateau in the early 1980s but ceased in 1985, when Korea started importing asbestos from overseas, especially Canada.

Pneumoconiosis among miners has been sporadically reported since the 1950s, but no notable action was taken by the government to prevent the health hazard due to asbestos. The first official action for this purpose was in 1984, 3 years after the Labor Safety and Health Law was enacted. The government mandated the monitoring of asbestos fibers in the air in workplaces. In practice, however, the monitoring was not well conducted, especially in small factories. Another shortcoming is that it was required only for the manufacturers of asbestos products, not for end users such as workers at construction sites and shipbuilders.

After 1990, the health hazards of asbestos were increasingly reported. The Labor Safety and Health Law was revised in 1991. The law required advance permission for the use of asbestos as an industrial material. Implemented in 1992, it provided for workers who had been engaged in asbestos handling for more than 3 years to undergo health checks, even after they left the company. In 1993, a case of mesothelioma in a female ex-worker was reported, and for the first time a worker received compensation for an asbestos-related illness. Around this time, when compensation for asbestos-related illnesses became official policy, environmental monitoring gradually increased, as did the implementation of preventive measures in the workplace.

After the mid-1990s, the asbestos companies in Korea moved to other countries, such as China and Indonesia, and the total consumption of asbestos, as well as the domestic manufacture of its products, gradually decreased. The use of crocidolite and amosite was prohibited in 1997, and a registry for mesothelioma was established in 2001. Concern about the health risks of asbestos spread among unions and citizens groups, and screening for asbestos-related disorders was conducted in many workplaces. Also, such disorders started to appear among people who lived near asbestos-handling factories. In 2003, compensation for asbestos-related illnesses was provided not only for employees of companies that manufactured asbestos products, but also to the users of these products and to persons exposed to asbestos while engaged in reconstruction or ablation.

Management of Food Risk: Cases of BSE-Related Human Risk Management

BSE is a cattle disease first identified in the UK in 1986. It was found to be a transmissible form of spongiform encephalopathy (TSE) caused by abnormal prions. The average incubation period of BSE is about 5 years and when clinically symptomatic, the disease presents abnormal neurological symptoms. In the 1990s, BSE was reported in many countries outside the UK. This cattle disease became a serious public concern in 1996, when the UK government and a committee of experts it formed acknowledged the possibility of BSE transmission to humans, causing a variant of Creutzfeldt-Jakob disease (Bradley et al. 2006; Department for Environment, Food and Rural Affairs 2005; Food Standards Agency 2007; McCallum et al. 2006).

BSE in Japan

In 1986, when the BSE cases were reported in the UK, the import of UK beef products, including meat and bone meal (MBM), was stopped, first voluntarily and then by the order of the Ministry of Agriculture, Forestry and Fisheries (MAFF), which subsequently banned the use of ruminants' organs for ruminants' feed. In 1997, when the spread of BSE to humans in the form of vCJD was confirmed, the Law to

Prevent the Infectious Diseases in Domestic Animals was revised, classifying BSE as a notifiable animal disease. Shortly thereafter, the WHO published its report on the risk of BSE risk in humans.

In response to the widespread incidence of BSE in Europe, the import of bovine brains and spinal nerves was voluntarily stopped, and in 2000 the MAFF banned the use of MBM for animal feed. In the following year, the MHW banned the import of European beef products, and the MAFF started a surveillance program to determine how many cows older than 24 months had abnormal neurological symptoms. The domestic beef industry launched a campaign called "Safe Domestic Beef," which aimed to reassure consumers and differentiate its products from imported ones.

The first BSE case in Japan was confirmed in August 2001. The MAFF immediately announced this fact and stopped the shipment of beef from the farm concerned. The Ministry soon conducted an emergency survey targeting all cattle and tightened BSE screening so that cows older than 24 months who showed abnormal neurological signs or were generally ill, as well as all cows older than 30 months, should be tested for BSE (the age limit of 24 months was adopted because the tests lacked the technology to detect the BSE prion). The ban on MBM imports was applied worldwide.

In October 2001, thorough testing of cattle, regardless of age, was introduced to reassure consumers, because exact cow ages were hard to determine. The following year, the Special Provisional Act for BSE Mitigation was adopted to protect both consumers and producers. Thus, the use of MBM for cattle feed was banned, notification of cattle deaths and BSE testing were made obligatory, BSE testing in slaughterhouses was introduced, compensation was provided to cattle farmers, and a cattle tracking system was established. With the adoption of the Food Safety Basic Law in 2003, the Food Safety Commission was created under the Prime Minister's office, and the MAFF was reformed so as to separate the consumer safety department from the department for industrial development.

When BSE cases were first reported in the US and Canada in 2003, a ban on imports of cattles and beef products from these countries was immediately enacted. As this ban soon became a trade issue, a US-Japan working group was created to address the issue, and the differences in how the two countries managed the risks of BSE was placed on the policy agenda. Triggered by this dispute, the Food Safety Commission started in 2004 to re-assess BSE-related policies in Japan. It concluded that the current complete cattle testing was hardly meaningful, because the BSE prions could not be detected. Testing cows older than 30 months, a cattle tracking system, the removal of specific risk materials (SRMs) for human consumption, and a ban on the use of MBM were cited as important for abolishing BSE and vCJD.

Around the end of 2006, the export of US beef to Japan was resumed on the condition that the beef export verification program guaranteed the safety of the exported beef. The program required SRM removal, verification of cows' ages, and BSE testing of cows older than 20 months. The US cattle and beef industry complained about the Japanese safety standards, especially the requirement for BSE testing of cows younger than 24 months (the international standard was 30 months). The trade dispute continues to this day.

BSE in the United Kingdom

In the UK, the first case of BSE was confirmed in 1985, it was announced in 1987 that the use of MBM caused the BSE epidemic and that the abnormal prions contributed to the disorder. In the following year, BSE was classified as a notifiable disease, and the industry was requested to voluntarily adopt a ban on the use of MBM and ruminant-originated animal feed, and the abandonment of the carcasses of affected cattle.

In response to the conclusions of the Southwood committee of experts in 1990 that the risk of BSE in humans is present but quite low, the UK government required the removal of SRMs from the food chain, but to soothe public concern it announced at the same time that UK beef is safe. However, concern about a BSE risk from beef products was not easily dismissed, as BSE infection was discovered in cats and pigs. Beef sales dropped by 25%, and many local governments banned the use of British beef for school meals. The national government therefore required the abandonment of cattle suffering from BSE, compensation payments to farmers, and a ban on the use of SRMs for animal feed.

After 1992, a few vCJD cases, probably resulting from the human transmission of BSE, were reported. In addition, despite the abandonment of BSE-infected cattle and the ban on MBM use, the BSE epidemic continued, inducing the public to consider the government policies ineffective. The use of ruminants' protein for animal feed was banned in 1994, and the Meat Hygiene Service was established in 1995 to more effectively enforce the policies already in place. Finally, in March 1996 the Southwood committee announced that the human transmission of BSE had been confirmed, which caused a much greater deterioration of public confidence in the safety of beef. The government response at that time was to ban the sale and use of ruminants' MBM.

The EU soon banned the import of cattle, beef products, and MBM from the UK. In 1997, it banned the use of SRMs, requested its member states to establish a cattle tracking system, set guidelines for BSE surveillance, and started a geographical (country-by-country) BSE risk assessment. The UK government, insisting that the EU claims were not backed by scientific evidence, established a cattle tracking system of its own, required the abolishment of BSE-infected cattle herds (herd culling), and necessitated the deboning of cow carcasses before shipment, as well as the incineration of SRMs and the carcasses of cows older than 30 months. Subsequently, in 2001, a BSE testing program was introduced for all cows older than 24 months that had died or presented abnormalities on the farm, and for the cows sampled from those older than 30 months (later changed to 24 months). After reviewing past BSE policies, the Food Safety Agency was established in 2002.

BSE in France

When the first case of BSE was found in the UK in 1985, only a few experts were interested in the issue. However, as the UK Southwood Committee did not exclude a risk to humans, and the use of SRMs had been banned in the UK, some in France became concerned that the risk to the French cattle industry was increased by the fact that UK cattle and MBM had been and were still being imported. It was estimated that the first French BSE case would appear as early as 1991. Accordingly, in 1989 the French Department of Agriculture issued a notice to French farmers that UK MBM should not be used for ruminants' feed.

Because the price of UK MBM dropped, and the ban on MBM use in France was only loosely enforced, the amount of beef and MBM imported from the UK increased. Reacting to requests by the French associations of MBM producers, and those dealing with animal feed, the government in 1990 enacted legislation banning the use of MBM (of any origin) as cattle feed. The ban only applied to cattle feed, because the feed industry and the Department of Agriculture were concerned about the possible economic distress that banning all animal feed might cause. With no domestic BSE cases reported, few people in France remained interested in the issue. In 1990, however, the government classified BSE as a notifiable disease and established a surveillance system. When the EU Council of Agricultural Ministers required the UK government to implement a set of effective preventive measures, more extensive than those recommended by the EU Veterinary Commission, the government decide to lift the import ban on UK beef products.

The first BSE case in France was reported in February 1991, and the number of cases gradually increased. This situation had been expected by the bureaucrats, who thought that sufficient measures had already been taken. BSE-infected cattle were being slaughtered and incinerated, but the incineration of the herds was not made obligatory until 1994, and the ban on the use of French SRMs was not introduced until 1996. Many regarded BSE as a UK problem, as the number of BSE cases in France was still quite small (6 cases by June 1993, 12 by May 1995). Although surveillance for vCJD was initiated in 1991 by the Department of Health, scientific research on BSE/vCJD did not advance substantially.

Immediately after the outbreak of vCJD was announced by the UK in March 1996, the French government banned the import of UK beef and set up an inter-agency committee on BSE. Domestic beef consumption dropped by 50% because of these BSE concerns, and the media criticized the government for not taking sufficient measures to benefit the agriculture industry. Parliament set up a committee to review past French BSE policies and eventually defended them. Nonetheless, reform of the Department of Agriculture was proposed, and the Food Sanitation Agency, established under the Ministry of Health, Agriculture, and Consumers, was assigned the tasks of assessing food risks, proposing policies, and supporting research.

In practice, the lifting of the import ban on UK beef was not implemented for a long time. There remained concerns on the French side about the risk of BSE in UK beef. It was argued that the UK had not adopted herd culling, that cattle from the same herd in which a BSE-infected cow was found could present a risk, and that the enforcement, and therefore the effectiveness, of the UK policy had not been verified. Only in October 2002, after the European Court of Justice levied fines on France for its ban of UK beef, was the ban finally lifted.

Meanwhile, in October 2000, a French meat producer sent to market a herd of cattle, but set aside one cow that was BSE-infected (the herd should have been culled). Following this event, concern about beef safety rose again among French citizens, beef consumption dropped by 35%, and criticism of the government's commitment re-emerged. When the issue appeared on the political agenda, a ban on MBM was adopted without waiting for the report of the Food Sanitation Agency. This decision was justified with the argument that the extent to which the current policies were being implemented could not be fully ascertained.

BSE in the United States

Following the 1985 report of BSE in the UK, an inter-agency working group was established by the US government in 1988 to study BSE. As a firewall, the United States Department of Agriculture (USDA) banned all imports of live cattle and other ruminants from the countries identified as presenting a high risk of BSE. In 1990, the USDA started its own BSE testing program, and 40 cattle brains were tested. When the Office International des Epizooties (OIE) created international standards for animal surveillance, the USDA instituted its own program in 1991 at a higher level. In 1997, the government extended the import ban on live cattle, cattle feed, and beef products to include all of Europe. The USDA for the first time banned the use of mammalian protein for animal feed.

In 2002, the BSE testing program expanded to nearly 20,000 cattle brains. This amounted to less than 1% of the cows slaughtered each year, and only cows that were disoriented or presented suspicious neurological symptoms were tested. The government reaffirmed its import restrictions and allowed the import of beef products only if the country of origin had been implementing adequate BSE controls since 1989 (Canada was the first one). When a BSE case was confirmed in Canada in 2003, the US immediately reinstituted the ban on the import of Canadian beef and cattle.

The first BSE case in the US was confirmed in December 2003, after which many countries banned the import of US beef. Although suspect cows were slaughtered and marketed before BSE test results were known, USDA officials stated that infectious animal parts such as the brain and spinal cord were removed before processing, and therefore the beef products being shipped were not so risky. Soon thereafter, the USDA ordered the removal of SRMs at the slaughterhouse from cows older than 30 months. In addition, it implemented new procedures to verify cows' ages, banned the use of advanced meat recovery and air injections, and introduced a test-and-hold rule for non-ambulatory cows.

Based on the results of the program to determine BSE rates, the USDA concluded in 2004 that it had been scientifically established that only cows over 30 months old were at risk of BSE. Thus, it only banned SRMs from cows whose beef was to be used for animal and pet food. In early 2004, a rapid screening procedure was adopted by the USDA, supplemented by an immunohistochemistry test for inclusive cases. When a second BSE case was found in June 2005, the Western blot test was added these latter cases.

Since 1996, when the human transmission of BSE was announced in the UK, the meat and agriculture industries in the US were concerned about the possibility that the government might take actions based on public pressure (responding to the irrational public food scare) rather than scientific evidence. So that the government and consumers would base their decisions and behavior on the scientific data, these industries expended a great deal of effort conducting their own consumer research and exchanging opinions with the media, policy makers, and consumers. In contrast to other countries, beef sales did not drop in the US, and the pressure for thorough cattle testing remained limited.

Comparative Study on Policy and Politics

In the management of (health) risks, an obvious early step is to protect against potential catastrophes. Aiming at preventing problems from occurring, devising protections that limit the damage (protection against severe risks). A second strategy is to proceed cautiously in protecting against potential catastrophes, assuming the worst rather than expect the likely (erring on the side of caution). A third strategy is to reduce uncertainty about the likelihood and magnitude of potential harm by testing whenever possible (advance testing, priority setting, and learning from error). If there has been no carryover of learning from one type of risk to another, there is little assurance that the next danger will be averted (Morone & Woodhouse 1986; Steinbruner 1974). How safe is safe enough? Implementation of safety precautions is a matter of degree, and each additional precaution typically involves increased cost.

Case studies presented in this book, from very different sectors, have a number of similarities with respect to the management and communication of risk. In much of the earlier periods of both, we can see a number of common features: Too close a relationship between the respective industry and the regulatory agencies, which were themselves underdeveloped; a denial of the real health risks or at least the extent to which they might arise; an increased recognition of the dangers once the risk to public health (as opposed to merely animal-based BSE or workplace-located asbestos) became so strong that the crisis turned into a medical one and a large number of professional and institutional actors had to be called upon to help handle it; a perception of the public as lacking the capacity to handle risk, and a strategy of reassurance that bore little relation to the evidence-based risks ahead.

A media that in the beginning assumed a relatively marginal role in the communication process but which gradually came to criticize the government's management of each case; a gradual strengthening and a distancing of the regulatory mechanisms from economic interests; the partial effect of a growing globalization and harmonization of standards, especially from within the European policy-making arena; a greater degree of accountability and a process of social learning whereby the government appears to have learned from its mistakes in earlier periods: this is especially so in respect to the recent outbreak of a different bovine disorder, foot-and-mouth disease, which was disastrously handled in 2004 but was contained much more rapidly and effectively in 2007.

Both the asbestos case and the BSE case show how a relatively clear set of risks can be badly managed – indeed, almost entirely ignored – when it is industry's responsibility. But even when the state institutions took on this role, and thus repositioned asbestos as a public health hazard and acknowledged that contaminated meat might be transmissible to humans as variant Creutzfeldt-Jakob Disease, the real responsibility for the long-term management of any illness fell both on the health care system and on individual patients and their families.

Again, science policy and risk management must address both the context of application and the context of implication, both immediately and in the long run. The case of asbestos and that of BSE provide very different examples of risk management and communication. Although it is true that the government has learned from both cases to be much more open and accountable, the long-term effects demand a recognition of responsibility long after the major crises have passed.

References

Adams, J. (1995). *Risk: Policy implications of risk compensation and plural rationalities*. London: UCL Press.

Almond, G. A., & Powell, G. B. (1966). *Comparative politics: A developmental approach.* Boston: Little Brown.

Altman, J. A., & Petkus, E., Jr. (1994). Toward a stakeholder-based policy process: An application of the social marketing perspective to environmental policy development. *Policy Sciences, 27*(1), 37–51.

Anderson, C. W. (1978). The logic of public problems: Evaluation in comparative policy research. In D. E. Ashford (Ed.), *Comparing public policies: New concepts and methods* (pp. 19–41). Beverley Hills, CA: Sage Publications.

Axelrod, R. (1973). Bureaucratic decision making in the military assistance program. In M. Halperin & A. Kanter (Eds.), *Readings in American foreign policy* (pp. 211–270). Boston: Little Brown.

Bauer, R. A., Pool, I. S., & Dexter, L. A. (1963). *American business and public policy: The politics of foreign trade.* New York: Atherton Press.

Bradley, R., Collee, J. G., & Liberski, P. P. (2006). Variant CJD (vCJD) and Bovine Spongiform Encephalopathy (BSE): 10 and 20 years on. *Folia Neuropathol, 44*(2): 93–101.

Chaffe, E. E. (1985). Three models of strategy. *Academy of Management Review 10*, 89–98.

Cobb, R. W., & Elder, C. (1972). *Participation in American politics: The dynamics of agenda building.* Boston: Allyn and Bacon.

Cobb, R., Ross, J., & Ross, M. H. (1976). Agenda-building as a comparative political process. *American Political Science Review, 70*(1), 126–138.

Cohen, M. D., March, J. G., & Olsen, J. P. (1972). A garbage can model of organizational choice. *Administrative Science Quarterly, 17*(1), 1–25.

Connolly, W. E. (1984). *Legitimacy and the state: Readings in social and political theory.* New York: New York University Press.

Crenson, M. A. (1971). *The unpolitics of air pollution.* Baltimore: Johns Hopkins University Press.

Department for Environment, Food and Rural Affairs (DEFRA). (2005). *Transmissible spongiform encephalopathies (TSE) in Great Britain 2005: A Progress Report*. London: Author. Retrieved, March 03, 2009, from http://www.defra.gov.uk/animalh/bse/pdf/tse-gb_progressreport12-05.pdf.

Dietz, T., & Rycroft, R. W. (1987). *The risk professionals*. New York: Russell Sage Foundation.

Edelman, M. (1971). *Politics as symbolic action: Mass arousal and quiescence*. Chicago: Markham.

Elder, C. E., & Cobb, R. W. (1983). *The political uses of symbols*. New York: Longman.

Etzioni, A. (1976). *Social problems*. Englewood Cliffs, NJ: Prentice-Hall.

Etzkowitz, H. (2002). Incubation of incubators: Innovation as a triple helix of university-industry-government networks. *Science and Public Policy, 29*(2), 115–128.

Farrar, E., DeSanctis, J. E., & Cohen, D. K. (1980). The lawn party: The evolution of federal programs in local settings. *Phi Delta Kappa, 62*(3), 167–171.

Fiorino, D. J. (1989). Technical and democratic values in risk analysis. *Risk Analysis, 9*(3), 293–299.

Food Standards Agency. (2007). BSE report archive. London: Food Standards Agency. Retrieved, March 3, 2007, from http://www.food.gov.uk/archive/bsearchive/aboutbsetse/aboutsbereviewarchive/bsereportarchive/.

Foyol, H. (1916). *Administration industrielle et generale. Prevoyance, organization, commandement, coordination, controle*. Bulletin de la Societe de l'Industrie Minerale. Geneva: International Management Institute.

Frantz, J. E., & Sato, H. (2005). The fertile soil of policy learning: Hansen's disease policy in US and Japan. *Policy Sciences 38*(2/3), 159–176.

Freedman, J. O. (1978). *Crisis and legitimacy: The administrative process and American government*. Cambridge, UK: Cambridge University Press.

Frewer, L. J., Howard, C., Hedderley, D., & Shepherd, R. (1996). What determines trust in information about food-related risks? Underlying psychological constructs. *Risk Analysis 16*: 473–486.

Frewer, L. J., Howard, C., & Shepherd, R. (1997). Public concerns about general and specific applications of genetic engineering: Risk, benefit and ethics. *Science, Technology and Human Values 22*: 98–124.

Frewer, L. J., Hunt, S., Miles, S., Brennan, M., Kuznesof, S., Ness, M., & Ritson, C. (2001). *Communicating risk uncertainty with the public: Final report project February 2001*. Newcastle: University of Newcastle.

Fuller, S. (2001). *Knowledge management foundations*. Oxford, UK: Butterworth-Heinemann.

Gibbons, M. (2000). Mode 2 society and the emergence of context-sensitive science. *Science and Public Policy, 27*(3), 159–163.

Graham, J. D., Green, L. C., & Roberts, M. J. (1988). *In search of safety: Chemicals and cancer risk*. Cambridge, MA: Harvard University Press.

Graham, P. J., & Dickinson, H. D. (2007). Knowledge-system theory in society: Charting the growth of knowledge-system models over a decade, 1994–2003. *Journal of the American Society for Information Science and Technology, 58*(14), 2372–2381.

Hall, P. M. (1973). A symbolic interactionist analysis of policies. In A. Effrat (Ed.), *Perspectives in political sociology* (pp. 35–75). New York: Bobbs-Merrill.

Hall, P. A. (1993). Policy paradigms, social learning, and the state: The case of economic policy-making in Britain. *Comparative Politics, 25*(3), 275–296.

Hinchliffe, S. (2001). Indeterminancy in decisions: Science, policy and politics in the BSE (bovine spongiform encephalopathy) crisis. *Transactions Institute of British Geographers NS, 26*, 182–204.

Jasanoff, S. (1990). *The fifth branch: Science advisers as policymakers*. Cambridge, MA: Harvard University Press.

Jenkins-Smith, H. (1991). Alternative theories of the policy process: Reflections on research strategy for the study of nuclear waste policy. *PS: Political Science & Politics, 24*(2), 157–166.

Jones, C. O. (1975). *Clean air*. Pittsburgh, PA: University of Pittsburgh Press.

Jones, C. O. (1978). *An introduction to the study of public policy* (2nd ed.). Boston: Duxbury Press.

Kasperson, R. E. (1983). Acceptability of human risk. *Environmental Health Perspectives, 52*, 15–20.

Kates, R. W., & Kasperson, J. X. (1983). Comparative risk analysis of technological hazards. *Proceedings of the National Academy of Sciences of the United States of America, 80*(22), 7027–7038.

Kazancigil, A. (1998). Governance and science: Market like modes of managing society and producing knowledge. *International Social Science Journal, 50*(155), 69–79.
Kingdon, J. W. (1973). *Congressmen's voting decisions.* New York: Harper and Row.
Kingdon, J. W. (1995). *Agendas, alternatives, and public policies* (2nd ed.) Boston: Little Brown and Company.
Lasswell, H. (1954). Key symbols, signs and icons. In L. Bryson & L. Finkelstein (Eds.), *Symbols and values: An initial study* (pp. 199–204). New York: Free Press.
Lasswell, H. D., & McDougal, M. S. (1992). *Jurisprudence for a free society: Studies in law, science and policy*. New Haven Press/Kluwer Law.
Lasswell, H. D., Lerner, D., & Montgomery, J. D. (1977). *Values and development: Appraising Asian experience*. Cambridge, MA: The MIT Press.
Levine, A. G. (1982). *Love canal: Science, politics and people.* Lexington, MA: D.C. Heath.
Light, P. C. (1982). *President's agenda.* Baltimore: Johns Hopkins University Press.
Lindblom, C. E. (1959). The science of muddling through. *Public Administration Review, 19*(2), 79–88.
Lindblom, C. E., & Woodhouse, E. J. (1993). *The policy making process* (3rd ed.). Englewood Cliffs, NJ: Prentice-Hall.
Lipset, S. M. (1981). *Political man: The social basis of politics.* Baltimore: Johns Hopkins University Press.
Majone, G. (1992). *Evidence, argument and persuasion in the policy process.* New Haven, NJ: Yale University Press.
Mannheim, K. (1936). *Ideology and utopia.* London: Routledge & Kegan Paul.
McCallum, M., Sutherns, R., & Haworth-Brockman, M. (2006). *Bovine Spongiform Encephalitis (BSE): An annotated review of international literature.* Winnipeg, Canada: Prairie Women's Health Centre of Excellence.
McClelland, G. H., Schultze, W. D., & Hurd, B. (1990). The effects of risk beliefs on property values: A case study of a hazardous waste site. *Risk Analysis, 10*(4), 485–497.
Mintzberg, H. (1988). Opening up the definition of strategy. In J. B. Quinn, H. Hintzberg, & R. M. James (Eds.), *The strategy process: Concept, context and cases* (pp. 13–20). Englewood Cliffs, NJ: Prentice Hall.
Moon, J. (2005). Innovative leadership and policy change: Lessons from Thatcher. *Governance, 8*(1), 1–25.
Morone, J. G., & Woodhouse, E. J. (1986). *Averting catastrophe: Strategies for regulating risky technologies.* Berkeley, CA: University of California Press.
Mucciaroni, G. (1992). The garbage can model and the study of policy making: A critique. *Polity, 24*(3), 459–482.
Newton, R. G. (1997). *The truth of science: Physical theories and reality.* Cambridge, MA: Harvard University Press.
Nowotney, H., Scott, P., & Gibbons, M. (2001). *Re-thinking science: Knowledge and the public in an age of uncertainty.* Cambridge, UK: Polity Press.
Pauchant, T. C., & Mitroff, I. I. (1992). *Transforming the crisis-prone organizations: Preventing individual, organizational, and environmental tragedies.* San Francisco, CA: Jossey-Bass Inc., Publishers.
Peters, R. G., Covello, V. T., & McCallum, D. B. (1997). The determinants of trust and credibility in environmental risk communication: An empirical study. *Risk Analysis, 17*(1), 43–54.
Pidgeon, N. (1999). Risk communication and the social amplification of risk: Theory, evidence and policy implications. *Risk, Decision and Policy, 4*(1), 1–15.
Porter, M. (1980). *Competitive strategy, techniques for analyzing industries and competitors.* New York, NY: Free Press.
Renn, O., Burns, W. J., Kasperson, J. X., & Slovic, P. (1992). The social amplification of risk: The theoretical foundations and empirical applications. *Journal of Social Issues, 48*(4), 137–160.
Rochefort, D. A., & Cobb, R. W. (1994). *The politics of problem definition: Shaping the policy agenda.* Lawrence, KS: University Press of Kansas.
Rogers, H. R., & Bullock, C. S. (1972). *Law and social change.* New York: McGraw-Hill.
Roux, D. J., Rogers, K. H., Biggs, H. C., Ashton, P. J., & Sergeant, A. (2006). Bridging the science-management divide: Moving from unidirectional knowledge transfer to knowledge interfacing

and sharing. *Ecology and Society, 11*(1), 4. Retrieved February 18, 2009, from http://www.ecologyandsociety.org/vol11/iss1/art4/.
Sabatier, P. A. (2007). *Theories of the policy process: Second edition*. Boulder, CO: Westview Press.
Sato, H. (2002). Abolition of leprosy isolation policy in Japan: Policy termination through leadership. *Policy Studies Journal 30*(1), 29–46.
Sato, H., & Frantz, J. E. (2005). Science, policy changes, and the garbage can model: Termination of the leprosy isolation policy in the US and Japan. *BMC International Health and Human Rights, 5*(3), 1–18.
Siddall, E. (1981). *Risk, Fear and Public Safety* (AECL-7404). Ontario, Canada: Atomic Energy of Canada Limited.
Siegrist, M., & Cvetkovich, G. (2000). Perception of hazards: The role of social trust and knowledge. *Risk Analysis, 20*(5), 713–719.
Slovic, P. (1987). Perception of risk. *Science, 236*(4799), 280–285.
Snare, C. E. (1995). Windows of opportunity: When and how can the policy analyst influence the policymaker during the policy process. *Policy Studies Review, 14*(3–4), 407–430.
Steinbruner, J. D. (1974). *The cybernetic theory of decision*. Princeton, NJ: Princeton University Press.
Thompson, F. J. (1981). *Health policy and the bureaucracy: Politics and implementation*. Cambridge, MA: MIT Press.
Thompson, F. J. (1984). Implementation of health policy: Politics and bureaucracy. In T. J. Litman & L. S. Robins (Eds.), *Health politics and policy* (Chapter 8). New York: John Wiley & Sons.
Tweedale, G. (2000). *Magic mineral to killer dust: Turner & Newwall and the asbestos hazard*. Oxford, UK: Oxford University Press.
Van Meter, D. S., & Van Horn, C. E. (1975). The Policy implementation process: A conceptual framework. *Administration & Society, 6*(4), 445–488.
Walker, J. W. (1981). The diffusion of knowledge, policy communities, and agenda setting: The relationship of knowledge and power. In J. Tropman, M. Dluhy, & R. Lind (Eds.), *New strategic perspectives on social policy* (pp. 75–96). New York: Pergamon Press.
Weiss, C. H. (1970). The politicization of evaluation research. *Journal of Social Issues, 26*(4), 57–68.
Weiss, J. A. (1989). The powers of problem definition: The case of government paperwork. *Policy Sciences, 22*(2), 97–121.
Wildavsky, A. (1988). *Searching for Safety*. New Brunswick and Oxford: Transaction Publishers.
Wise, C. R. (1991). *The dynamics of legislation: Leadership and policy change in the Congressional process*. San Francisco, CA: Jossey-Bass Inc. Publishing.
Wittrock, B. (1989). Social science and state development: Transformations of the discourse of modernity. *International Social Science Journal, 41*(122), 497–507.

Part II
Management of environmental risk: Cases of asbestos

Chapter 2
Development of Asbestos Regulation in Japan: Incremental Policy Making and Crisis Politics

Hajime Sato

Introduction

Asbestos is the fibrous form of mineral silicate. It belongs to rock-forming minerals of the serpentine group, i.e., chrysotile (white asbestos), and of the amphibole group, i.e., actinolite amosite (brown asbestos, cummingtonite-grunerite), anthophyllite, crocidolite (blue asbestos), tremolite, or any combination of these. The natural resistance of asbestos to heat and acid, its tensile strength, and its remarkable thermal, electrical and sound insulating properties (Sawyer 1977) have led to its use in over 3,000 applications, including floor and ceiling tiles, boiler and pipe insulation (sprayed or troweled on), wall board, roofing materials, brake linings, and cement water pipes (Rom 1992). These properties have been well known since ancient times. The name asbestos comes from the ancient Greek *asbestos*, meaning non-extinguishable or non-perishing, the medieval Latin *asbeston*, and the middle English word for mineral, *albeston*, which apparently means non-extinguishable.

Exposure to some forms of asbestos in highly contaminated environments can lead to asbestosis, lung cancer, and mesothelioma. Causal linkages of asbestos with these diseases gradually became evident in the early twentieth century (Navarro 2003; Wikeley 1992). At the beginning of the century, knowledge about the dangers of asbestos was limited. After the first decade, however, concern over the impact of asbestos on human health arose as a result of the inordinately high rate of disease discovered among employees handling the material in mining, manufacturing, and construction. During the 1950s and 1960s, clinicians and researchers reviewed the medical records of asbestos workers and found positive correlations between the number of people who had been exposed to extraordinarily high concentrations of asbestos in the air and specific diseases such as asbestosis, bronchogenic carcinoma, and mesothelioma.

H. Sato
Department of Public Health, Graduate School of Medicine, The University of Tokyo, Tokyo, Japan
e-mail: hsato-tky@umin.net; hsato@post.harvard.edu

H. Sato (ed.), *Management of Health Risks from Environment and Food*, Alliance for Global Sustainability Bookseries 16,
DOI 10.1007/978-90-481-3028-3_2, © Springer Science+Business Media B.V. 2010

By the early 1970s, these associations had been confirmed in animal experiments. Since the 1980s, additional concerns have arisen about the very low levels of exposure to asbestos found in non-occupational settings (Boden et al. 1988).

In response to the emerging, and sometimes widespread, health hazards from asbestos, many countries introduced regulations for the safe use of asbestos products, and they gradually banned their production, sale, transfer, and use. International organizations, such as the World Health Organization (WHO), the International Agency for Research on Cancer (IARC), the International Labor Organization (ILO), and the European Commission (EC), have occasionally published scientific reports and issued policy recommendations summarizing the accumulating evidence (Tossavainen 1997). Though the hazards of asbestos are scientifically evident, countries have nonetheless adopted a range of public policies on the asbestos problem. Many economically developed countries have already implemented some sort of ban on asbestos usage, resulting in a rapid decline in its consumption. However, asbestos use is rapidly increasing in many developing countries that have no strict safety standards for preventing the associated health hazards.

The regulation of asbestos is at the intersection of several policy domains, including health, the environment, industry, commerce, and trade. This fact has complicated the situation, making it difficult to regulate or ban asbestos use. Asbestos has been so widely used that the cost of its regulation was substantial by the time the issue appeared on the government's and public agenda. From a political perspective, it is sometimes difficult to regulate the use of a product that was widely used before its dangers had become known and/or scientifically proven. This is especially true when the products are of natural origin and in use for a long time. Many societies have experienced far more difficulty in regulating asbestos products than artificial products recently introduced in the market: Tobacco is a good example.

A number of factors have provided for a plethora of scientific conclusions, value judgments, policy discussions, and political action. They include the nature of the asbestos problem itself, the ambiguity of the scientific evidence about the extent of asbestos-related health hazards, the long latency before symptoms appear, the lack of adequate information about the hazards, the number of victims, the technical difficulties in controlling asbestos exposure, and the uneven distribution of economic benefits and health costs. Even in the 2000s, well after the health hazards of asbestos had been scientifically established, different policy positions led to an international trade dispute: Canada brought the French ban on asbestos products before a panel of the World Trade Organization (WTO).

If we must always wait for a health disaster before taking action, there is no way to effectively prevent the emergence of future health hazards, although it might be possible to lessen their magnitude; precautionary measures would have no place in policy. In this context, it is important to examine the politics of asbestos regulation. In this chapter, the history of asbestos regulation in Japan is examined, with a focus on the political aspects of policy-making, especially the political leadership, a topic that has been on the public agenda and hyped by the media. The possibilities of, and obstacles to, the establishment of effective measures to prevent risks to health and to protect the environment are also discussed.

Materials and Methods

To examine policy-making concerning the regulation of asbestos, a comprehensive search of historical documents was conducted. After thoroughly searching the archives of public records, both printed and on-line, relevant documents were collected from libraries, governmental and non-governmental organizations in Japan, and international organizations. The specific sources include the National Diet Library (Tokyo) and academic databases such as the Social Science Citation Index, the Science Citation Index, MEDLINE, and ICHUSHI. Archives of Office of Prime Minister were throughly examined (Office of Prime Minister, website). Other official archives on asbestos issues and policies were carefully scrutinized (Ministry of Economy, Trade and Industry; Ministry of Education, Culture, Sports, Science and Technology; Ministry of Internal Affairs and Communications; Ministry of the Environment; Ministry of Health, Labor and Welfare; Ministry of Education; Ministry of Land, Infrastructure, Transport and Tourism). Other types of documents and books, such as theses, essays and recollections, were also gathered. During 2005–2007, supplementary interviews were conducted with bureaucrats, politicians, researchers, and some members of non-governmental organizations.

To observe the political process, a set of models, from studies in different policy domains, were consulted to find components that can explain the dynamics of policy making. Representative examples were found in both classical and contemporary textbooks on policy and politics, including linear and rational policy making (Anderson 1996), bounded rationality modeling (Simon 1991; Gigerenzer & Selten 2002), bureaucratic incrementalism (Aberbach et al. 1981), political leadership (Heifetz 1994), and the garbage-can model (Kingdon 1995). The models were used to analyze the sequence of events so that the important contributions and obstacles to policy making would not be missed.

Asbestos-Related Health Hazards and Regulatory Policies in Japan

Early Scientific Reports and Policies Regarding Asbestosis

The records of asbestos use in Japan date back to ancient times. A ninth-century fairy tale, "Taketori monogatari," describes incombustible clothes, which are now considered to have been made of asbestos (Sugiyama 1934). The industrial use of asbestos started in the late nineteenth century. Raw asbestos and asbestos products were exported from overseas to Japan in the 1880s. The first asbestos factory, which manufactured asbestos packing and other insulation materials, was founded in Osaka in 1886. Asbestos textile factories that produced the soles of Japanese socks started operating in 1907. Soon a variety of asbestos-containing textiles were being manufactured in the Osaka area (Environmental Agency 1987). At the beginning of twentieth century, the use of asbestos had extended to the construction of battleships, the installation of military facilities, and the manufacture of underground water pipes, boilers, and brake linings.

In the 1910s and 1920s, the risk of pneumoconiosis caused by asbestos inhalation was frequently reported in Europe, the US, and Canada. In 1924, the term "asbestos lung" was coined to name the pneumoconiosis caused specifically by asbestos (Cooke 1924). The UK Ministry of Interior reported that about one-fourth of the workers in their asbestos textile factories had abnormalities in their lungs. Then in 1931, the UK government required ventilation fans to be installed in asbestos-handling factories. Shortly thereafter, the US government launched a series of epidemiological studies. In 1935, the first cases of lung cancer were reported in asbestos workers in the UK and the US. In 1938, the US Public Health Service (PHS) officially confirmed the health risk to workers in asbestos textile plants. Many medical reports linking asbestos with long cancer were published in the 1940s (Cook 1942; Greenberg 1999).

These events alerted local Japanese health officials and researchers, who then conducted surveillance in the areas where asbestos was widely processed for industrial use. Consequently, in 1929 the first case of asbestosis in Japan was uncovered and reported in a medical journal. In a study conducted from 1937–1940 in Osaka by Dr. Sukegawa and colleagues, asbestosis was found in 65 of 251 workers in factories that manufactured asbestos (Mizuno 2007). As raw asbestos could not be imported during World War II, many small asbestos mines were opened nationwide, but the limited amount of asbestos ore they produced was of inferior quality. The mines were closed soon after the war, when imports resumed. However, no actions were taken by the Japanese government until the end of World War II.

After the late 1950s, both the quantity and the use of asbestos expanded remarkably. It was used in shipbuilding, the manufacturing of chemicals, and the plants for power generation, because of its superior durability and chemical stability. In addition to water pipes, asbestos-containing slate boards and tiles became quite popular in construction, as did roof boards and wall boards. Under the new Constitution, a series of acts and regulations protected the health and safety of workers. These included the Labor Standards Law of 1947, the Compensation Law for Workers' Accidents of 1947, and the Workplace Safety and Health Regulations of 1947. The latter, though not specifically and exclusively targeting pneumoconiosis, implicitly had its prevention as one of its aims.

In 1954, the Tokyo Labor Standards Office consulted with the Ministry of Labor (MOL) concerning the official recognition of asbestosis as an occupational disease covered by the Compensation Law. Whenever impairments possibly resulting from occupational hazards were not explicitly covered by the law, administrative adjudications were sought on a case-by-case basis. The first recognized asbestos case occurred in the following year. A director at the Osaka Kosei-en National Sanatorium conducted a new health-checkup study of asbestos-handling workers in the Osaka area (Horai et al. 1957). He reported a high incidence of lung disease (especially asbestosis) among the workers.

The bureaucrats did not move quickly to protect workers. In 1955, when "the special act to prevent the impairment by silicosis (stoneman's disease)" was placed on the agenda, the explicit inclusion of asbestosis under the law was discussed. Eventually, asbestos was excluded, because it was "too early." Silica was better known at the time as a cause of pneumoconiosis than was asbestos. However, in 1956, the MOL started providing research grants to study asbestos-related health hazards, and in 1958 to create diagnostic standards for asbestosis. A 1957 study measuring airborne asbestos fibers in four factory workplaces in Sennai (Osaka) revealed that the fiber concentration exceeded 1,000 fiber/mL in the mixing process in 3 of the mills, with no official order or sanction (Horai 1959; Sera et al. 1960). Instead, the MOL Labor Standards Bureau required employers to implement a special health-checkup program for asbestos-handling workers. In 1958, the construction industry began using asbestos spraying for insulation. It did not take long for this to become very popular in Japan.

In 1960, the "Pneumoconiosis Law of 1960" was adopted. Among other things, it identified asbestos as a cause of pneumoconiosis, and it set rules and regulations regarding environmental safety in the workplace, employers' obligation to provide health checkups for their workers, and compensation for the afflicted. This was the first time that asbestos was explicitly incorporated in the list of diseases and injuries targeted by law.

Reports on Asbestosis-Induced Cancer and Associated Policies

After the initial case reports of lung cancer in asbestos-handling workers during the mid-1930s, more epidemiological studies were conducted in the UK, the US, and Europe. In the 1950s, an increased incidence of lung cancer in asbestos workers was established by research. Above all, Dr. Doll's 1955 report confirmed the elevated cancer risk from asbestos. In the meantime, a special type of lung cancer, mesothelioma, was reported among workers in the UK in 1935, in Canada in 1952, and in West Germany in 1953. In 1960, mesothelioma cases were reported in the US and South Africa. A landmark article, "Asbestos Exposure and Neoplasia," by Selikoff et al. (1964), was published in *JAMA* in 1964. This article thoroughly established the carcinogenic potential of asbestos. It also shows the greatly increased mortality of insulation workers exposed to asbestos, and it makes clear that an epidemic of occupational and environmental cancer was underway. In this year, the ILO held the Employment Injury Benefits Convention (C121), which required asbestosis to be compensated as an occupational injury (International Labor Organization 1964).

In the meantime, mesothelioma cases, as well as cases representing specific forms of cancer in the pleura, peritoneum, and pericardium associated almost exclusively with asbestos exposure, were sporadically reported in many countries, for example, the UK in 1935, Canada in 1952, Germany in 1953, and the US and South Africa in 1960. Asbestos as a cause of mesothelioma was confirmed by animal experiments in 1969. Furthermore, it gradually became evident that the health hazards of asbestos were not confined to workers but were also present in family members and people living near asbestos factories (LaDou 2004; Newhouse et al. 1972). In 1968, the UK Association of Industrial Hygienists revised the environmental standards for asbestos use. At the same time, 35 years after the first case was reported, the UK government launched a disease registration system for mesothelioma. Two years after that, in 1970, the government called for the voluntary cessation of crosidolite imports, while the US introduced a set of federal regulations on the environmental monitoring of asbestos, the ventilation in the workplace, the use of personal protection devices, and health checkups for asbestos.

In 1960, the first case of lung cancer in asbestos workers was reported in Japan (Sera et al. 1960). Asbestos consumption was rapidly increasing. Spraying asbestos on ceilings, walls, and iron frames for construction purposes had begun in 1957 and soon became very popular. Also, the production of asbestos-containing building materials remarkably increased after 1965. Both chrysotile and amosite were used for slates and boards. The Japanese government continued to focus its efforts on preventing asbestos (pneumoconiosis). In 1968, the Labor Standards Bureau of the MOL added workplaces in which asbestos was handled to the list of venues covered by the occupational safety law. The health regulations required the installation of local ventilation facilities to limit dust concentrations. The production of asbestos-containing construction materials rose steadily and remarkably from 1965 to 1975.

In 1970, an official inspection of the release into the environment of 46 hazardous substances revealed that only 70.3% of the 150 asbestos-handling workplaces surveyed had installed an appropriate air ventilation system. These results led the MOL to convene an expert panel. Based on the panel's report, "the Ordinance on the Prevention of Hazards due to Specified Chemical Substances (SCS) of 1971 (SCS Ordinance of 1971, [Tokutei Kagaku-busshitsu tou Shougai Yobou Kisoku])" was enacted. Though asbestos had not been listed as a carcinogen in the preceding report, the Ordinance included the detailed safety measures for asbestos processing, a requirement for ventilation, environmental standards for workplaces, compulsory health checkups for workers, and official inspections for manufacturers' compliance with the other provisions of the ordinance. The local ambient concentration allowed for hazardous substances was defined as the permissible concentration recommended by either the Japan Society for Occupational Health (JSOH) or the American Conference of Governmental Industrial Hygienists (ACGIH). For asbestos, the adopted concentration was 2 mg/m^3 (33 $fibers/cm^3$).

During the discussion of this regulation in 1970, it was reported that eight asbestos-handling workers had died of lung cancer in the preceding 11 years. This report attracted the attention of experts to the risk of lung cancer caused by asbestos (Sera 1971; Sera et al. 1973). With reference to the SCS Ordinance of 1971, an MOL officer explained that asbestosis was as severe an impairment as silicosis and that a certain kind of asbestos can cause lung cancer and malignant mesothelioma in the pleura of the lung. Although the scientific evidence was yet to be forthcoming, asbestos was included in the regulations for the prevention of the associated health hazards. As described above, at this time asbestos was not officially treated as carcinogenic, although some considered it to be just as hazardous as silica.

However, no action was taken in response to these reports, especially as they applied to neighborhoods and the general population. The ordinance was reformulated only as an administrative order, in compliance with the new Occupational Safety and Health Law of 1972.

In 1972, the ILO and the WHO/IARC convened a meeting of experts in Lyon, entitled "Evaluation of the Carcinogenic Risk of Chemicals to Man." The experts listed asbestos as a carcinogen, and several European countries introduced stricter regulations: Iceland in 1973 and Norway in 1974 banned the use of all kinds of asbestos; Sweden in 1975 prohibited the use of crocidolie (blue asbestos), and in 1972 the UK banned its import; Denmark in 1972 and the US in 1973 prohibited the spraying of asbestos. Canadian industry introduced a voluntary cessation of asbestos spraying in 1973, and at the same time the EC adopted a directive that prohibited the distribution and use of crocidolite. Finally, the ILO in 1973 adopted the Occupational Cancer Convention (C139), which recommended international standards for protection against carcinogenic substances, including asbestos; it took effect in 1974.

In Japan, some experts, bureaucrats, and politicians were already aware of the carcinogenic risk caused by various kinds of exposure to asbestos, including occupational, para-occupational (workers' family), and environmental (residents in the factories' neighborhood) exposure (Konami et al. 1974; Shishido 1986). They were also aware of the actions taken by other countries and the WHO in this regard. Research supported by the Environmental Agency (EA) summarized the scientific reports from foreign countries. Furthermore, another EA-granted research confirmed the elevated incidence of lung cancer among asbestos workers. In 1972, a politician from the Japan Communist Party raised this issue in the Diet, and a Director of the Public Health Bureau of the Ministry of Health and Welfare (MHW) stated that health checkups might be necessary if asbestos were shown to have detrimental health effects on neighborhood residents. In 1973, a study of residents in an area of Osaka revealed 10 lung-cancer cases among asbestosis patients, as well as the first case of peritoneal mesothelioma (Sera et al. 1973). Cases of pleural mesothelioma were increasingly publicized during the following year (Kishimoto et al. 2003). The JSOH responded by further tightening its asbestos regulations.

Shortly after the ILO convention of 1972, the MOL revised the Ordinance of 1971, setting the permissible asbestos concentration in the local air as 5 fibers/mL. The Bureau Director mentioned that the revision was necessary because asbestos was now known to cause malignant neoplasms such as lung cancer and mesothelioma, and several countries were tightening their environmental standards for asbestos dust. An expert panel was again called to discuss the hazards and regulation of asbestos. In 1975, the SCS Ordinance was again revised. It banned the spraying of materials in which the asbestos concentration exceeded 5%, encouraged the use of asbestos substitutes, designated other control measures, and explicitly branded asbestos as a carcinogen.

In 1976, the Labor Standards Bureau of the MOL circulated information on asbestos hazards, including a foreign report noting an increase in the incidence of lung cancer among workers and neighborhood residents and calling for better compliance with the revised Ordinance. The Working Environment Measurement Law of 1976 further specified that asbestos exposure levels should be reduced to less than 2 fibers/mL. Coincidentally, in the same year, the results of a local occupational health survey revealed that 4 relatives of workers at asbestos-handling brake factories, and 11 people living nearby, had died of lung cancer (Ebihara 1981). The IARC released a monograph documenting the risk of lung cancer and mesothelioma caused by crocidolite, amosite, and chrysotile (International Agency for Research on Cancer 1977). In 1978, 18 years after the first case report of cancer in Japan, and 5 years after the first mesothelioma case was reported there, the official standards for workers' compensation for these cancers were institutionalized: Asbestos-induced lung cancer and mesothelioma were identified as occupational cancers by administrative order; more than 10 years' employment was required for lung cancer and more than 5 years for mesothelioma.

For many years, crocidolite had been used in Japan for making asbestos cement pipes and amosite had been used for making boards; both had been used for spraying. After 1976, most of the asbestos was being used in the manufacture of boards and slates. As a result of the SCS Ordinance, more artificial (man-made) fibers, such as rock wool, were being used for spraying (rock wool containing less than 5% asbestos continued to be used for spraying until 1980). On the whole, the substitution of substances for asbestos in fact proceeded quite sluggishly, as the industry saw the substitutes as inferior in terms of their properties and costs. Also, many insisted that asbestos could be used safely if the users took sufficient caution. The major small- and medium-sized companies continued to produce asbestos-containing construction materials, because they could not afford to invest in asbestos-free products (National Diet Library 2005: p. 3). Asbestos consumption reached its peak in the mid to late 1970s (Fig. 2.1).

In Europe, West Germany banned asbestos spraying in 1979, and Norway and Denmark in principle prohibited the use of asbestos in 1980. Also in 1980, the ILO added lung cancer and mesothelioma, along with asbestosis, to its list of occupational accidents and diseases eligible for compensation (Nevitt et al. 2007). Japan ratified this convention a year later.

Possible Environmental Hazards and Policies for Reassurance

Soon after the release of the IARC monograph in 1977, the US National Cancer Institute (NCI) sent warning letters on asbestos risks to all physicians nationwide. All asbestos spraying was banned in 1979 (Nicholson et al. 1979; Baldwin et al. 1982; Stavisky 1982). The risks in school buildings appeared on the public agenda, and some states, such as New York, enacted laws to remove asbestos surfacing materials. The Environmental Protection Agency (EPA) set guidelines for clearing asbestos from school buildings. In 1980, Congress passed the Asbestos School Hazard Detection and Control Act, which required the

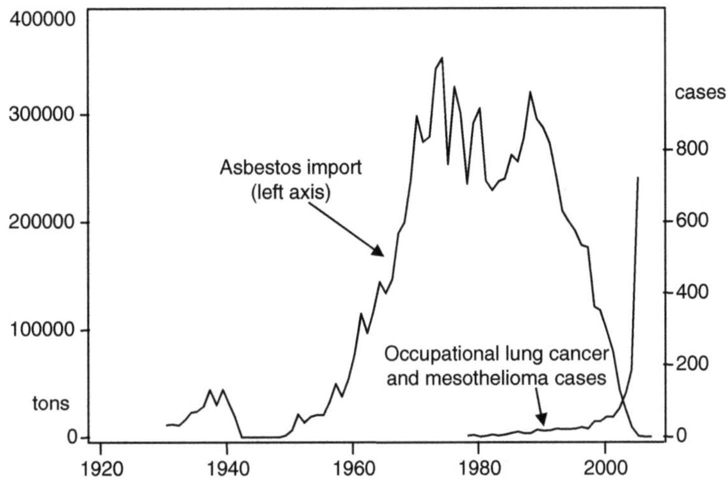

Fig. 2.1 Asbestos import to Japan, lung cancer and mesothelioma

investigation of asbestos risks in schools, based on the premise that there was no safe level of asbestos exposure (Mossman et al. 1990). The ACGIH revised its permissible concentration level for asbestos (chrysotile 2 fibers/cm^3, amosite 0.5, crocidolite 0.2).

In Japan, in 1979, soon after the institutionalization of compensation for asbestos-induced cancers, the EA began collecting scientific papers on the risks of airborne carcinogens (including asbestos). Some of these papers documented an increased incidence of cancer in workers' families and among people living near the factories. Others noted the EC's recent determination that there was no firm evidence about the health risks of asbestos in the air. Based on these papers and the recommendation of the expert panel, the EA started an environmental measurement program in 1980. The media and politicians occasionally stressed the possible health hazards of non-occupational exposure to asbestos, such as in residences located near asbestos factories.

The EA justified its inaction by stating that it remained to be seen whether all the cancer cases with no history of occupational exposure to asbestos could or should be attributed to environmental pollution, but it reaffirmed its commitment to further research. In 1981, the Agency formed an expert panel to study factory emissions of asbestos, its concentration in the environment, the methods used to measure and control it, and viable alternatives. The EA released the panel's report to the national press early in 1985. It concluded that asbestos concentration in the ambient air was far below that at worksites and therefore the risk of airborne asbestos in the general environment was minimal. The panel also noted that the persistence and accumulation of asbestos in the human environment would be a problem in the future. They concluded that, therefore, the development of alternative materials was important, as was the development of technology to minimize the release of asbestos into the air, especially during demolition and waste disposal.

Regarding occupational health, the German government and asbestos industry reached a voluntary agreement in 1982 that prohibited asbestos in construction materials until 1990. In 1983, after the IARC report on the carcinogenicity of asbestos, the ILO adopted a code of practice entitled "Safety in the Use of Asbestos" that stated the obligations of both the government and employers to protect workers from asbestos (International Agency for Research on Cancer 1982). Also in 1983, the EC issued a set of directives (83/477–478/EEC) that prohibited the sale and use of crocidolite before March 1986, and the spraying of asbestos before January 1987.

Meanwhile in Japan, 11 of 427 factories nationwide were processing crocidolite and 52 were processing amosite. In 1982, doctors at a hospital in Yokosuka city began studying the pathological anatomy of 848 patients, mostly workers at military bases or engaged in shipbuilding, who had died in the previous 5 years. They found that one-third of them died from lung cancer caused by asbestos exposure (Kazan-Allen 2003; Morinaga 1989; Morinaga et al. 1982). Several clinical and epidemiological studies were further conducted (Kikuchi and Hiraga 1983; Miyazaki 1983). Also in 1982, the JSOH recommended a lower permissible concentration level for crocidolite (0.2 fibers/cm^3). The MOL did not comply with this recommendation but instead issued a set of notices that urged local governments and employers to adopt the recommended measures for improving the work environment. In September 1986, after the EA report, the MOL issued a notice that laborers working in demolition and refurbishment must be appropriately protected from the health hazards of asbestos dust (Gunji 1987). In 1985, it was reported that 43 cases of lung cancer and 9 of the approximately 500 cases of mesothelioma deaths in Japan were compensated through the injury benefits scheme. Thirty thousand workers in 3,000 workplaces enrolled in special health-checkup programs for asbestos-related diseases. The MOL further reported that its inspection uncovered no asbestos factories that were processing crocidolite.

In Europe, Switzerland banned the use of asbestos. The UK banned the import of products containing crocidocite and amosite, and it followed the EC in prohibiting the spraying of asbestos. In 1986, the WHO's Environmental Health Criteria (EHC53) were released. This document summarized the possible health hazards of asbestos for workers and nearby residents, but not for the general environment (World Health Organization 1986). The Japanese mass media and politicians picked up on the issue only sporadically. In July, the Japanese government ratified the ILO Asbestos Convention of 1964. When a nationwide investigation was called for, the EA argued that no epidemiological survey could detect the real extent of the health hazards to the general environment.

In 1985, the 72nd ILO Convention put the safe use of asbestos on its agenda. In June 1986, the ILO General Conference held the "Convention concerning Safety in the Use of Asbestos (C162)" and adopted the "Recommendation concerning Safety in the Use of Asbestos (R172)," which listed the main policy instruments for the safe handling and uses of asbestos, including the option of banning its use

(International Labor Organization 1986a, 1986b). At the Convention, there was a dispute over the safe use or banning of asbestos, and how these processes should be measured. The Japanese government protested the asbestos ban proposed by the EU and recommended by the ILO, arguing that asbestos could be used safely. The Convention also recommended a ban on crocidolite use and asbestos spraying; the government ratified this recommendation, but only after two decades.

The asbestos issue returned to the public agenda in October 1986, when it was reported that 1.5 tons of asbestos from the renovation of the *USS Midway* at the Yokosuka Navy Base had been illegally dumped. The matter was discussed by the Yokosuka city assembly. A notice was delivered to the municipal governments concerned, and the waste was sent to the US. Both the mass media and the elected officials closely watched the course of events. Shortly thereafter, asbestos use in school buildings was taken up by the Environmental Committee of the House of Representatives in the National Diet. In February 1987, the crumbling of old asbestos in a national university building was reported, and stories of deaths from mesothelioma of people living near asbestos factories were featured in the press. Newspapers and TV programs started to refer to "the asbestos issue." An IARC report entitled "Evidence for Carcinogenicity to Humans," which clearly confirmed the hazards of asbestos, was also released, although the content of the report was not new (International Agency for Research on Cancer 1987).

Asbestos was a hot public and political topic for several years. Researchers, the mass media, and the public became concerned about asbestos in talcum powder for babies and in water pipes. There was a scientific report of 12 mesothelioma cases among former shipbuilders living in the Yokosuka area (Miura et al. 1986). Between 1984 and 1986, it was reported that 12 dockyard workers had contracted pleural mesothelioma following definite asbestos exposure during World War II (Kishimoto et al. 1989). Since the 1980s, several cases of lung cancer were reported among shipbuilders in Kure and Yokosuka (Kishimoto & Okada 1987; Katayama 1993; Miura 1982). The incidence of mesothelioma from 1977 to 1984 increased steadily (Morinaga 1988). Soon, in response to all this, several citizens' groups formed to protest the use of asbestos, and the construction workers' unions formed a national congress that petitioned for stricter government regulations. Many insurance companies in Japan adopted a policy that exempted them from the responsibility to compensate victims of asbestos exposure.

Since 1986, several ministries have taken action. The EA set up a series of expert panels and publicized their reports on carcinogenic chemicals (including asbestos) and the health effects of asbestos, but they denied the risks of asbestos in the general environment (Kohyama 1989; Koike 1992). The MOL issued a guideline on asbestos measurement (Ministry of Labor 1986). In response to the discussions in the Diet and to mass media reports, the Ministry of Education (MEd) in 1987 conducted a quick national survey, which disclosed that asbestos had been sprayed in 1,337 Japanese public schools (3.3% of the total). This report was publicized by the media, which called it "a school asbestos panic." Soon thereafter, the MEd announced its program to partially subsidize the remodeling of public school buildings to remove old asbestos-containing construction materials. The Ministry of

Construction investigated asbestos spraying in governmental buildings and later in public housing.

Since 1987, the Ban Asbestos Network Japan (BANJAN, Sekimen Taisaku Zenkoku Renraku Kaigi) has been working to raise awareness of the hazards of asbestos, supporting asbestos victims and their families, and campaigning for stricter regulations (Kazan-Allen 2003). They also took a critical look at the government's action (or inaction). The MEd survey was limited in scope: it excluded from inspection corridors, meal rooms, broadcast rooms, and utility rooms, and it reviewed only three of the materials used for spraying. The survey targeted only buildings constructed before 1976, even though up to 1980 asbestos-containing spraying materials had been used in about 5% of them. The Ministry explained retrospectively that schools should take appropriate countermeasures against asbestos hazards for themselves, and that the survey was only intended to provide a cursory snapshot of the general state of affairs (National Diet Library 2005: p. 10). No follow-up study of the measures taken by them was conducted for another 20 years.

The Ministry of International Trade and Industry (MITI) started providing subsidies to improve the working conditions in asbestos-handling small- and medium-sized factories. The MHW issued a notice that asbestos-free talc must be used in making baby powder, and it requested surveys to investigate the use of asbestos-containing construction materials in social welfare facilities. It also launched research into the status and management of asbestos dust that might affect health indoors. The MOL supported research on the health hazards of asbestos alternatives. In 1987, when the IARC again confirmed the carcinogenicity of asbestos to humans (International Agency for Research on Cancer 1987), the Japan Asbestos Association prohibited the use of crocidolite.

Early in 1988, the Environmental Health Bureau of the MHW and the Air Quality Bureau of the EA jointly announced that the condition of asbestos in the buildings was generally stable and there was no significant risk. Only deteriorated or worn-out surfaces of asbestos-sprayed or asbestos-containing construction materials posed a risk of releasing asbestos into the air. Study on the asbestos risks from tap water was also officially conducted (Ministry of Health and Welfare 1989). Though the WHO/IARC released a report classifying several artificial mineral fibers, such as glass wool, rock wool, and ceramic fibers, all of which are used as asbestos alternatives, as possibly carcinogenic (class 2B; reassessed in 2001), the EA issued an interim report that asbestos-free construction materials had been developed and were available, although they cost 20–50% more. The MOL repeatedly sent out notices alerting the public about the health hazards of asbestos and calling for safer procedures for construction, remodeling, and demolition.

While denying the direct evidence of real asbestos hazards in ambient air, the EA in 1989 revised the Air Pollution Control Law and regulated airborne asbestos concentrations in close proximity to factory buildings (factory borders). The MHW started an investigation of the asbestos control procedures used in the factories. The MITI set up an expert panel of academics, industry representatives, and workers to provide research grants for the study of asbestos substitutes and the control of asbestos dust. Many bureaucrats later confessed that around this time they were well aware of and concerned about the increasing number of asbestos-related lawsuits in the US. In 1989, a WHO report stated that asbestos was so hazardous that there was no safe minimum exposure.

Product Regulation Following International Trends

In the early 1970s, when the carcinogenicity of asbestos became evident, a ban on its use was proposed in several European countries. The ban gradually spread from crocidolite and amosite to chrysotile, especially in the 1990s and 2000s. Specific uses, such as spraying for construction, were early targets of these bans, and also of lesser regulations. Later in the 1990s, the use of asbestos was banned totally. Austria banned its use in 1990, and the Netherlands did so in 1991. Italy and Finland followed in 1992, and Germany banned the manufacturing and sales of crocidolite and amosite in 1993. The EU adopted a directive (EC 91/656/EEC) to ban the sale and use of crocidolite and amosite in 2005.

In Japan, politics and policy proceeded as usual. Incremental administrative decision making took place, although not without political input. In 1991, with support from the Japan Socialist Party (JSP), citizen groups circulated petitions signed by 630,000 people and visited the MOL, demanding a total ban on asbestos use. The Ministry officials responded that the ban could not be adopted instantly, as most of the asbestos substitutes were also classified as carcinogenic. In 1992, the JSP and the Socialist Democratic Federation introduced the "Asbestos Control Act," which had originally been drafted by the BANJAN. This bill was rejected by the majority party (the Liberal Democratic Party), the asbestos industry, and the workers' unions. It was killed without even going to committee for review. Absent political support, it was never resubmitted.

MOL officials claimed that, considering the fate of the bill above, a total ban on asbestos, including chrysotile, was neither appropriate nor feasible. Instead, they held a liaison conference with officials from other agencies. The MOL continued to provide research grants on the health effects of asbestos and its substitutes, and they watched the actions taken overseas. In 1994, France banned the sale of all asbestos except chrysotile (i.e., crocidolite and amosite). In the meantime, the Japan Asbestos Association announced that amosite was no longer to be used in Japan. However, the Japan Automobile Manufacturers Association (JAMA) reported to the Ministry of Construction that the development of asbestos substitutes was not yet complete, implicitly arguing against an immediate total ban.

Scientists continued to publish reports on asbestos-related health hazards in Japan. From the late 1980s through the early 1990s, several epidemiological studies were conducted on asbestos-related malignancies (lung cancer and mesothelioma). Several case-control studies showed an increased risk of lung cancer among those exposed to asbestos (Kawami et al. 1991; Kishizuchi et al. 1996), as well as an increase in lung cancer deaths among workers in asbestos factories and people living near them (Higashi & Tsuchiya 1991; National Cancer Center 1991). Since the early 1980s, cohort studies have revealed increases in the standardized mortality rate for lung cancer in asbestos workers. Similar reports on mesothelioma in asbestosis patients were also being compiled (Ebihara et al. 1999; Sakai et al. 1994;). The upshot of all this is that asbestos has been shown to be the leading cause of occupational respiratory cancer since 1997.

Finally in 1995, the MOL revised the enforcement regulations in the Occupational Health and Safety Law: the manufacture, import, sale, and use (provision) of crocidolite and amosite were prohibited; the permissible amount of asbestos in sprays was lowered from 5% to 1%; and dust control was tightened. When the law was still being negotiated, the MITI asked the MOL to postpone the enforcement date for 6 months, in consideration of its economic effects on small- and medium-sized firms. The Ministry of Construction requested that asbestos-containing molding materials be exempted from regulation on the premise that proper handling of these materials could be assured. The request was rejected by the MOL, which argued that publicizing the law revision in advance of its implementation was sufficient to avert any social and economic turmoil. It was also confirmed that no domestic manufacturer had any crocidolite or amosite in stock.

The regulations for the management of asbestos-related diseases were revised in 1996. All asbestos-handling workers (not just those with signs of pneumoconiosis) became the targets of a special health-checkup program. Reports of pending asbestos abatement had to be officially filed in advance, and the control of dust was to be regulated by the Air Pollution Control Law. The Ministry of Transport decided to ban asbestos in automobile brakes in 1998, 2 years after the JAMA declared that some other material replace asbestos in automobile parts. The Law Concerning the Monitoring and Control of the Environmental Discharge of Specified Chemicals was enacted in 1999, in compliance with the Organization for Economic Co-operation and Development (OECD) admonition to introduce the PRTR (Pollutant Release and Transfer Register) principle.

Despite the emerging concern about possible environmental hazards, asbestos in the workplace remained the government's focus. It was the non-government organizations that actively campaigned for the elimination of the health hazard for the entire population. The BANJAN works closely with the Japan Occupational Safety and Health Resource Center (JOSHRC) (Furuya 1993) and the Japan Citizen's Network for Wiping out Asbestos (ASNET) (Furuya 2001). These organizations assist victims, monitor environmental contamination, and generate public awareness. The International Expert Meeting on Asbestos, Asbestosis and Cancer, convened in Helsinki in 1997, used the term "global asbestos epidemic" to characterize asbestos-related health issues on a global scale. Inspired by the meeting, the BANJAN convened the International Conference on the Total Elimination of Asbestos Use in 1998. After the meeting, and following adoption of version 10 of the International Classification of Diseases (ICD), mortality statistics on mesothelioma in Japan became officially available for the first time.

Comprehensive Remedies, but No Clear Liability

In 1999, the EU issued a directive (1999/77/EC) mandating that all types of asbestos be banned by 2005 (2008 for the diaphragms used in chloride plants). The UK and Germany soon tightened their regulations on asbestos use, thereby reducing the number of items exempted from the ban. In 2001, the WHO and the IARC revised their hazard assessments and identified the major asbestos substitutes (e.g., glass wool and rock wool); these substitutes were reclassified as non-carcinogenic. In the same year, Chile and Argentina banned asbestos use. A year after that, following the WTO settlement between France and Canada, France introduced a total ban on asbestos products.

In Japan, workers' compensation was granted to patients who had worked in an environment where asbestos had been used for at least 10 years, provided that the severity of asbestosis was PR 1/0 or higher and pleural plaques were confirmed by chest radiography, or that asbestos bodies were pathologically found in the lung tissues (Ishii 2002). Although the researchers reported that the health risks in asbestos-handling factories were well-controlled (Yoshizumi et al. 2001), the annual number of compensated lung cancers due to asbestos exposure in the workplace has been rising: Up to the end of March 2000, 162 cases of malignant mesothelioma and 197 cases of lung cancer were compensated (Morinaga 2000; Morinaga et al. 2001). These figures sometimes appeared in newspaper reports.

Inspired by newspaper articles reporting an expected surge in the incidence of mesothelioma (*Asahi Shimbun*, April 2, 2002; *Maichichi Shimbun*, April 28, 2002), and under pressure from the BANJAN, Nakamura, a member of the Diet, submitted an official query to the chairman of the House of Representatives, asking about the government's position on the total ban of asbestos. About a month later, the Minister of State, Fukuda, on behalf of the Prime Minister, defended the government's current policies, citing its past efforts to overcome the occupational hazard. Specifically, they pointed to a relative reduction in the amount of chrysotile available for use, the superior chemical and physical properties of asbestos, and its usefulness compared to possible substitutes, which they noted were all acceptable. Nevertheless, further investigation was promised to examine the substitutability of asbestos and to consider more carefully a more extensive ban on asbestos use. The Minister of Health and Welfare, Sakaguchi, reiterated these arguments to the media in June. It was then reported that more than 90% of the asbestos in Japan was being used for the manufacture of construction materials (about half of that for roofing materials).

In August of the same year, the Labor Standards Bureau of the Ministry of Health, Labor and Welfare (MHLW, created in 2001 as a merger of the MHW and the MOL) initiated a questionnaire survey on the substitutability of asbestos products, targeting 26 asbestos product manufacturers, 10 asbestos industry associations, 19 asbestos-using industry associations, and nine importers. They concluded that asbestos is indispensable and could not be replaced with other materials in the manufacture of 43 out of 107 construction products examined (about 40%).

In response, the MHLW set up a panel to study the substitutability of asbestos. The panel included representatives from the Ministry of Economy, Trade, and Industry (METI), the Ministry of Land, Infrastructure, and Transport (MLIT), the EA, the MEd, and the National Defense Agency. It held a series of hearings in which manufacturers and users of asbestos testified. Finally in December, the panel concluded that asbestos, contrary to prior industry claims, was not indispensable. From the technical standpoint, they determined that appropriate substitutes were available for the manufacture of many materials, including all kinds of construction materials. The exceptions included a few products used in some chemical and nuclear plants. Later, the panel further concluded that asbestos was not indispensable for automobile brakes (asbestos was no longer used for them anyway by 1996) and for heat insulation adhesives.

When the use of crocidolite and amosite was prohibited in 1995, and the carcinogenicity of asbestos had become known, its import gradually decreased. However, as late as 2003, about 110,000 tons of asbestos were imported, primarily for use in construction and the automobile industry (Murai 1998). On October 16, 2003, a revision of the Occupational Health and Safety Law was passed: With three exceptions, the manufacture, import, transfer, provision, and use of all asbestos, including chrysotile, were banned. Also, the inclusion of any asbestos in construction, friction, and adhesive materials (which accounted for 90% of the asbestos used in Japan) was prohibited.

Letters protesting the ban were delivered to the Minister of Health, Labor and Welfare by the Canadian ambassador to Japan, but they did not affect overall support for the revised law. At around the same time, the official standards for asbestos-related occupational diseases were agreed upon, allowing more people harmed by asbestos to be compensated. Groups consisting of patients with asbestos-related diseases and their families were formed early in 2004. These groups visited the MHLW, pleading for support. Construction workers' unions and the Association of National Health Insurance Organizations committed themselves to helping ill workers who were eligible for compensation. In November 2004, citizens groups convened the Global Asbestos Conference in Tokyo, inviting many scholars, activists, and representatives of other citizens groups from all over the world. The Japanese government, however, did not respond: Politics and policy proceeded as usual.

The asbestos issue next landed on the public agenda on June 29, 2005, when the Kubota Corporation disclosed that between 1978 and 2004, 79 of its workers had died of lung cancer or mesothelioma, and three non-employees living within half a kilometer of the factory had died of mesothelioma. Later updates reported 31 mesothelioma deaths among workers' families and nearby residents. This corporation, a major machinery manufacturer established in 1890, has been the largest asbestos user in Japan. Its factory in Kanzaki, Hyogo Prefecture, used crocidolite in the manufacture of water pipes and construction materials between 1954 and 1975, and chrysotile until it closed down in 1995. This disclosure was widely covered by the mass media and attracted public attention. Within a week, two other asbestos-handling companies, the Nichias Corporation (formerly Nihon Asbestos, established in 1896) and the

Taiheiyo Cement Corporation (originally founded in 1881 and the largest cement company in Japan after its merger with other companies) followed suit.

While the MHLW repeatedly urged the labor bureaus of local governments nationwide to adopt appropriate control measures for asbestos, such as inspections, workers' health checkups, and notification of rules and regulations, the METI and the MLIT requested the industry associations under their jurisdiction to report the number of registered asbestos-related disease cases. By July 15, 89 companies under the jurisdiction of the METI reported 127 mesothelioma cases and 207 pneumoconiosis cases (deceased or living). The shipbuilding companies under the jurisdiction of the MLIT reported 85 deaths from asbestos-related diseases. The Chief Cabinet Secretary and the Vice-Minister of the MHLW stated in public that asbestos use should have been banned earlier. In the Diet, the asbestos issue was debated more frequently than ever. The MHLW issued a notice that all asbestos-containing products produced prior to the ban on their manufacture and import must be discarded.

On July 29, the cabinet ministers announced a set of pro-tempore solutions consisting of anti-scatter measures, requests to cease the manufacture and use of asbestos-containing materials whenever possible, the collection and disclosure of pertinent information, and the swift provision of injury benefits to people with asbestos-related diseases. In consideration of the fact that people living near the factories, in addition to workers within the factories, had suffered serious health damage, it was announced that asbestos substitution should be hastened (Ministerial Meeting on Asbestos Issues 2005). The intent was to enforce a complete asbestos ban starting no later than 2008. A comprehensive policy for asbestos-related injuries was adopted as a special act, and shortly thereafter, in August, the cabinet ratified the ILO's Asbestos Convention (C162, adopted in 1986) (International Labor Organization 1986a).

A month later, the MHLW released a list of 415 factories and offices in which some of the workers had been registered during the preceding 6 years as suffering from asbestos-related occupational injuries. The Ministry also provided the public with health-consultation services, examined how asbestos was being used in building public hospitals, requested manufacturers to disclose how they had used asbestos-containing materials for construction, listed all the medical and cosmetic products containing asbestos, and convened an expert panel on asbestos substitution as a way to the eliminate its use. The MEd resurveyed asbestos use in public schools. Every ministry was requested, whenever applicable, to examine the occurrence of asbestos-related diseases in the industries under its jurisdiction (e.g., the shipbuilding industry under the MLIT), and, at the same time, to review its own policies on asbestos, assessing how they had thus far tackled the issue.

On August 26, the government released its report "Assessment of the Past Governmental Policies concerning the Asbestos Issue," which was based on reports submitted by several ministries (Ministerial Meeting on Asbestos Issues 2005). The report recapitulated past and present policies and the results of interviews and surveys of ex-officials. Also, comparisons were made with policies overseas. The relevant agencies, most notably the MHLW (formerly the Ministry of Health and

Welfare plus the Ministry of Labor), the MITI/ METI, and the MLIT, defended their positions, stating that they had taken the recommended and necessary steps and complied with international standards. The EA added that there was no demonstrable threat to the general environment, noted that the precautionary principle had not been widely accepted until recently, and confirmed that the agency had committed itself to monitoring the areas in close proximity to the factories, as indicated by the thrust of the end-of-pipe regulation. A month later, on September 29, an additional report was approved by the cabinet panel. The MHLW revised the Occupational Health and Safety Law and banned the manufacture, import, transfer, provision, and use of all the asbestos-containing products by 2006, with just a few exceptions (e.g., joint-sealing materials used in petrochemical plants).

On February 3, 2006, the Relief Act for Health Victims of Asbestos was established (drafted in November 2005, endorsed in January 2006 by the cabinet). The Environmental Restoration and Conservation Agency (ERCA) created the Relief Fund for Asbestos Victims, to which the national and local governments, companies with workmen's compensation insurance, and ship owners annually contribute. A separate ordinance requires contributions from companies that have consumed large amounts of asbestos and/or have had certain rates of asbestos-related diseases among their employees. Upon notification, the ERCA provides money from the Fund to help victims and their families with medical care expenditures, funeral costs, and condolence money.

The Act went into effect on April 1, 2007. It enables the families and neighbors of factory workers, as well as self-employed workers, to apply for injury compensation. A worker's descendants can also apply provided that they could have had done so within 5 years after the worker's death. The governments are required to contribute to the Fund because many asbestos-using companies have gone out of business or gone bankrupt, not because of government negligence. It was urged as the government's duty that the public, physicians, and employers be informed about the program, as many eligible workers had not been told that they were eligible to apply for compensation. (Less than 5% of mesothelioma deaths between 1995 and 2003 were covered by occupational compensation insurance.)

In February 2006, following the recommendation of the expert panel, it was decided that the products exempt from the 2005 ban would be banned by 2008. However, construction materials with an asbestos content of less than 1% remained permissible. The MHLW revised its official standards to compensate for the occupational injuries caused by asbestos. Compensation was also provided for those diagnosed with mesothelioma (no other condition required), lung cancer (with the presence of a certain amount of asbestos fibers and bodies pathologically proven), and diffuse pleural thickening. Six months later, the Ministry announced programs to provide the public with asbestos-related information and consultation services, substantiate the prevention and management of occupational exposures, and initiate education for physicians.

In April 2007, at the annual conference of the JSOH, mesothelioma was reported in three ex-teachers with no clear history of asbestos exposure other than working in school buildings.

Discussion

Bureaucratic Incrementalism and Crisis Politics

The selection of policy issues for the agenda, as well as the development, selection, implementation, and appraisal of policies, is not always rational. Decisions about which issues to address are not always based entirely on scientific, comprehensive judgments about their importance as social issues. Likewise, solutions are not necessarily adopted exclusively on the basis of their prospective technical effectiveness and efficiency.

According to classical linear policy making models, the ideal is for the socially more important issues to be selected, the most effective and efficient policies to be devised, adopted, and implemented. These policies then must be properly appraised, and perhaps adjusted or terminated, by technically competent bureaucrats (Anderson 1996). However, the results of several studies suggested that we cannot expect this level of competence from any policy maker. In the real world, rationality is limited (bounded). The abilities of policy makers can be restricted because of insufficient time, energy, and information, and/or because the political leadership cannot always be counted on to mobilize and support them (Simon 1991). Furthermore, individuals and social groups differ in their perceptions of issues, and politically powerful special interests might even oppose government involvement with an issue, as well as its authority to (re-)allocate the costs and benefits associated with its actions.

The problem is even more complicated when scientific knowledge concerning the potential, latent, and emerging hazards (such as asbestos), including their causal interrelationships, as well as the feasibility and effectiveness of possible policies, is incomplete (or not "fully" developed). Searching for complete knowledge is, in most cases, too much of a burden or simply impractical. The consideration or reconsideration of policy issues is necessarily punctuated.[1] In such cases, therefore, satisficing might be the most realistic alternative. By taking this shortcut, aspirations can be adjusted and the search for alternatives ended as soon as some possible policy action is encountered that exceeds one's aspiration level (Simon 1983).

In Japan, bureaucrats, in cooperation with expert panel members, have occasionally searched for domestic and foreign scientific reports and standard policy guidelines related to previous and current asbestos regulations. In one way or another, the information derived therefrom has led to the revision of official regulations. Permissible exposure limits have been lowered time and again. The JSOH, as an

[1]Argumentation is the key process through which citizens and policymakers arrive at moral judgments and policy choices. The argument is the link that connects data and information with the conclusions of an analytic study. However, the structure of an analytic argument is typically a complex blend of factual propositions, logical deductions, evaluations, and recommendations. In this context, evidence, when used in the argumentative process, is intrinsically instrumental. In other words, because perfect data are impossible, the standards of acceptance must be based on craft judgments of what is good enough for the functions the data perform in a particular problem. Evidence is information selected from the available stock and introduced at a specific point in an argument to persuade the mind that a given factual proposition is true or false (Majone 1989).

academic society, was sometimes consulted by bureaucratic policy makers, who, in turn, drove the regulatory changes. International conventions, guidelines, and policy trends certainly alerted both agency officials and scientists to the need for a reconsideration of the appropriateness of existing policies. They then had to justify their actions in scientific terms. If this justification was made in a clear and manifest way, it could have revealed how much of a health risk was assumed as acceptable.

Thus, bureaucrats have continually sought technical rationality, but for a long time they failed to exert sufficient effort to detect the latent health risks of asbestos to non-workers, or even to establish an information system to check the trends. Bureaucrats' decision making was inevitably dependent on their environment, including their jurisdiction, motivation, the availability of information, support, and other resources, and their incremental decision making – incremental in terms of its accordance with scientific information updates, and international regulatory trends, and of changing policy measures adopted – had to be a satisficing one (Byron 2004; Simon 1983).

The EA justified its inaction through an audit, stating that its authority had long been confined to issues that could be handled by the "end-of-pipe regulation," and that there had been little awareness or acceptance of the "precautionary principle" by either policy makers or the society at large. The MHLW (formerly the MHW) defended itself by saying that it had taken the necessary measures based on their scientific assessments, and that the Ministry could not move faster without clear evidence of real health hazards. Bureaucratic turf wars and competition also hindered swift and concerted government action.

As a result, the "precautionary principle," which was included in the 1992 Rio Declaration on Environment and Development to justify early efforts at prevention even without full scientific certainty about the risks, has never been the core ideology driving asbestos regulation. In Japan, the Basic Environmental Plan of 1994, enacted after the Basic Environmental Law of 1993, incorporated the concept behind the precautionary principle, but this concept was never applied. During the discussion on the asbestos issue, the idea of precaution was only alluded to, although it was stated that it would be desirable to decrease (or at least, not to increase) the total amount of asbestos in society to prevent future health hazards.

This incremental policy making was interwoven with policy changes resulting from political initiative. Public concern about asbestos hazards was stirred by media reports on asbestos use in public schools in the 1980s. The media were obviously aware of the former social and political debates on asbestos in the US, and their attention was aroused again in 2005 by the corporation's disclosure of an increase in the incidence of cancer in its neighborhood. This disclosure also put asbestos on the agenda of the Diet. In the 1980s, the debate in the Diet mobilized various administrative agencies, e.g., the MHW and the MEd, to act; in the 2000s, they led to the enactment of the "Relief Act."

This alternating mode of policy making has been modeled by Grindle (1991). Differences in how issues are perceived have major impact on the process and possibly even the results of policy making. The perception of either a crisis or politics-as-usual alters the dynamics of decision making, because it affects the political stakes for policy makers, altering their identity and their place in the hierarchy as well as changing their time constraints.[2] Whereas policy changes in the context of politics-as-usual are handled mainly with narrow clienteles or by taking into account "micro-political" relations within the bureaucracy, crises induce policy makers to consider "macro-political" effects (e.g., the impact upon their legitimacy and authority) and therefore to make a greater commitment to resolve the issues. Policy makers usually sense that their capacities are limited. They are constrained by societal pressures and interests, lack of political stability and support, sociohistorical factors, and bureaucratic incapacity. However, they usually continue to occupy a broad policy space, which allows them to exercise discretion in policy making, and/or gives them the ability to maneuver the policy space with the help of the technical, economic, political, and bureaucratic resources available to them.

A rise in public concern signifies a kind of political crisis. Failing to make deliberate responses can be detrimental to the legitimacy of elected officials as well as administrative agencies. When the problem was limited to occupational hazards, bureaucratic policy making could take place: the risks were implicitly accepted, although it began to be argued that they had no threshold limit. When the problem extended to the health of the general public, on the other hand, efforts to reduce public concern were called for, even if that meant employing the precautionary principle. Especially in the 2000s, a large policy window allowed policy makers, highly inspired and mobilized by the threat to their legitimacy, to achieve major policy changes. Political mobilization helped the MHLW and the EA override bureaucratic turf wars and overcome resistance by the industry. Consequently, a new paradigm of legislative enactment was created that involved innovative methods of financing, led by the Cabinet Office.

From this vantage point, both modes of policy making required a long time to introduce policies to detect and prevent the health hazards caused by asbestos exposure and to compensate the victims. Although health checkups for workers had been introduced, they did not seem to be entirely successful, especially because the disease registry for mesothelioma in Japan was established quite late (the late 1990s). Surveillance of workers' families, people living near asbestos-handling factories, and the general population were never complete. Consequently, policy makers had to rely on a weak information base. Furthermore, bureaucrats and scientists had to wait for "sufficient" scientific evidence and/or authoritative recommendations from international organizations to guarantee their legitimacy.

On the other hand, politicians could act only when very large policy windows opened. Both in the 1980s and the 2000s, the media pushed asbestos onto the

---

[2]When dominated by a single interest, the subsystem is best thought of as a policy monopoly. A policy monopoly has a definable institutional structure responsible for policy making in an issue domain, and its responsibility is supported by some powerful idea or image. Punctuated-equilibrium theory includes periods of equilibrium or near-stasis, when an issue is captured by a subsystem, and periods of disequilibrium, when the issue is forced onto the macropolitical agenda. When issues (areas) are on the macropolitical agenda, small changes in the objective circumstances can cause large changes in policy (True et al. 1999). When a policy shifts to the macropolitical institutions for serial processing, it generally does so in an environment of changing issue definitions and heightened attentiveness by the media and the broader public (Jones 1994).

agenda. By expanding the circle of asbestos victims from workers, a distinctive and limited population, to a broader population including school children, former patients, the workers' families, and other people living near the factories, they made the problem worthy of political attention. Policy precedents were already available from abroad: laws on asbestos abatement in school buildings had been enacted in the US before 1980, and product bans had been gradually introduced in Europe before 2000. In Japan, a disease registry for mesothelioma that was established in 1997 disclosed a remarkably high incidence of asbestos poisoning among workers. As the garbage-can model suggests, the convergence of these factors finally bore fruit a decade later in the Relief Act of 2006 (Kingdon 1995).

Policies for Safety and Reassuring the Public

Cases of para-occupational (familial) exposure have been reported since the 1970s. Numerous persons who have been exposed to asbestos fibers as a result of living in the same household as an asbestos-exposed worker have contracted mesothelioma (Anderson et al. 1979; Kilburn et al. 1985). Furthermore, abnormally high disease rates have been associated with people who lived near asbestos mines and factories in the 1970s and the 1980s. However, the scientific evidence on the true health hazards of environmental exposure to asbestos in the general population is not convincing. In general, fiber concentrations in the atmosphere seem to be on the order of 1,000 times less than in the typical working environment. The EEC report concludes that air concentrations of this magnitude do represent a sufficient safety margin (Zielhuis 1977).

In the decades since then, the available data have not indicated that malignancies or functional impairments result from exposure to most airborne concentrations of asbestos in buildings. The models used to estimate cancer risk assume no particular threshold for cancer and suggest that any exposure at all is carcinogenic. In the absence of information on the risks of low-level asbestos exposure, the epidemiological findings from occupational studies have been used to extrapolate the risk in the general environment. The asbestos levels measured in indoor air cast some doubt on whether asbestos-containing materials make a significant contribution to the asbestos levels occupants are actually exposed to, even when some of the material is friable or in bad condition (Whysner et al. 1994)

Nonetheless, asbestos easily engenders fear and even panic in society. Asbestos has many attributes, called "outrage factors," which sometimes cause media hype and defy rational calculations of risks and their acceptance (Covello et al. 1989; Sandman 1993). Asbestos exposure is neither voluntary nor natural. It can result in dreadful cancers. Distribution of the benefits and costs, concerning the use of asbestos-containing products, is uneven across society. Political theories also suggest that clear causal stories, namely, the explanations of events that the causes are thought to be purposeful human actions, are likely to induce political mobilization (Stone 1998).

This was the case in Japan. When asbestos became a big social issue in the 1980s and 2000s, it posed a political challenge. In the former case, the policy clientele expanded from a chosen occupational group to the general public, who were characterized as the guardians of children, and the issue was pushed onto the political agenda. Elected officials and bureaucrats tried to soothe public (and, of course, parental) anxiety by introducing asbestos abatement programs for public buildings. In the latter case, the political spectacle itself was triggered by the company's disclosure in anticipation of possible damage to its corporate reputation. Other companies followed suit, and the resulting increase in media hype, which was considerable beforehand, drove legislators to act.[3]

This remarkable increase in public concern, which at its height could be considered a panic, indicates the presence of a social problem to be solved, whether it was real or a phantom. This panic had the potential to endanger social stability and cause additional problems. Furthermore, panic can result in the inefficient use of limited resources, as it can impair rational decision making and efficient implementation of programs to resolve the problem. Additional programs may also be necessary to soothe public anxiety, even though this anxiety is sometimes disproportional to the actual threat as determined by technical judgments. Thus, the persistence of a social panic is a reflection of government incompetence, and therefore it is a genuine threat to the political legitimacy of policy makers, such as legislators and bureaucrats, as well as to the social and corporate legitimacy of organized interests such as the asbestos industry. In such a case, all the parties must make a commitment to extinguish the fire, to protect themselves by fending off public criticism.

This was also the case in the US in the 1980s. The recognition that asbestos-containing materials had been used in schools, governmental offices, and hospitals generated public concern about children's health. The resulting political mobilization led to the enactment of the Asbestos Hazard Emergency Response Act of 1986, a mandate from the EPA that requires inspection of the nation's public and private schools for asbestos. Around the time of the legislative debate, the Environmental Health and Safety Council of the American Health Foundation discovered that even complete removal of asbestos from all the buildings would provide no measurable benefits for public health (Whysner et al. 1994). The testimony at an EPA public hearing indicated that the claim of a huge risk to school children was not supported by the scientific evidence (Jones et al. 1996). These scientific arguments were overwhelmed by political considerations.

In Japan, the ban on asbestos products in the 2000s was justified by the argument that it is important to keep possible exposure levels everywhere as low as possible by stopping the production of asbestos so its use would not increase. Major policy changes, including the ban on asbestos use as a precautionary (preventive) measure,

[3]The increasing recourse to individual litigation against environmental risks poses real dangers for companies that previously might have had to contend only with the demands of state regulation. The dominance of corporate, professional, and governmental defendants in the use of civil damage judgments as a policy instrument suggests that "toxic tort" is an extremely inefficient and inequitable approach to environmental regulation (Galandter 1994).

were possible only when there was sufficient political leadership. At the insistence of the MHLW, the legislators in the Diet backed a total ban on asbestos, overriding industry resistance. This action certainly increased the symbolic appeal of the policy to the public by creating the impression that the government was taking decisive measures to forestall further hazards, and that domestic policy was catching up with the policies of other countries and the recommendations of international organizations. However, it is not clear how much a "complete" ban can really contribute to decreasing the hazards. Attempts to deal with concerns about the health risks associated with asbestos exposure have been based on a mixture of scientific knowledge and social judgments, and activism has played an increasingly large role as well.

Thus, the government's policies were intended not only to attain certain technical objectives (i.e., practical solutions to social issues, which were now recognized as worthy of attention), but also to accomplish political objectives (i.e., increasing political pressure, earning credits, and protecting legitimacy). Consequently, the policies in both periods were a hodgepodge of rational, scientifically justified decisions and unscientific palliatives intended to provide symbolic assurance (Edelman 1967).

A number of issues remain. They include the consistency of the policy debates, how scientific evidence is to be employed (including judgments on the validity and clarity of the available information), the degree of hazard and the socially acceptable margin of safety, the costs and benefits for both individuals and society, and, finally, how the above considerations can be integrated (Majone 1985). If scientific arguments were to be employed to soothe public anxiety and obscure the extent of the real hazards, it might be difficult to scientifically justify the introduction of arguably mitigating measures on other occasions, unless sufficiently persuasive new evidence is presented. In other words, if scientific arguments were to be used to justify policy measures adopted for public reassurance but lacking a rigorous scientific foundation (including an appeal to the precautionary principle), it would be hard to find support for changes in the program without sufficient new scientific evidence on product safety. This would be true even if the revised policies seemed to be more rational in the sense that they seemed to be based on a scientific rationale. The role of science in justifying policy should be carefully revised and a new public image adopted.

An important question from the perspective of democracy is whether efforts have been made to build a social consensus thorough deliberation (Sato & Akabayashi 2005). A technical search for a consistently rational policy can be accomplished by using, for example, a cost-benefit analysis. Other methods, such as opinion polls and public hearings, can also be employed. In the case of asbestos, there were no such efforts until the last stage of product regulation, hazard mitigation, and victim compensation, namely, the debates concerning the Rescue Act of 2006.

The Nature of Policy: Styles of Regulation

Policy and politics set the goals of the government action (and inaction), allocating their costs and benefits to different social groups and individuals (Montgomery & Rondinelli 1995). In some cases, the government imposes its policies on industry to promote public welfare. In other cases, the organized interests effectively negotiate with the government to alter policy, thereby influencing the costs and benefits to themselves. There are many cases, from economic and industrial policies to health and environmental policies, in which it is difficult to judge if the regulator led the regulation or vice versa, in terms of either the power relationships involved in determining a policy or the consequences of implementing the policy. Samuels (1987) explains that policy making based on mutual consensus between the state and organized private interests, a procedure that has been widely adopted in Japan, reflects no domination by either party.

In regulating asbestos, either its usage per se or its presence in other products, the Japanese government occasionally enacted regulations after the industry had acted voluntarily and/or ceased to produce and/or use asbestos products. The government also enacted policies to establish the pace of government intervention, sometimes after consultation with the industry. As was reported elsewhere (Bernstein 1995), it can be inferred that the industry's voluntary actions helped to postpone or even fend off official government action. Through their respective actions, the industry and government decreased the adverse effects of regulation. For example, the asbestos manufacturers were able to exhaust their stocks of asbestos and switch to substitutes well before the use of asbestos became officially prohibited. Moreover, the industry could avert possible lawsuits and the government could avoid criticism of its negligence or inaction. Sanctions were rarely required or imposed. Consequently, a large amount of discretion was allowed to the industry in terms of how quickly it must take action, make changes, or comply with regulations. The effect was to reduce the economic burden of the regulations.

When considering the legitimacy of regulatory interventions in a market economy in a democratic political system, it is sometimes important to examine the role of the government in overcoming (possible) market failures, such as inadequate information and externality, and to determine if the policies are likely to promote the political and economic equity of individuals (Krugman & Wells 2006). The long latency period for the appearance of asbestos-related symptoms, and the dispersed, unorganized, and resource-poor status of factory workers, their families, and nearby residents, not to mention the end-users of asbestos-containing products, were the obvious causes of these market failures.

In retrospect, it appears that the past policies in Japan were intended either to inform asbestos users (e.g., manufacturers and construction companies) and their employees about the need to change their behavior or to inform local governments about the need for them to provide oversight. Later, it was argued, these mandates were extended to the managers of public buildings, such as schools, government offices, and hospitals. Except for targeting the people in those settings, however, the policies have never effectively conveyed the possible health hazards of asbestos. In addition to workers' families and those living near the factories, the general public as end-users have never been properly warned about the hazards of asbestos, either for themselves or for the environment. Nor have they participated in the policy-making process as informed and competent actors. In summary, there were no clear efforts to communicate with or empower the public to make informed choices about asbestos-containing products, both at the time specific products were selected and the time the corresponding regulations were adopted.

Later, the Rescue Act of 2006 at least partially transferred the costs for compensating people with asbestos-related diseases from the industry (manufacturers and heavy users) to the government, but it did not clearly assign liability. It is arguable that all the individuals and groups in society, including the end-users of asbestos products, in one way or another benefited from the use of these products. It is evident, however, that the economic benefits of this usage were concentrated more in the industry than among the end-users, and that the former had better information about asbestos hazards than the latter. Because of the large gap between the time the products were marketed and the time their hazardous nature was recognized, as well as the fact that some companies had gone out of business, it was practically impossible to track all the transactions. As a result, the industry was eventually exempted from full responsibility. The inability to redress the market failures helped to perpetuate and exacerbate similar problems related to moral hazard.

International Trends and Domestic Policies

In any case, the sentinel case reports did not promptly alert policy makers and society to the need to adopt policies that would effectively prevent an epidemic and at the same time compensate all the victims.

In European countries and the US, it took 20–30 years after the sentinel cases were reported for epidemiological studies establishing the causal link between asbestos and diseases to accumulate.[4] It took an additional 10–30 years for preventive measures (such as ventilation) to be adopted and compensation programs to

[4]There was a 20-year gap between the first identification of asbestos as the cause of fibrotic disease and the general acceptance of asbestos dust as a health hazard. A causal link between exposure to asbestos and asbestosis with lung cancer was suspected in the 1930s and 1940s, but it was not established until the mid-1950s. The relationship between exposure to asbestos dust and the development of malignant mesothelioma was demonstrated in 1960. To complicate the matter, the latency period for cancer was longer than for asbestosis, and it was found that lower levels of asbestos dust could cause cancer. A latency of 20 or more years from onset of exposure to diagnosis has also been shown for asbestos-related lung cancer.

The first rigorous epidemiological study of asbestos poisoning appeared in 1955 when Sir Richard Doll (1955) documented a tenfold increase in the risk of lung cancer in a group of men employed 20 or more years at an asbestos textile plant in northern England. The association of mesothelioma with asbestos exposure was first described in 1960 in the northwest Cape area of South Africa, where long, thin crocidolite fibers were mined. The seminal paper published by Peto in *The Lancet* about mesothelioma mortality in Great Britain raised public awareness of the hazards of asbestos exposure throughout Europe. By 1973, the full scope of the asbestos danger was apparent (Hammons & Huff 1974; Huff et al. 1974).

become institutionalized by law. There was a 10–35 years gap between the publication of the sentinel reports on lung cancer and mesothelioma and the enactment of legislation banning the manufacture of asbestos products (e.g., crocidolite) and their use (e.g., spraying). It also took more than 30 years for scientific knowledge about the non-occupational hazards of asbestos to be translated into clear policy action (i.e., a total ban on both occupational and non-occupational asbestos use). It remains for society to address two issues related to the adverse effects of asbestos on health: (1) how to deal equitably with existing diseases and (2) what safeguards should be implemented to prevent the continuation of asbestos-associated health effects (Galandter 1994). In many countries, the last step in the process, a product ban, is hardly enough to overcome the criticism, as the widespread use of abundant amounts of asbestos was already in play.

Asbestos consumption in Japan tended to lag behind that in many European countries and the US by 20–30 years. Whereas heavy consumption of asbestos started in the late 1940s in the EU countries and the US, it started in the late 1960s in Japan. Correspondingly, the sentinel case reports on asbestosis, lung cancer, and mesothelioma appeared somewhat later in Japan, again lagging behind the EU and the US (asbestosis: 1906–1918 overseas and 1929 in Japan; lung cancer: 1935–1940s overseas and 1960 in Japan; mesothelioma: 1930s–1960 overseas and 1970s in Japan). In Japan, the epidemiological studies, which necessarily depended on the accumulation of cases even if the motivation was there to conduct them beforehand, followed comparable studies overseas by about 10–30 years. When a clear relationship was reported between mesothelioma incidence/mortality and the preceding per-capita asbestos consumption rate (Takahashi et al. 1999), there was already, without question, a huge amount of asbestos consumption.

In retrospect, scientific knowledge about asbestos-related health hazards and the idea of policy alternatives had certainly spread beyond national borders. Scientists and the bureaucrats in charge of the issue were well aware of these international trends. As a result, the introduction of preventive measures in workplaces, such as the requirement of air ventilation and the ban on spraying, took place almost simultaneously (i.e., the ban on spraying was implemented in the 1970s, both overseas and in Japan). A ban on crocidolite and amosite was introduced in many countries between the late 1980s and the early 1990s. A complete ban on crocidolite products in the UK was implemented in 1986, and a ban on the principal asbestos-containing products was introduced in Germany, France, and Japan in 1986, 1988, and 1995, respectively (the discontinuation of asbestos use in Japan was confirmed in 1986). Many of these actions took place several years after the ILO recommendations of 1986. The regulation of amosite-containing products varied from country to country. The UK adopted a total ban on amosite in 1986. The corresponding WHO recommendation came in 1989, and the ban was introduced in Germany, France, and Japan in 1993, 1994, and 1995, respectively. The US and Canada have not banned amosite.

As a result of delayed history of asbestos consumption, Japan could have benefited from the already accumulated scientific evidence and the experiences of other countries. However, because heavy use of asbestos in Japan was not deterred, the potential advantage of this history for preventing asbestos-related health hazards was not fully exploited. In particular, Japan was about a decade behind (2006–2008) in introducing its primary ban on asbestos use; many EU countries abandoned it in the 1990s. In Japan, most of the asbestos in the latter years was used for construction materials, although it was eventually proved not to be indispensable. Historians have addressed both the social and the political determinants of environmental and occupational safety and health regulations (Corn 1992; Rosner & Markowitz 1991). According to Corn and Corn (1995), broad social and political issues arise in part because of the uncertainties and complexities involved in scientific measurement, the extent of the risk, and the degree of control sought.

Absent a record of manifest value judgments, it cannot be determined whether the prolonged and loosely regulated use of asbestos resulted from the implicit decision to accept the possible and emerging health hazards resulting from its use (and the necessary development costs for the state and/or the industry) or from the incapacity of the policy makers (and society at large), given that the required intelligence and rationality could not escape the limitations imposed by time. Contingency plans about possible courses of actions and possible outcomes have never been discussed in public.

Conclusions

Incremental bureaucratic policymaking has dominated past asbestos regulation in Japan. When the issue appeared on the public agenda, accompanied by media hype, elected officials were mobilized to exercise leadership. Bureaucratic officials continually sought a scientific basis for regulation. Because of incomplete data and unresolved scientific issues, the scientific evidence provided at best a weak background for conclusive policy debates. Politicians, on the other hand, adopted policies to reassure the public, even when its anxiety was not rational and the necessity and effectiveness of such measures was not scientifically proven. Thus, past asbestos regulations were intended to provide either effective technical solutions or symbolic assurances, depending on the current political environment. These alternate policy objectives were never made explicit or communicated to the public.

Although the technical solutions implicitly accepted a certain degree of hazard, the intent being to reduce the hazard to an acceptable level rather than to completely overcome it, the political solutions reflected the more ambitious goal of eliminating the hazard entirely. They were adopted, in response to the public pressure to demonstrate its political commitment. On such occasions, politicians employed scientific arguments, more or less as a tactic to convince the public of the desirability of their decision as preferable precautionary measures. The advent of the precautionary principle did not per se induce legislation to completely ban asbestos products. Neither precautionary nor preventive measures are possible unless sufficient political leadership is exercised. Precautionary arguments were employed instrumentally to justify their decisions.

Besides the difficulty in scientifically assessing the health hazards, in retrospect it has proven difficult to regulate the use of asbestos, especially since its use had been widespread, and it existed for a long time before the agenda was set. Until the final push by the health ministry, the industry was reluctant to replace asbestos with substitutes in producing its products. Although a series of policies were implemented to improve the working conditions at asbestos-handling sites, those efforts did little to enlighten the public; the biased information about asbestos risks was never substantially addressed, which adversely affected the ability of the end-users to make rational decisions.

When a political spectacle is created, a window opens for policy change. Though constrained by the political environment, this window can provide a valuable opportunity to adopt policy measures that could not be passed otherwise. They can be either palliatives employed to reassure the public or genuine preventive measures, even precautionary ones. It can also be an opportunity for players other than policy makers, such as the mass media, organized private interests and other non-governmental organizations, and individuals, to create a public forum to promote understanding of the issues and to deliberate on a rational approach to risk management.

References

Aberbach, J., Putnam, R., & Rockman, B. (1981). *Bureaucrats and politicians in western democracies*. Cambridge, MA: Harvard University Press.

Anderson, H. A., Lilis, R., Daum, S. M., & Selikoff, I. J. (1979). Asbestosis among household contacts of asbestos factory workers. *Annals of New York Academy of Science, 330*, 387–399.

Anderson, J. (1996). *Public policy making: An introduction*. Boston, MA: Houghton Mifflin College Division.

Baldwin, C. A., Beaulieu, H. J., Buchan, R. M., & Johnson, H. H. (1982). Asbestos in Colorado schools. *Public Health Report, 97*(4), 325–331.

Bernstein, M. H. (1955). *Regulating business by independent commission*. Princeton, NJ: Princeton University Press.

Boden, L. I., Miyares, J. R., & Ozonoff, D. (1988). Science and persuasion: environmental disease in U.S. courts. *Social Science & Medicine, 27*(10), 1019–1029.

Byron, M. (2004). *Satisficing and maximizing: Moral theorists on practical reason*. Cambridge, UK: Cambridge University Press.

Cook, W. A. (1942). The occupational disease hazard: evaluation in the field. *American Industrial Hygiene Association Quarterly, 3*(2), 193–197.

Cooke, W.E. (1924). Fibrosis of the lungs due to the inhalation of asbestos dust. *British Medical Journal, 2*, 147–152.

Corn, J. (1992). *Response to occupational health hazards: a historical perspective*. New York: Van Nostrand Reinhold.

Corn, J. K., & Corn, M. (1995). Changing approaches to assessment of environmental inhalation risk: a case study. *The Milbank Quarterly, 73*(1), 97–119.

Covello, V. T., McCallum, D. B., & Pavlova, M. T. (1989). *Effective risk communication: the role and responsibility of government and nongovernment organizations*. New York, NY: Plenum Press.

Doll, R. (1955). Mortality from lung cancer in asbestos workers. *British Medical Journal, 12*: 81–86.

Ebihara, I. (1981). Asbestos-related-respiratory impairment in Japan: From occupational exposure to exposure at home – expanding pollusion. *Journal of Science of Labor [Roudou Kagaku], 57*(8), 363–396.

Ebihara, I., Fujii, M., & Kawami, M. (1999). Asbestos-related occupational diseases among construction workers in Japan. *Journal of Science of Labor [Roudou Kagaku], 75*(3), 87–114.

Edelman, M. (1967). *The symbolic uses of politics*. Champaign, IL: University of Illinois Press.

Environmental Agency. (1987). *Sekimen, zeolite no subete* [All about asbestos and zeilite]. Kawasaki, Japan: Nihon Kankyo Ensei Center.

Furuya, S. (1993). Asbestos issues in Japan: working environment and pollution problems, number 4. Tokyo: JOSHRC.

Furuya, S. (2001). Ban Asbestos Network Japan (BANJAN) and Japan's Citizen's Network for Wiping Out Asbestos (ASNET). Annals of the Global Asbestos Congress (CD), 3, 2. London: International Ban Asbestos Secretariat (IBAS).

Galandter, M. (1994). The transnational traffic in legal remedies. In: S. Jasanoff (Ed.), *Learning from disaster: Risk management after Bhopal* (pp. 133–157). Philadelphia: University of Pennsylvania Press.

Gigerenzer, G., & Selten, R. (2002). *Bounded rationality*.Cambridge: The MIT Press.

Greenberg, M. (1999). A study of lung cancer mortality in asbestos workers: Doll, 1955. *American Journal of Industrial Medicine, 36*(3), 33–347.

Grindle, M. S. (1991). The new political economy: Positive economics and negative politics. In G. M. Meier (Ed.), *Politics and policy making in developing countries: Perspectives on the new political economy* (pp. 41–67). San Francisco: ICS Press.

Gunji, N. (1987). Sekimen ni yoru kenko shogai no yobo ni kansuru hourei oyobi kanren-tsutatsu ni tuite. *Kuuki Seijo, 24*(5), 2–13.

Hammons, A. S., Huff, J. E. (1974). Asbestos: World concern, involvement and culpability. *International Journal of Environmental Studies, 6*(4), 247–251.

Heifetz, R. A. (1994). *Leadership without easy answers*. Cambridge, MA: Belknap Press.

Higashi, T., & Tsuchiya, K. (1991). Asbestos research and regulation in Japan. *International Asbestos Medical Research, 6*, 207–262.

Horai, Y. (1959). Clinical study on asbestosis. *Hai (Tokyo), 6*, 294–304.

Horai, Y., Tsujimoto, H., Uejima, R., & Sano, H. (1957). Study on asbestos lung. *Journal of Nara Medical Association, 9*, 48–57.

Huff, J. E., Dinger, C. Y., Kline, B. W., Whitfield, B. L., & Hammons, A. S. (1974). A health view of asbestos: an annotated literature collection – 1960–1974. *Environmental Health Perspectives, 9*:341–462.

International Agency for Research on Cancer. (1977). *Asbestos: IARC summary and evaluation, volume 14*. Lyon: IARC.

International Agency for Research on Cancer (1982). *Monographs on the evaluation on the carcinogenic risk of chemicals to humans: Supplement 4*. Lyon: IARC.

International Agency for Research on Cancer. (1987). *Asbestos: Actinolite, amosite, anthophyllite, chrysotile, crocidolite, tremolite*. Lyon: IARC. Retrieved February 28, 2009, from http://www.inchem.org/documents/iarc/suppl7/asbestos.html

International Labor Organization. (1964). *Convention concerning benefits in the case of employment injury* (C121). Retrieved February 18, 2009, from http://www.ilo.org/ilolex/english/convdisp1.htm

International Labor Organization. (1986a). *Convention concerning safety in the use of asbestos* (C162). Retrieved February 18, 2009, from http://www.ilo.org/ilolex/english/convdisp1.htm

International Labor Organization. (1986b). *Recommendation concerning safety in the use of asbestos* (R172). Retrieved February 18, 2009, from http://www.ilo.org/ilolex/english/recdisp1.htm

Ishii, Y. (2002). Acknowledgment for lung cancer induced by asbestos. In: K. Morinaga (Ed.), *Occupational asbestos exposure and asbestos-related diseases* (pp. 24–245). Tokyo: Sanshin-tosho.

Jones, K. (1994). *The making of social policy in Britain 1930–1990: Second edition*. London: Athlone Press.

Jones, R. N., Hughes, J. M., & Weill, H. (1996). Asbestos exposure, asbestosis, and asbestos-attributable lung cancer. *Thorax, 51*(suppl 2), S9–S15.

Katayama, S. (1993). Asbestos exposure and cancer development. *Iryo, 47*, 661–666.

Kawami, M., Ebihara, I., Kawami, M., & Shinokawa, E. (1991). Asbestos exposure and occupational background: Evidence from asbestos fibre and ferruginous body concentrations in the lungs. *Journal of Science of Labor [Roudou Kagaku], 67*(10), 469–480

Kazan-Allen, L. (2003). The asbestos war. *International Journal of Occupational Environmental Health, 9*(3), 173–193.

Kikuchi, K., & Hiraga, Y. (1983). Clinical and epidemiologic investigation of asbestos exposed workers. *Sapporo Medical Journal, 52*(6), 599–612.

Kilburn, K. H., Lilis, R., Anderson, H. A., Boylen, C. T., Einstein, H. E., Johnson, S. J., & Warshaw, R. (1985). Asbestos disease in family contacts of shipyard workers. *American Journal of Public Health, 75*(6), 615–617.

Kingdon, J. W. (1995). *Agendas, alternatives, and public policies: Second edition*. Boston, MA: Little Brown and Company.

Kishimoto, T., Ohnishi, K., & Saito, Y. (2003). Clinical study of asbestos-related lung cancer. *Industrial Health, 41*, 94–100.

Kishimoto, T., & Okada, K. (1987). The relationship between lung cancer and asbestos exposure. *Chest 94*, 486–490.

Kishimoto, T., Okada, K., Sato, T., Ono, T., & Ito, H. (1989). Evaluation of the pleural malignant mesothelioma patients with the relation of asbestos exposure. *Environmental Research 48*, 42–48.

Kishizuchi, K., Takeshima, Y., Kitaguchi, S., Nishida, T., & Inai, K. (1996). Comparison of asbestos fibers deposited in lung tissue between pleural plaque cases and malignant mesothelioma cases. *Pathology and Clinical Medicine, 14*, 783–790.

Koike, S. (1992). Health effects of non-occupational exposure to asbestos. *Japanese Journal of Industrial Medicine, 34*, 205–215.

Kohyama, N. (1989). Airborne asbestos levels in non-occupational environments in Japan. *IARC Science Publication, 90*, 262–272.

Konami, S., Uematsu, M., Okamoto, S., Kameda, Y., Kawai H., & Harada T. (1974). Clinicopathological study on the asbestos bodies found in the 830 autopsy cases. *Haigan (Tokyo), 14*, 320–338.

Krugman, P. & Wells, R. (2006). *Economics*. New York, NY: Worth Publishers.

LaDou, J. (2004). The asbestos cancer epidemic. *Environmental Health Perspectives, 112*(3), 285–290.

Majone, G. (1985). *Guidance, control and evaluation in the public sector*. Berlin: Walter De Gruyter & Company.

Majone, G. (1989). *Evidence, argument, and persuasion in the policy process*. New Haven, CT: Yale University Press.

Ministerial Meeting on Asbestos Issues. (2005). Records of the ministerial meetings on asbestos issue. Tokyo: Office of Prime Minister. Retrieved February 28, 2009, from http://www.kantei.go.jp/jp/singi/asbestos/index.html

Ministry of Economy, Trade and Industry. *Asbestos kanren* [Information on asbestos]. Retrieved (last access), March 10, 2009, from http://www.meti.go.jp/press

Ministry of Education, Ministry of Education, Culture, Sports, Science and Technology. *Asbestos taisaku eno torikumi* [Actions for asbestos issues]. Retrieved (last access), March 10, 2009, from http://www.mext.go.jp/submenu/05101301.htm

Ministry of Internal Affairs and Communications. *Asbestos mondai eno taiou ni tsuite* [Policies for asbestos issues]. Retrieved (last access), March 10, 2009, from http://www.soumu.go.jp/menu_04/asbest/index.html

Ministry of the Environment. Sekimen/asbestos mondai eno torikumi [Information on asbestos issue and related policies]. Retrieved (last access), March 10, 2009, from http://www.env.go.jp/air/asbestos/index_link.html

Ministry of Health, Labor and Welfare. *Asbestos/sekimen joho* [Information on asbestos]. Retrieved (last access), March 10, 2009, from http://www.mhlw.go.jp/new-info/kobetu/roudou/sekimen/other.html

Ministry of Health and Welfare. (1989). *Tap water and asbestos*. Tokyo: Nihon Suido Kyokai.

Ministry of Land, Infrastructure, Transport and Tourism. *Asbestos mondaieno taiou ni tduite* [Policy for asbestos issues]. Retrieved (last access), March 10, 2009, from http://www.mlit.go.jp/sogoseisaku/asubesuto/top.html

Ministry of Labor. (1986). Working environment measurement law. In Industrial Health Division, Department of Industrial Safety and Health, Ministry of Labor (Ed.), *Working environment system in Japan*. Tokyo: Japan Association for Working Environment Measurement.

Miura, H. (1982). A study on the relationship between the occurrence of lung cancer and asbestos exposure. *Lung Cancer 22*, 283.

Miura, H., Takayama, S., Nakayama, M., Yoshizawa, M., Natori, Y., & Kimura, Y. (1986). An analysis of 12 cases of malignant pleural mesothelioma. *Lung Cancer, 26*, 467.

Miyazaki, R. (1983). Study of causal factors of lung cancer associated with pulmonary asbestosis. *Journal of Nara Medical Association, 34*(5), 473–476.

Mizuno, H. (2007). Showa zenhan-ki no roudou eisei to Sukegawa sensei [Industrial hygiene in the early Showa era]. *Sangyo Eiseigaku Zasshi, 49*(6), 240.

Montgomery, J.D., & Rondinelli, D.A. (1995). *Great policies: Strategic innovations in Asia and the Pacific Basin*. Westport, CT: Praeger.

Morinaga, K. (1988). The present status and future trend of asbestos-related cancers. *Japanese Journal of Trauma and Occupational Medicine, 36*, 361–365.

Morinaga, K. (1989). Wagakuni ni okeru sekimen-kanren shikkan no ekigaku-teki chiken. *Byori to Rinsho, 7*, 686–694.

Morinaga, K. (2000). Dust and respiratory cancers. *Japanese Journal of Occupational Medicine and Traumatology, 48*(3), 385–390.

Morinaga, K., Yokoyama, K., Hara, I., Tsukuma, H., Hiyama, T., Yutani, S., Oshima, A., Fujimoto, I., & Sera, Y. (1982). Lung cancer mortality among asbestos workers: A retrospective cohort study in Sennan district. *Proceedings of Japanese Cancer Association, 41*, 423.

Morinaga, K., Kishimoto, T., Sakatani, M., Akira, M., Yokoyama, K., & Sera, Y. (2001). Asbestos-related lung cancer and mesothelioma in Japan. *Industrial Health, 39*, 65–74.

Mossman, B. T., Bignon, J., Corn, M., Seaton, A., & Gee, J. B. L. (1990). Asbestos: Scientific developments and implications for public policy. *Science, 247*, 294–301.

Murai, Y. (1998). Asbestosis in the general public. *Toyama Medical and Pharmaceutical Journal, 11*, 8–11.

Navarro, A. M. (2003). Shaping industrial health: The debate on asbestos dust hazards in the UK, 1928–39. In E. Rodriguez-Ocana (Ed.), The Politics of the healthy life: An international perspective (pp. 69–87). EAHMH Publications.

National Cancer Center. (1991). Annual report of the cancer research, Ministry of Health and Welfare 1990 (pp. 608–610). Tokyo: National Cancer Center.

National Diet Library. (2005). Asbestos mondai to sono taiousaku [Asbestos issue and policies]: National Diet Library Issue Brief, number 495. Tokyo: National Diet Library.

Nevitt, C., Daniell, W., & Rosenstock, L. (2007). Workers' compensation for nonmalignant asbestos-related lung disease. *American Journal of Industrial Medicine, 26*(6), 821–830.

Newhouse, M. L., Berry, G., Wagner, J. C., & Turok, M. E. (1972). A study of the mortality of female asbestos workers. *British Journal of Industrial Medicine, 29*(2), 134–141.

Office of Prime Minister. *Asbestos mondai* [Asbestos issues]. Retrieved (last access), March 10, 2009, from http://www.kantei.go.jp/jp/asubesto/

Rom, W. N. (1992). Accelerated loss of lung function and alveolitis in a longitudinal study of non-smoking individuals with occupational exposure to asbestos. *American Journal of Industrial Medicine, 21*(6):835–844.

Rosner, D., & Markowitz, G. (1991). *Deadly dust: Silicosis and the politics of occupational disease in twentieth-century America*. Princeton, NJ: Princeton University Press.

Sakai, K., Hisanaga, N., Huang, J., Shibata, E., Ono, Y., Aoki, T., Takagi, H., Ando, T., Yokoi, T., & Takeuchi, Y. (1994). Asbestos and nonasbestos fiber content in lung tissue of Japanese patients with malignant mesothelioma. *Cancer, 73*, 1825–1835.

Samuels, R. J. (1987). *The business of the Japanese state: Energy markets in comparative and historical perspective*. Ithaca, NY: Cornell University Press.

Sandman, P. M. (1993). *Responding to community outrage: Strategies for effective risk communication*. Fairfax, VA: American Industrial Hygiene Association.

Sato, H., & Akabayashi, A. (2005). Bioethical policymaking for advanced medical technologies: institutional characteristics and citizen participation in eight OECD countries. *Review of Policy Research, 22(4)*, 571–587.

Sawyer, R. N. (1977). Asbestos exposure in a Yale building: Analysis and resolution. *Environmental Research 13*, 146–169.

Selikoff, I. J., Churg, J., & Hammond, E. C. (1964). Asbestos exposure and neoplasia. *JAMA, 188*, 22–26.

Sera, Y., Yokoyama, K., & Tanaka, S. (1960). An autopsy case of asbestosis with lung cancer. *Japanese Journal of Occupational Health, 2*, 326.

Sera, Y. (1971). Asbestos-handling works and lung diseases. *Roudo No Kagaku, 26*, 4–12.

Sera, Y., Kang, K-Y., & Yokoyama, K. (1973). Asbestosis and lung cancer in Osaka Sennan district. *Gann, 64*, 313–316.

Shishido, S. (1986). Studies of asbestos pollution in the lungs of the populae: Trend of the incidence of feeruginous bodies during 45 years. *Journal of Nara Medical Association, 37*, 214–223.

Simon, H. A. (1983). *Reason in human affairs*. Stanford, CA: Stanford University Press.

Simon, H. A. (1991). Bounded rationality and organizational learning. *Organization Science* 2(1): 125–134.

Stavisky, L. P. (1982). State responsibility for the control of asbestos in the schools. *The Journal of School Health, 52*(8), 358–364.

Stone, D. A. (1998). Policy paradox and political reason: The art of political decision making. New York, NY: W. W. Norton & Company.

Sugiyama, A. (1934). *Sekimen [Asbestos]*.Tokyo: Kouseikai Shuppan-bu.

Takahashi, K., Huuskonen, M. S., Tossavainen, A., Higashi, T., Okubo, T., & Rantanen, J. (1999). Ecological relationship between mesothelioma incidence/mortality and asbestos consumption in ten western countries and Japan. *Journal of Occupational Health, 41*(1), 8–11.

Tossavainen, A. (1997). Consensus report: Asbestos, asbestosis, and cancer – the Helsinki criteria for diagnosis and attribution. *Scandinavian Journal of Work, Environment, and Health, 23*(4), 311–316.

True, J. L., Jones, B. D., & Baumgartner, F. R. (1999). Punctuated-equilibrium theory. In: P. A. Sabatier (Ed.), *Theories of the policy process* (pp. 97–116). Boulder, CO: Westview Press.

Whysner, J., Covello, V. T., Kuschner, M., Rifkind, A. B., Rozman, K. K., Trichopoulos, D., et al. (1994). Asbestos in the air of public buildings: a public health risk? *Preventive Medicine, 23*(1), 119–125.

Wikeley, N. (1992). The asbestos regulations 1931: A license to kill? *Journal of Law and Society, 19*(3), 365–384.

Nicholson, W. J., Swoszowski, E. J. Jr., Rohl, A. N., Todaro, J. D., & Adams, A. (1979). Asbestos contamination in United States schools from use of asbestos surfacing materials. *Annals of the New York Academy of Sciences, 330*: 587–596.

World Health Organization. (1986). *Environmental health criteria 53: Asbestos and other natural mineral fibers*. Geneva: World Health Organization. Retrieved February 28, 2009, from http://www.inchem.org/documents/ehc/ehc/ehc53.htm

Yoshizumi, K., Hori, H., Satoh, T., & Higashi, T. (2001). The trend in airborne asbestos concentrations at plants manufacturing asbestos-containing products in Japan. *Industrial Health, 39*, 127–131.

Zielhuis, R. L. (1977). *Public health risks of exposure to asbestos: Report of a working group of experts prepared for the Commission of the European Communities, Directorate-General for Social Affairs, Health and Safety Directorate*. Oxford: Pergamon Press.

Chapter 3
Emergence of Asbestos-related Health Issues and Development of Regulatory Policy in the UK

Andrew Webster, Conor M.W. Douglas, and Hajime Sato

Introduction

Asbestos is a naturally occurring silicate mineral with long, thin fibrous crystals. It became increasingly popular among manufacturers and builders during the industrial revolution and thereafter in the mid and late nineteenth century, due to its resistance to heat, electricity and chemicals. Asbestos, however, can be hazardous. When asbestos fibers are inhaled or ingested, they can cause serious health damages, including malignant mesothelioma, lung cancer, and asbestosis (Agency for Toxic Substances and Disease Registry 2001). As early as 1898, the Chief Inspector of Factories in the UK reported that asbestos had "easily demonstrated" health risks. In the early 1900s, researchers began to notice a large number of early deaths and lung problems in asbestos mining towns. By the 1930s, England regulated ventilation and made asbestosis an excusable work related disease (Johnston & McIvor 2000; Tweedale 2001).

This chapter examines the emergence of asbestos-related health hazards, the development of asbestos regulations, and the socio-political context surrounding them in the UK, where, as noted above, health hazards of asbestos were noticed quite early. Our study found three distinct periods in the UK history of asbestos-regulations: 1924–1974, 1974–1999, and 1999–2006. The key events, major actors, and socio-political characteristics of asbestos-related risk management in each period will be presented.

A. Webster and C.M. Douglas
Science and Technology Studies Unit (SATSU), Department of Sociology, University of York, UK

H. Sato (✉)
Department of Public Health, Graduate School of Medicine, The University of Tokyo, Tokyo, Japan
e-mail: hsato-tky@umin.net; hsato@post.harvard.edu

H. Sato (ed.), *Management of Health Risks from Environment and Food*, Aliance for Global Sustainability Bookseries 16,
DOI 10.1007/978-90-481-3028-3_3, © Springer Science+Business Media B.V. 2010

Period 1 (1924 to 1974): The Pre-risk Management Era of Asbestos Production and Regulation

Characterization of the Period for Its Approach to the Management of Asbestos Risk

The use of and research on asbestos in the UK began much earlier than 1924. Alleman and Mossman (1997, p. 70), for example, discuss how the mineral was a topic of exploration within the Royal Society almost from the Society's inception in 1660. Geoffrey Tweedale (2001) provides a detailed account of asbestos production before and after 1924. This date of 1924 is significant for our purposes, because it is the year that saw William Cooke publish the first medical paper reporting a death caused by fibrosis of the lungs.[1] The papers Cooke published on the death of Nellie Kershaw were significant because they made a direct link between fibrosis of the lungs and the asbestos fibers that Nellie inhaled while working in the spinning room of an asbestos factory in Rochdale, England (Bartrip 2004, p. 72; Selikoff & Greenberg 1991, p. 898). From that moment on, asbestos was considered a health hazard requiring some kind of management.

The 50 years of asbestos production and management in question can be largely characterized by industrial expansion and influence, as well as quasi-self-regulation.

Industrial Expansion

The use of asbestos boomed with the expansion of industry and the wars that were prevalent during the 1930s, 1940s, and 1950s, with world production jumping from 339,000 tons in 1930 to 1.2 million tons in 1950 (Jeremy 1995, p. 255). Demand for the mineral was high in the UK, and when imported it was transf ormed locally to support the electrical and engineering sectors, the motor car industry, and shipbuilding after the rearmament following World War I (Jeremy 1995, p. 255). Rearmament was good for the asbestos industry, but World War II was especially significant, as the mineral came to be seen and treated as an essential component of the war effort. After the bombings of World War II, manufacturers were producing asbestos during the rebuilding process to such an extent that capacity had to be expanded (Monopolies Commission 1973, p. 12). The largest British player in asbestos, Turner & Newall, saw their workforce explode from 5,000 in 1926 to 40,000 in 1961 (half were based in the UK and the other half abroad), and sales were over £300 million by 1958[2] (Warren 1997, p. 53). This early era of asbestos use was one of growth and large profits.

[1]Later, in 1927, Cooke termed this fibrosis of the lungs, pulmonary asbestosis (Borron et al. 1999).

[2]At some of the times between 1926 and 1980, T&N returned to its shareholders a 16% yield on their investments (Warren 1997, p. 53).

Self-Regulation

With the health risks of asbestos achieving widespread recognition only within the medical profession in 1927/8 (Jeremy 1995, p. 257), any control of asbestos dust up to that point was at the whim of industry. The regulations concerning dust content in factories' air was enforced only minimally, if enforced at all, on the moral basis of providing the workers with comfortable working conditions. In 1926, when Turner and Newall had 5,000 employees (Warren 1997, p. 53), with the exception of the very general Factories Acts there was not a single piece of legislation or regulations that required industry to manage in any way the asbestos dust that these 5,000 workers were breathing.

Even with the advent of the first Asbestos Regulations in 1931 (which are dealt with in more detail below), the inspections and fines that these regulations demanded for the management of asbestos health hazards were very ineffective. There were a grand total of only two convictions (with fines) handed down as a result of the 1931 Regulations between the years 1931 and 1969, the year when new regulations were being drafted. There was only one conviction in the following years of 1970 and 1971. Even in relative terms, the fines that were levied for these violations averaged just £12 in the cases from 1931 to 1969, and £25 for the fine in 1971 (Tweedale 1999; Tweedale 2001, p. 212). With the extreme rarity of convictions based on the Asbestos Regulations of 1931, and the small amounts of the fines,[3] any actions taken by the industry to manage the health risks associated with asbestos would have had to be almost entirely voluntary.

The asbestos industry in Britain from 1924 until 1969 was self-regulated. This self-regulation worked in dangerous conjunction with another characteristic of this "pre-risk management" era, namely, that research on hazards was exclusively the province of industry. This situation tended to ensure that the potential harm of asbestos would be understated (Jeremy 1995; Tweedale 2000) (this industry-sponsored science and the denial of the scope of the asbestos hazard is addressed in more detail in later sections). Given these characterizing factors, it is clear that 1924 to 1969 was a period in which "risk management" as we know it today scarcely existed, at least to the extent that we can call it a "pre-risk management era."

Relevant Actors in the Use and Management of Asbestos, and Key Events

Given the fact that asbestos use in the UK was self-regulated during the first half of the twentieth century, it is important that we provide the reader with a profile of the major industrial players who determined the conditions that allowed the health-related

---

[3]For instance, £12 in 1945 would in 2005 be £344.95 (using the retail price index), £366.89 (using the GDP deflator), £1,086.32 (using average earnings), £1,211.68 (using per capita GDP), £1,483.31 (using the GDP); £25 in 1971 would in 2005 be £236.71 (using the retail price index), £242.31 (using the GDP deflator), £419.62 (using average earnings), £494.89 (using per capita GDP), £532.77 (using the GDP) (MeasuringWorth 2007).

hazards to manifest. It is also important that the reader understand the nature of the asbestos industry prior to 1974, because this industry financed most of the research studies on the potential health impact of asbestos, research that was supposed to guide the development of the asbestos regulations.

Industry

Of the three major firms that manufactured asbestos-related products, Turner and Newall (T&N) was by far the largest (Monopolies Commission 1973, p. 12; Tweedale 2001, p. 4; Warren 1997, p. 53). The company was established in 1871 as Turner Brothers and in 1907 was renamed Turner Brother Asbestos (Tweedale 2001, pp. 2, 5). This company was actually merged with three other companies, which propelled it to the status of a major asbestos manufacture. As Warren describes it:

> Turner and Newall (T&N) which was formed in 1920 by merging four other firms: Turner Brothers Asbestos, The Washington Chemical Company, Newalls Insulation, and J.W. Roberts. T&N was floated on the London Stock Exchange in 1925 and acquired Ferodo Ltd a brake linings manufacturer soon after, making it the largest vertically integrated asbestos based business in the UK (Warren 1997, p. 53).

The other two major industrial players in the UK were The Cape Asbestos Company Ltd. and British Belting & Asbestos (BBA). As the Monopolies Commission described in detail, these "leading competitors" were very much a post-war development, following on the coattails of T&N, at least insofar as "the history of the [asbestos] industry in the United Kingdom in the 1920s and 1930s was in effect the history of T&N; to some extent it has remained so to the present day" (Monopolies Commission 1973, p. 12).

Government

Despite the characteristically industry-driven approach to asbestos management that dominated the decades prior to 1974, the British government was technically involved with the safety of all factory workers. More specifically, the 1931 Asbestos Regulations meant that there was some kind of recognition of the potentially hazardous nature of the mineral. Lucy Deane, the first female Inspector of Factories, warned in 1989 (i.e., prior to the 1931 Regulations) that asbestos was dangerous, "on account of [its] easily demonstrated danger to the health of workers and because of ascertained cases of injury to bronchial tubes and lungs medically attributed to the employment of the sufferer … the evil effects of asbestos dust have also instigated a microscopic examination of the mineral dust by HM Medical Inspector. Clearly revealed was the sharp glass-like jagged nature of the particles, and where they are allowed to rise and to remain suspended in the air of the room in any quantity, the effects have been found to be injurious as might have been expected." (Deane 1898; Gee & Greenberg 2001).

Although this report was later confirmed by similar reports in 1909 and 1910, and irrespective of whether these earlier reports "appeared in the annual reports of HM

Chief Inspector of Factories, which were widely circulated amongst policy-makers and politicians" (Gee & Greenberg 2001), the asbestos industry was largely its own regulator when it came to taking any action against asbestos hazards.

The position of the British government in the first half of the twentieth century in regard to asbestos was further complicated by its view of the perceived necessity of the mineral in "the war effort." As the Monopolies Commission (1973, p. 12) observed: During the war T&N and its major subsidiaries, and other manufacturers of asbestos products in the United Kingdom, were declared controlled undertakings by the Ministry of Supply under the Defence Regulations and their activities were almost wholly directed towards furthering the war effort.

Given the primacy of the war in the 1930s and 1940s, and the acknowledgement of the importance of asbestos by the British government, which classified the industry's activities as "controlled undertakings by the Ministry of Supply under the Defence Regulations," it is not surprising that the 1931 Asbestos Regulations were not enforced more stringently nor reformed for nearly 40 years.

Key Events

At the outset of this section, we identified the death of Nelly Kershaw and the subsequent publication of William Cooke's paper, which identified asbestosis and linked it to asbestos exposure, as the key moment at which the mineral came to be considered a hazard in need of management. A series of other medical discoveries and publications proved equally important during this period. The disheartening fact is that these scientific papers are **not** key events because of their impact on the way the asbestos industry chose to operate, but rather because these discoveries and papers were ignored. As we shall see below, the asbestos industry, as well as the British government, had access to medical information that linked asbestos to a host of deadly diseases. Rather than accepting the dangers of asbestos exposed by this work, the industry chose to conduct its own research into the problem and for the most part continued with "business as usual."

Along with Cooke's early work on asbestosis, Dr. Edward Merewether reported a statistical analysis in the *Annual Report of the Chief Inspector of Factories* for 1947. In his report, he stated that "cancer of the lungs or pleura was present in 13.2% of cases [of asbestosis death from 1924 to 1946], compared to an incidence of 1.32% in silicotics and a similar figure in the general population" (Jeremy 1995, p. 258). In essence, these statistics demonstrate a link between asbestos and lung cancer. In 1955 this link was confirmed by Dr. Richard Doll's work on the risk of cancer in asbestos workers (Doll 1955).

The third major killer disease associated with asbestos exposure did not come to light until 1960, when Wagner et al. (1960) published their first paper indicating a relationship between pleural mesothelioma and asbestos exposure. Mesothelioma, a type of cancer almost uniquely associated with asbestos, takes hold when malignant cells develop in the mesothelium, a protective lining that covers most of the body's internal organs.

Its most common site is the pleura, the outer lining of the lungs and chest cavity, but it may also occur in the peritoneum, the lining of the abdominal cavity, or the pericardium, a sac that surrounds the heart (Mossman et al. 1990).

As the number of cases of asbestosis, lung cancer, and mesothelioma increased, so did the visibility of the asbestos issue. "Between 1964 and 1967, stories about the health hazards of asbestos appeared in such national newspapers as *The Times, The Sunday Times, The Daily Herald, The Guardian, The Daily Telegraph, The Morning Star, The New York Times, and The Wall Street Journal*, as well as in local and regional papers. In January 1967, the BBC broadcast a film on the subject on its early evening news program, 24 Hours. Thereafter, asbestos health hazards regularly featured in newspaper and television reports." (Bartrip 2004, p. 73).

Table 3.1, taken from Bartrip (2004, pp. 74–75), lists the key medical events and publications during this period of asbestos use. Other key legislative and institutional events are discussed later.

The Institutional Framework for the Management of Asbestos Risk

The small number of state-run agencies responsible for the protection of factory workers, the pressure of the wartime economy, and the toothless regulations that were supposed to manage asbestos hazards together meant that the government

Table 3.1 List of key medical events and publications in the history of asbestos

Key events	
1924	W. E. Cooke publishes the first paper on asbestos related disease
1925	Thomas Oliver coins the term "asbestosis"
1930	Edward Merewether confirms that inhalation of asbestos dust can cause a fatal disease
1935	Kenneth M. Lynch and W. Atmar Smith identify a "possible relationship" between pulmonary asbestosis and carcinoma of the lung
1955	Richard Doll finds that certain asbestos workers face a "notably higher risk" of contracting lung cancer than the rest of the population
1960	Wagner, Sleggs, and Marchand publish their first paper indicating a relationship between pleural mesothelioma and asbestos exposure
1964	Selikoff, Churg, and Hammond demonstrate that insulation contract workers face a health hazard resulting from asbestos exposure
Key publications	
1930	Merewether E. R. A., & Price, C. W. *Report on effects of asbestos dust on the lungs and dust suppression in the asbestos industry.* London: Her Majesty's Stationary Office
1955	Doll, R. Mortlity from lung cancer in asbestos workers. *British Journal of Industrial Medicine, 12*(2), 81–86
1960	Wagner J. C., Sleggs, C. A., & Marchand, P. Diffuse pleural mesothelioma and asbestos exposure in the North Western Cape province. *British Journal of Industrial Medicine, 17*(4), 260–271
1964	Selikoff, I. J., Hammond, E. C., & Churg, J. Asbestos exposure and neoplasia. *JAMA, 188*, 22–26

paid little attention to the management of asbestos. As mentioned above, this lassitude resulted in just two convictions based on the 1931 Asbestos Regulations from their inception until 1969 (Tweedale 2001, p. 212).

The release of the first report by the British Occupation Hygiene Society (BOHS)[4] on chrysotile asbestos standards in 1968 signaled that this and other professional organizations would ultimately come to play a role in asbestos management. However, the BOHS was established in 1953 (Ogden 2003, p. 3), which means that it was largely silent and ineffective regarding the asbestos issue for a decade and a half.

What can explain the paucity of convictions by the government inspectorates and the silence of a significant scientific/professional body such as the BOHS for so many years? Given that these organizations were not highly involved with the management of asbestos, what other organizations or institutional frameworks were in place to take up the slack, and how did this situation affect the management of asbestos?

Industry-University-Research Complex – The Asbestos Research Council

The laissez-faire position of most government institutions and significant professional bodies coincided dangerously with the emergence of a pseudo "Industry-University-Research Complex," which Geoffrey Tweedale (2000, 2001) has efficiently recorded in his history of the Asbestos Research Council.

In the 1950s, the industry was skeptical about the purported health hazards associated with asbestos.[5] It was afraid that its wartime profits would be reduced by workers' compensation claims, let alone the increasing regulation of the manufacture of asbestos-related products. As a result, the "Big Three" industry players (T&N, Cape Ltd., and BBA) began funding university academics to conduct research on asbestos. Instead of individual research projects being funded here and there – such research had been conducted in the 1950s at the University of Reading with support from the Big Three – the Asbestos Research Council (ARC) was established in 1957 to formalize the research enterprise. Prior to the establishment of the ARC, Turner Brother Asbestos – T&N's main factory – had established the Asbestosis Research Committee (ASR) in 1942. The ASR would eventually be the model on which the ARC was based (Tweedale 2000, pp. 724–726). Although this research was "in-house" by nature, it did take place at the British Postgraduate Medical

---

[4]Perhaps an explanation at this point of what the BOHS is would be good?

[5]Recall the following quotes from T&N in Jeremy (1995, p. 254): "We repudiate the term 'Asbestos Poisoning.' Asbestos is not poisonous and no definition or knowledge of such a disease exists" (T&N's Turner Brother Asbestos board view, 1922). "Disease associated with asbestos is rare. The general public is not at risk, and very few workers are. The whole subject has been sensationalised because some recent medical research is of a kind which easily attracts headlines, and because asbestos dust can, in a minority of cases, lead indirectly to cancer, which is always a 'scare' word" (draft by UK Asbestos Information Committee 1967).

School in London, and it tended to focus on dust-counting technology and animal/tissue-culture experiments.

When the ARC was formally established in 1957, the academics from Reading University were joined on the Council by academics from the University of Cambridge. In 1961, their participation resulted in the first appointment of an ARC research fellow at Cambridge. The organizational structure of the ARC is illustrated in Fig. 3.1 and is described by Tweedale (2000, p. 724) as follows:

> … the ARC functioned through a management and research committee, linked by a joint-secretary. The management committee was to meet twice a year; the research committee (which was to include one medical and one technical member, and an independent expert) was to meet three times each year. The ARC's initial budget was a little over £4000 (this at a time when T&N annual profits after tax were over £6 million) (Asbestos Research Council 1966, 1987).

However, as Tweedale shows, by the 1970s the ARC lost any kind of respect that it may have had as a research body, because it ultimately was the asbestos company that set the research agenda and determined the direction and public visibility of the research. The research strategy was set by the management committee, which in turn responded to the wishes of the sponsoring directors. As should be clear by now, these men did not see the ARC as fundamentally a council for scientific research. Ultimately, its activities were revealed as simply an attempt to capture the scientific agenda and influence public policy (Tweedale 2000, p. 732) (Fig. 3.1).

Tweedale notes that "[t]he ARC stated publicly that it intended to concentrate on the prevention, diagnosis, and treatment of asbestosis." With that in mind, the ARC decided internally that the measurement of dust suppression should be separated from the biological and health research. Like the work that took place in the 1940s, before the establishment of the Council, the more recent research tended to focus on animals and the chemical analysis of fibers. In fact, over 50 of the 69 pre-1976 ARC

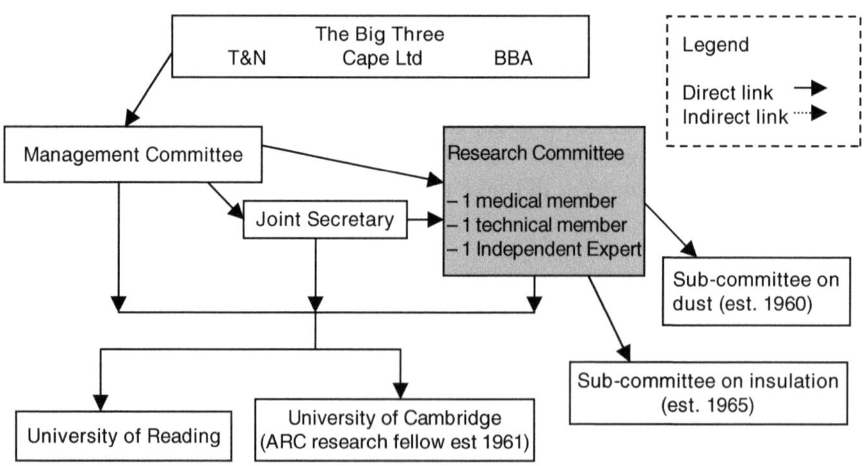

Fig. 3.1 Structure of the Asbestos Research Council (ARC) 1957–1968

publications that Tweedale (2000, pp. 724–726) reviewed were concerned with these topics; only seven were focused on dust-fiber counting and dust control. In the animal studies, it is not surprising that rodents injected with asbestos developed asbestosis. The studies involving the chemical analysis of fibers were concerned with determining how asbestos could continue to be used, in addition to the prevention of asbestosis.

Given that T&N alone was employing 40,000 people by 1961 (Warren 1997, p. 53) and also that there was strong medical evidence – if not proof – that asbestos was linked not just to asbestosis but also to lung cancer and mesothelioma, it is somewhat surprising that the ARC did not undertake epidemiological studies aimed at identifying the precise risks associated with various levels of exposure. This relative lack of research is explained by the fact that epidemiological studies were excluded from the ARC's remit. By 1965, neither industry nor the government had conducted large-scale population studies on the health risks associated with asbestos (Tweedale 2000, p. 725).

By 1968/9 the ARC had shifted its research from Reading and Cambridge to the University of Edinburgh. This seems to have been an opportune decision, as the National Coal Board had previously established a charitable foundation at Edinburgh called the Institute of Occupational Medicine (IOM). With the decline of the coal industry, the IOM was looking for fresh streams of funding comparable to those previously obtained from the ARC. In return, the IOM could provide expertise for industry-based occupational research. This process culminated in 1971 with the ARC establishing the Asbestosis Research Foundation (ARF) within the IOM. This change from the early institutional arrangements of the ARC is illustrated in Fig. 3.2.

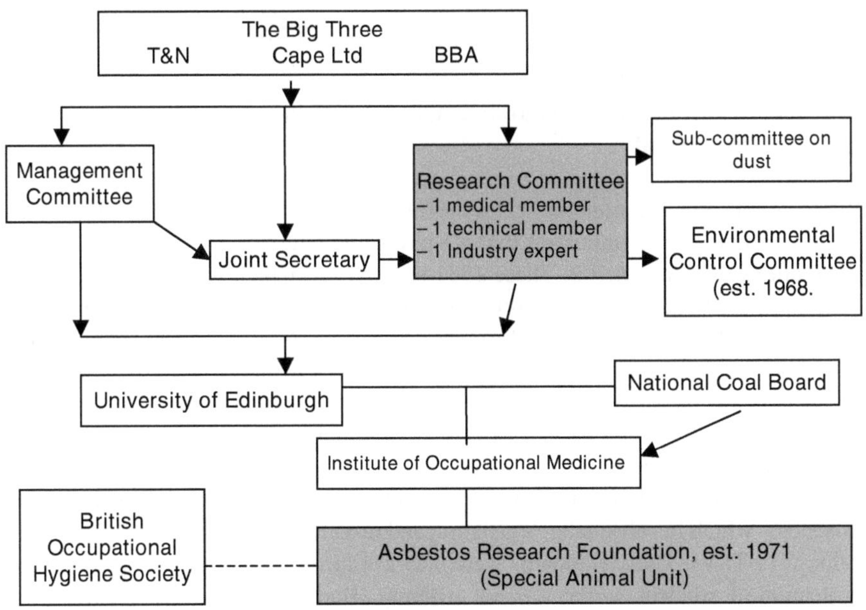

Fig. 3.2 Organizational structure of the Asbestos Research Council (ARC) 1968–1990

For 10 years, money was invested by the ARC in the research at Edinburgh. The ARC's total budget jumped from £4,000 in 1957 to £30,000 in 1970, and a few years after that it more than doubled to £80,000. By the end of the 1970s, the budget exceeded £2,000,000 (Tweedale 2000, p. 725).

Did this massive increase in the ARC's budget mean that they were conducting ground-breaking research? If not, was the money being spent in other ways (on public relations, perhaps)? Animal experiments continued until 1978, and it was only then that the funds were considered to have been exhausted to the point that the amount of research had to be reduced (Tweedale 2000, p. 726). Nonetheless, by the end of the 1980s the ARC was responsible for approximately 200 published papers; unfortunately, in 1971 its Research Committee had to admit that most of these experiments (the animal studies, in particular) "were of no value in improving the lot of the asbestos worker" (Asbestos Research Council 1973).

If the ARC was largely ineffectual in conducting research that would improve the health plight of asbestos factory workers, what did their work amount to? From its inception in 1952, the British Occupational Hygiene Society (BOHS) was silent on the asbestos issue, but when the government sought in 1968/9 to reform the 1931 Asbestos Regulations (a matter discussed in more detail below) the BOHS was called upon to advise the government as to the safe threshold for asbestos dust. Throughout this process of "reform," the ARC was able to exert influence and affect how management strategies for controlling asbestos would be applied to the industry (Tweedale 2000, p. 725).

Policies, Regulations, and Legislative Acts That Emerged

1931 Asbestos Industry Regulations, Asbestosis Scheme, and Asbestosis Fund

Before exploring the content of the 1969 Asbestos Regulations and how they came into being, it is important that we pay some attention to the 1931 Asbestos Regulations that preceded it. These 1931 regulations, which were not fully implemented until 1933, were the product of negotiations between the senior members of the Home Office, the Factory Inspectorate, selected union representatives, and, of course, the major industry players (Bartrip 2004, p. 73). The purpose of these regulations was the suppression of asbestos dust in factories. The industry apparently thought this task was important, "not to protect the health of their workers, but [to protect themselves] against legal action from them." The industry's response to the proposal for "dust suppression" was to collectively set "standards" for exposure to asbestos – which in subsequent legislation and research were called "threshold limit values," or TLVs. These standards were later criticized as totally inadequate for the protection of asbestos workers, if not arbitrary (Budgen 2004, p. S78). They were modified in the 1969 Asbestos Regulations.

During this early period of asbestos use, it is not only the research into very specific aspects of the asbestos problem that can be characterized as "in-house," or industry-controlled, but also the attempts to manage the problem that followed from the 1931 Regulations. These regulations were the dust-control arm of a tripartite approach to deal with the asbestos problem. The remaining two parts were a medical monitoring arm, the government-run Medical Arrangement Scheme, and a compensatory arm, the Asbestosis Scheme.

As we have already described in detail, the attempts to achieve "dust control" through inspections and fines, as mandated by the Asbestos Regulations of 1931, was largely ineffective in managing the health risks caused by asbestos. "Medical monitoring" was just that – monitoring – and the most that the government-run Medical Board could do was to perform medical examinations and thereby assess the degree of disability, which in turn would affect the amount of compensation[6] (Tweedale 2001, p. 73). This was essentially a reactive approach to dealing with asbestos health hazards. The "compensation" arm was called the Asbestosis Scheme, through which employers were instructed to provide avenues for workers to seek damages for the ill health they incurred while working with asbestos.

This, then, was the industry, "in-house" component of asbestos management. Although the Asbestosis Scheme was a government initiative established in 1931, the onus was on industry to provide the financing and administer the claims once their validity had been assessed by the state-run Medical Board. Technically, this meant that the industry could choose to use outside insurance companies, but the majority decided to deal with compensation themselves. For T&N, the impetus provided by the Asbestosis Scheme culminated in the creation of the Asbestosis Fund in 1931 (Tweedale 2001, p. 70). The establishment of this fund would prove to be a significant development, as the final decisions regarding compensation were now made by the Board of Directors, and all the finances had to be cleared with T&N, or more precisely, its main factory, Tuner Brothers Asbestos (TBA). This compensatory fund was accessible only to those who worked in the main preparatory phases of asbestos production, and then, to the dismay of many workers, only if they were hired on or after May 1, 1931. In other words, if they left the industry before then, or they later developed asbestosis, they were not covered. Moreover, they could only claim compensation for up to 3 years after leaving the industry. This is not enough time, because the biochemical processes associated with the development of asbestosis take longer than 3 years to manifest (Tweedale 2001, p. 70).

[6] "Partial incapacity brought a payment of half the difference between a worker's present wage and his pre-suspension earnings; total compensation for complete incapacity could not exceed 50 percent of the worker's average weekly earnings" (Tweedale 2001, p. 73).

In the case of T&N, the cash to support this fund came from the individual subsidiaries of the company, and it was split into two funds: "Into Fund A, subsidiaries subscribed 7.5% of the total wages paid by the company to employees who came within the Medical Arrangements Scheme. This covered scheduled employees who were subject to periodic Medical Board examinations. Fund B was established to meet the claims of workers who came within the Compensation Act, but were not covered by the Medical Arrangements Scheme. These workers might develop asbestosis and then make a successful claim against a Turner & Newall company. Subscription to the second fund were originally at the rate of 2.5% of the total wage bill paid to such employee" (Tweedale 2001, p. 71).

Whereas the Asbestosis Fund was financed and cleared internally, a process that stands in a negative light with regard to health-risk management, it was administered externally by the Manchester branch of the Commercial Union Assurance Company. This allowed T&N to wash its hands of dealing with the claimants, and it allowed the Commercial Union to make settlements on a 75% liability basis (Tweedale 2001, pp. 72–73).

The British Occupational Hygiene Society and the 1968 Threshold Limit Values

Perhaps one of the reasons why so few convictions and fines were handed out to the asbestos industry from 1931 to 1974 is that the standards that were supposed to be upheld were unclear. Then again, perhaps it was because the government had no interest in regulating asbestos during the World War II, or maybe even that there was no evidence that asbestos was a health risk. By 1966, neither of these last two explanations held any water, because asbestos had been directly linked to asbestosis, lung cancer and mesothelioma (Bartrip 2004, pp. 74–75). Finally, after nearly a decade and a half of operations, the British Occupational Hygiene Society took action in the case of chrysotile asbestos (Greenberg 2004, p. 538). The idea that a "threshold limit value" (TLV), or an acceptable limit of exposure, could be established for asbestos happened to mesh with the interests of industry and with the research that the ARC had been carrying out.

The ARC believed that short asbestos fibers (i.e., under 5 μm) had no harmful effects. Therefore, they argued that no TLV should be set for a concentration of dust of fibers shorter than 5 μm (Tweedale 2000, p. 727). The TLV value that the BOHS ultimately chose was determined by the following steps. (1) X-ray data (but not the x-rays themselves) were provided by TBA for 290 of their workers, along with dust measurements taken during their employment. (2) Risk was then calculated from the observable relationship between dust exposure and the incidences of the disease in the men, who had worked in the factory for 10 years or more since 1933. (3) A TBA/ARC/BOHS member reported that although there were 70 to 80 workers who had been operating in conditions in which the dust concentration was 3–4 fibers per cubic centimeter (f/cc), there was only one abnormal x-ray in the group (Tweedale 2000, 2001).

(4) The key finding was that the risk of [the asbestosis symptom] basal rates would be <1% for a cumulative exposure of ~100 fibre year/mL measured by the membrane filter method, equivalent, for example, to 2 fibres/mL for a 50 year working lifetime. The fine detail of the report stated that to be better than 95% certain that the risk was <1%, the level would have to be <1 fibre/mL The fibre/mL limit would apply to an average over 3 months, determined either by continuous sampling or by random samples, taken at times decided in advance, in such a way that the upper 90% confidence limit of the average exposure was <2 fibres/mL. So the committee's recommendation was not a 4 or 8 h limit of 2 fibres/mL, but a sampling strategy which gave a 90% confidence that the long-term average was <2 fibres/mL (Ogden 2003, p. 4). (5) It was concluded [by the BOHS] that after 50 years of exposure to an average atmosphere of 2 fibres/mL of chrysotile over 40 h working weeks, fewer than 2% of workers would have developed early minor signs of lung disease (Greenberg 2004, p. 538).

The hygienic standard TLV of 2 fibers of chrysotile per asbestos/cubic centimeter was published by BOHS in 1968 (Ogden 2003, p. 3), and it was immediately "adopted" by the government (Tweedale 2001, p. 200).

Asbestos Regulations – 1969

Although the government appeared to accept the BOHS's TLV in its 1969 Asbestos Regulations, the regulations themselves did not actually take effect for the industry until May1970. This delay gave T&N and the other players time to adjust their practices and redirect their policies. Significantly, the "Industry" section of the 1969 Regulations had subsequently been dropped, meaning not only that the restrictions now applied to all work sites, but also that asbestos was now recognized as a public hazard.

The TLV of 2 fibers/cubic centimeter that the BOHS had determined applied only to chrysotile asbestos. The TLV of 0.2 f/cc set for crocidolite (blue asbestos) was so stringent that the use of this mineral was in effect abandoned (Bartrip 2004, p. 74). Tweedale argues that this voluntary compliance by industry to a ban on blue asbestos was strategic, because this substance was seen as particularly dangerous by regulators, and voluntary compliance would help them to achieve (if not through explicit negotiation) more lenient rules for chrysotile (white asbestos) and amosite (brown asbestos). Tweedale argues that this strategy was "substituting one hazardous material with another" (Tweedale 2001, p. 207).

It turned out that despite the efforts of the BOHS to establish a TLV, the fibers/millilitre regulatory ratios were listed as **guidelines only** (Ogden 2003, p. 5). The 1969 Regulations themselves referred only to dust "consisting of or containing asbestos to such an extent as is liable to cause danger to the health" (Department of Employment and Productivity [DEP] 1970), and the BOHS's TLVs were seen as a "goal to be aimed at eventually rather than a standard to be rigidly enforced" (statement of S. Holmes, at US Department of Labor, who was a member of the ARC and the BOHS and was on the T&N/TBA payroll). The details of the specific chrysotile measurements and enforcement procedures[7] were presented only in *Technical Data Note 13*, which the government issued in March 1970. This document, which contained the details of the TLVs, was meant to be read along with the Regulations, as these details were absent from the Regulations themselves (Ogden 2003, p. 5; Tweedale 2001, p. 207).

The Regulations themselves hampered any attempts the Factory Inspectorate might have made to manage asbestos risks. *Technical Data Note 13* stated that exposure should never exceed 2 fibers/mL averaged over a 4-hour period; but at the same time, medical experts were unwilling to claim a "danger to health" based on 50-year exposure estimates that themselves were based on only a couple measurements during the 4-hour periods in which the fiber/mL threshold was exceeded (Ogden 2003, p. 5). As Table 3.2 illustrates, although the number of convictions by the Factory Inspectorate increased dramatically, the average fines remained small. Further, the 1969 Regulations were later found to be "very optimistic," were severely criticized, and were ultimately overhauled in 1987.

Table 3.2 Factory Inspectorate convictions and fines under the asbestos regulations from 1931 to 1977

Year(s)	Convictions	Average fine
1931–1969	2	£12
1970	0	–
1971	1	£25
1972	40	Not available
1973	15	Not available
1974	39	Not available
1975	23	£79
1976	23	£43
1977	84	£182

Source: Chief Medical Inspector of Factories, 1931–1978, Annual Reports (London 1931–1978).

[7]The Inspectorate, at a meeting in April 1968, was to decide how the forthcoming Asbestos Regulations would be applied. The view was expressed "that there was hardly any sound biological evidence to justify the fibers per cc shortly to be adopted by the British Occupational Hygiene Society", but "Even the best textile factories in the United Kingdom had fiber counts higher than 2 and some as high as 8." A final outcome of these deliberations was that anytime the concentration over any 10 min period was <2 fibers/mL, the Inspectorate would not seek to enforce the regulatory requirements for exhaust ventilation and respiratory protection. If the 10 minutes concentration was between 2 and 12 fibers/mL, further sampling would determine if the average concentration over 4 hour was >2 fibers/mL. If so, further control would be required (DEP 1970). So 2 fibers/mL measured over a short period became the benchmark of good control. These guidelines evidently were never written down by any of the Inspectorate's experts: the version of the membrane filter they used was incapable of measuring 2 fibers/mL over 10 minutes (Ogden 2003, p. 5).

Perceived Failures of These Frameworks and Resultant Policies, Regulations, and/or Legislative Acts

As we have just alluded to, this "pre-risk" period in the history of asbestos use in the UK is marred by many more failures than successes, and it is worthwhile if not mandatory that the foundations for these failures in the management of asbestos are dealt with fully. The very nature of the present risks of asbestos management in contemporary Britain is rooted in these failures, which will be dealt with in turn.

Policy – Successful or Not?

It is difficult to identify the highlights of this era of industrial self-regulation and denial vis-à-vis asbestos. One notable fact, perhaps, is that the UK was the first country in the world to implement regulations on asbestos (i.e., the 1931 Asbestos Regulations). This achievement, however, is tarnished by the government's failure to uphold the regulations, inspect the factories, and issue fines to those in violation.

Another potential success gone astray was the close cooperation between players from the asbestos companies T&N, Cape Ltd. and BBA, and universities such as Cambridge, Edinburgh, and Reading. Together, these players formed the membership of the Asbestos Research Council (ARC). Academics such as Leydesdorff and Etzkowitz (1998) advocated close relationships between industry and universities in technological innovation, so that the (basic) research developed in academia could create technologies for use in the market or by the state. An argument can be made that risk management might benefit from a similar productive relationship in which the knowledge developed in academia on risk assessment and management could be transferred to the industry, thereby creating safer working conditions and better services. However, the Leydesdorff & Etzkowitz model for technological innovation is not restricted to a productive relationship between industry and universities; rather, these authors proposed a "triple-helix" model that crucially includes a role for government or the state. The institutional framework of the ARC and the policies (or lack thereof) that emerged from this framework did not include the third leg, which was necessary if a meaningful and effective management of asbestos was to stand.

For the most part, commentators have characterized this period as one in which the early warnings about the health hazards of asbestos were ignored (Gee & Greenberg 2001; Selikoff & Greenberg 1991). Therefore, it is hard to find any success stories from this period. In fact, the nature of the asbestos risk in the UK, as it stands today, is largely the result of the appalling factory conditions that workers during this period experienced, combined with the dismissive and indifferent approach of the industry to the risks of asbestos.

Failures to Uphold the 1931 Regulations

The problems associated with the 1931 Asbestos Industry Regulations have already been described. As we have seen, numerous factors led to their failure as a management tool for asbestos risks. One of these factors was the geo-political context in which asbestos was used in this early period and the central role it played in the British war effort. The perceived necessity of the mineral probably deterred regulators from clamping down on its production. The lack of standards which the Factory Inspectorate could measure the process against did not help either, nor did the absence of any convictions during the entire period in which the 1931 Regulations were in effect.

In the view of most analysts of the period covered in this chapter (Bartrip 2004; Budgen 2004; Gee & Greenberg 2001; Jeremy 1995; Tweedale 2000, 2001; Tweedale & Jeremy 1999; Warren 1997), the most significant cause of asbestos risks and the most significant contribution to the failure of management were the strong relationships between the industry, the regulations that were to keep the industry in check, and the research that was meant to underpin these regulations. Not only was the industry directly patronizing research through the ARC, but the results that the ARC were producing led to confusion about the grounds the policy was supposed to be based on.[8] As Tweedale notes:

> The ARC created uncertainty and suggested that various "problems" needed to be resolved by time-consuming research (rather than by common sense) ... much more investigation is necessary before firm conclusions can be reached. First to be clarified is whether the nature of the cancerous change might be related either to a specific type of asbestos or to some accompanying impurity contained in the asbestos fibre ...The disentangling of these and other questions will take much time and effort (Asbestos Research Council 1966; Tweedale 2000, p. 729).

Not only was the ARC creating an aura of uncertainty about the evidence linking asbestos to cancer, but certain ARC committees simply "vetted and approved" the research as suitable for publication (Tweedale 2000, p. 728). If a finding was too threatening to the asbestos industry, the research was simply shelved.

> The ARC's files, however, show that controversial subject-matter was delayed or amended [Asbestos Research Council 1974]. In 1967, Cape chemist Dr. Richard Gaze (who later died of mesothelioma) wanted to block publication of a paper on "Asbestos Fibers in the Air of Towns," because it was such "an intensely important and provocative subject," and because he felt that a proper study should be more comprehensive [Gaze 1967]. The ARC, however, never conducted one.

[8] It should be stressed that neither complex scientific research nor laboratory work has ever been necessary to understand the dangers of asbestos. The hazards could be appreciated simply by looking at the impact of the dust on the workforce and by examining the pathology data (Egilman & Reinhart 1995). Edward Merewether, a Medical Inspector of Factories, had shown the merits of this approach in the late 1920s. His sample survey of asbestos workers demonstrated the catastrophic health effects of inhaling asbestos and paved the way for government regulation (Merewether & Price 1930). For 1949, the pneumoconiosis statistics published routinely by the government provided further evidence for the association between asbestos and lung cancer (Ministry of Labour & Factory Inspectorate 1949). This association was confirmed again 6 years later by the publication of Richard Doll's study of lung cancer mortality at "a large asbestos works" (Doll 1955). Later epidemiological studies demonstrate the prevalence of mesotheliomas in asbestos workers and also show that even individuals not exposed to asbestos at work could develop the disease (Wagner et al. 1960; Selikoff et al. 1964; Newhouse & Thompson 1965; Tweedale 2000, p. 725).

Further, the ARC simply did not engage in epidemiological studies, which is odd given the size of the workforce exposed to asbestos and the nature of asbestos and cancer. One of the rare occasions for such research arose in 1952, when Dr. John Knox asked Dr. Richard Doll to study the mortality rates among textile workers who were exposed to asbestos at Turner Brothers' Rochdale factory. Doll showed an increased risk of lung cancer for the cohort. However, this report too was blocked by T&N (Tweedale 2000, p. 726). When Doll pressed ahead, the industry contemplated legal action and tried to intimidate the editor of the medical journal that eventually carried the piece (Greenberg 1999).

In 1966, when a reformulation of the 1931 Regulations was inevitable and the BOHS finally stepped in to contribute to the scientific edifice for the new regulations, the major players in the asbestos industry – by way of the ARC – were still able to exert influence. In fact, we have already shown how members of the ARC – who would have been funded directly or indirectly by companies like T&N – actually sat on the BOHS subcommittee that was responsible for developing the new asbestos thresholds that took effect in 1969/70.

Critique of the BOHS TLV

It is bad enough that the BOHS was heavily influenced by the ARC (and thus the industry) and that the BOHS subcommittee responsible for developing the TLVs contained two members from the TBA, one member from Cape Ltd, and one member from the BBA (Tweedale 2001, p. 199). But to make matters worse, the TLVs were subsequently criticized severely by some of the very same people responsible for creating them. Berry, the statistician responsible for the TLV calculations, later published a paper arguing that the TLVs were flawed in the following ways: (1) they were based on current employees only (i.e., the sick and the dead were discounted); (2) they took no account of the progression that might occur in the future, even if there was no future exposure (it had long been appreciated that asbestosis often developed after the cessation of the exposure, and dust measurements were not available for the early years of exposure, requiring that the lower limits be estimated); (3) the measurement of the health effects of asbestos was dependent on a single observer (the techniques available at the time for reducing observer bias and standardizing the reporting of qualitative and quantitative effects were not employed) (Berry 1978).

Greenberg (1997a) pointed out that the TLVs were created with the relationship of asbestos and asbestosis in mind, and "the greatest weakness was surely that the risk estimates took no account of cancer." This was because, as the report said, "the quantitative relationship between asbestos and cancer risk is not known," although the risk was expected to be less at lower exposures (Ogden 2003, p. 4).

Improvements and Reform of the Perceived Policy Failures

The 1970s brought some light to the dark period of asbestos use and its (lack of) management. This relief came largely in the form of the 1974 Health and Safety at

Work Act (which is dealt with in the next section). Nonetheless, there was a gradual decline of the asbestos industry during this period.

The Health and Safety at Work Act ultimately established more effective state-run forms of institutionalized risk management. Unfortunately, the seriously flawed 1969 Asbestos Regulations remained intact until they underwent a major overhaul in 1983–1984. This overhaul followed in the wake of the establishment of the government Advisory Committee on Asbestos in 1979 and the European Community's directives in 1983 (all of which are dealt with in the next section).

In the meantime, the 1960s saw the rise of litigation against the asbestos industry by its workers. In 1963, the case that would come to be known as *Smith v. Central Asbestos Company* was launched. In this case, the claims of seven individual workers were jointly tried in the civil courts. After an original ruling in favor of the claimants and an appeal by the Central Asbestos Company on May 26, 1971, the men were finally awarded damages for negligent exposure to asbestos – the first ruling of its kind (Budgen 2004, p. S78). Whereas this landmark ruling was a bittersweet victory for the asbestos workers, it was also a sign of the near-total failure of the policies for managing asbestos risk. Court rulings and litigation are not meant to supplement policy, but rather to be a last resort for righting injustices. It was the job of government – not the courts – to create and uphold regulations based on existing scientific and lay evidence and that were supposed to protect the citizenry. The government simply failed to do this, and thus in 1963, when the courts were forced to step in and hand down judgments for damages, it was a sign that the existing regulatory policies had failed to protect workers, and that self-regulation was likewise an inadequate approach to the management of asbestos hazards.

Period 2 (1974 to 1999): Establishing an Institutional Risk Framework for Asbestos

The Context – Institutionalization of Risk and the "Controlled Use" of Asbestos

In the 1970s, the unwillingness of the asbestos industry to regulate itself was recognized and the ineptitude of attempts to regulate and manage asbestos was accepted. Legislative acts and the bureaucratic arrangements that followed from them were intended to compensate for these failings. The management of health risks became much more strategic during this period. In general, they can be characterized as the institutionalization of risk assessment and calculation, a concurrent overload of policy innovations to deal with these risks, an expanded participation by various relevant actors in discussions of asbestos risk, increased media pressure and press exposure of the risks, and a certain amount of scientific controversy about the threat to **public health** (rather than simply to industry) posed by white (chrysotile) asbestos. All these factors together have led commentators to see this era in the history of asbestos as one of "controlled use" of the substance, particularly when juxtaposed with the free-for-all period that it followed.

The rationale behind this segmentation (i.e., 1974 to 1999) of the history of asbestos use and management is straightforward. In 1974, the Health and Safety at Work Act (HSWA) was passed. It created mechanisms whereby government institutions could monitor conditions in the workplace and protect workers who might be exposed to asbestos (Health and Safety Commission 2002, p. 4). The period ended in 1999, when the UK "Europeanized" its approach to asbestos risk management by implementing an EU ban on all remaining chrysotile (Bartrip 2004, p. 74). The significant developments that took place during these 25 years have come to characterize the period. Many of these developments created legacies that can be seen in current approaches to asbestos risk management. It is necessary, therefore, to examine these developments in detail.

The Legislation – Health and Safety at Work Act (HSWA) – 1974

The laws concerning health and safety at work began to be reviewed as early as 1967. In 1970, a parliamentary committee was convened to review these specific statutory and voluntary arrangements (Simpson 1973, p. 192). The committee was headed by Lord Robens (who learnt his name to the committee – the Robens Committee) and it filed its report in 1972. One of the Committee's conclusions was that "the most important single reason for accidents at work is apathy" (Robens Committee 1972, para. 13). In an attempt to respond to this apathy, as well as to a number of other problems raised in the Robens Report, the Health and Safety at Work Act (HSWA) was passed in 1974 (Health and Safety Executive 2004, p. 3).

Robens criticized the overly detailed and incomprehensible nature of existing work safety legislation, as well as the fragmentation of administrative jurisdiction. This fragmentation ultimately led to "situations in which some establishments fell outside of all legislation, different parts of others fall under different Acts, and it is impossible for any single department to take decisive initiatives" (Robens Committee 1972, chap. 1). The HSWA basically embodied the recommendations of the Robens Report: "In place of existing detailed and prescriptive industry regulations, it created a flexible system whereby regulations express goals and principles, and are supported by codes of practice and guidance." (Health and Safety Executive 2004, p. 3).

Given the particular history of how asbestos was "managed" in the early twentieth century, it is somewhat surprising to discover that the Robens Report and the HSWA operated on the premise that "those that create risk are best placed to manage it" (Health and Safety Executive 2004, p. 3). The crucial difference between the industrial self-regulation of asbestos in the period from 1924 to 1974 and the HSWA proposition that "those that create risk are best placed to manage it" is that the HSWA provided for a state-run institutional apparatus to ensure that this "self-management" actually took place. This institutional apparatus took the form of the Health and Safety Commission (HSC) and the Health and Safety Executive (HSE). Very much in line with Robens' criticisms of the previous work safety apparatus, the goals of the HSWA – and its subsequent institutional manifestations – were (1) to centralize the responsibility for health and safety in the workplace in one legislative act and institution,

(2) to establish clear regulations and responsibilities that employers could follow, and (3) to set out general duties that were to apply across industries or sector-specific responsibilities to be stated in other regulations (Health and Safety Executive 2001, p. 7).

The Institution – Establishment of the HSC and HSE in 1974

The two new institutions established by the Health and Safety at Work Act (HSWA) of 1974 were the Health and Safety Commission (HSC) and the Health and Safety Executive (HSE). The HSC is responsible for making the "arrangements to secure the health, safety and welfare of people at work, and the public" (Health and Safety Commission 2002, p. 4). This means primarily that the HSC proposes laws and standards concerning health and safety at work, conducts research, and provides information and advice (particularly on occupational health matters) (Health and Safety Commission 2002, p. 4). The HSE, on the other hand, is responsible for upholding the laws, regulations, and standards that the HSC creates. Its staff is made up of inspectors, policy advisors, technologists, scientists, and medical experts. These two bodies work closely together (Health and Safety Commission 2002, p. 4).

The main idea behind the HSWA, the HSC, and the HSE was "preventative enforcement," which means that workers are protected better by dealing with a hazard before it causes a problem than by retrospectively punishing a company for violating the laws and standards (Health and Safety Executive 2004, p. 8). This goal was accomplished by issuing notices to employers who were operating risky work situations or were in direct violation of a standard. In 1976, the HSE sent out "7,334 notices and instituted 1,200 prosecutions. By 2002/03, the number of notices issued had risen to 13,263, while prosecutions were taken for 1,688 separate alleged offences" (Health and Safety Executive 2004, p. 8). These notices effectively alerted the employer that the company's working conditions needed improvement. The relatively slow growth from 1976 to 2003 in prosecutions compared to notices indicates that this approach was functioning as an institutional instruction to make changes rather than as a form of punishment. According to the HSC:

> The philosophy underpinning the HSW Act is that prevention of harm is the primary aim and enforcement has never been regarded as the only or even the main way of securing compliance. Over the years, HSC has remained wedded to the principle of a discretionary approach to enforcement and to obtaining an appropriate balance between enforcement and advice. HSC's current enforcement policy stresses that any action should be: proportionate to any risks to health and safety and to the seriousness of any breach; targeted primarily on those whose activities give rise to the most serious risks or where hazards are least well controlled; consistent; and transparent (Health and Safety Executive 2004, p. 8).

More details on the makeup and institutional structure of the HSC and HSE are provided in the later section, where the current strategic management and communication framework for asbestos risk in the UK are outlined. What is important to note now is the timing and context for the establishment of these institutions, as well as their "preventative enforcement" approach and the role this approach subsequently played in asbestos regulations.

The 1976 Advisory Committee on Asbestos (the Simpson Committee)

The numerous health hazards associated with asbestos finally came to the forefront in a 1976 report by the parliamentary and health service to the Ombudsman, Sir Alan Marre.[9] This report concerned the Cape's Acre Mill asbestos factory in Hebden Bridge. Marre discovered that since the factory closed in 1970, more than 10% of the 2,000+ workers contracted devastating asbestosis. These numbers exceeded what the previous TLV and the previous understanding of exposure-risk (dose-response) relationships estimated (London Hazards Centre 1995, chap. 10).

The government ordered a major inquiry. It was led by Sir William (Bill) Simpson, chair of the Health and Safety Commission and an ex-trade unionist with a strong Labor Party affiliation (Tweedale 2001, pp. 244–251). The committee conducting the inquiry became known as the Advisory Committee on Asbestos (ACA), or simply the Simpson Committee. It reviewed not only the implications of Marre's Ombudsman report but, more generally, the 1969 Asbestos Regulations (Greenberg 1997b; Greenberg 2004, p. 539). See Table 3.3 for a complete list of the ACA members, who were intended to represent a cross-section of interests and expertise within and surrounding the asbestos regulations.

Table 3.3 Members of the Advisory Committee on Asbestos (ACA), the Simpson Committee

The following membership represents the balanced set of 'interests' deemed appropriate after the 1974 Health and Safety at Work Act (HSWA)	
W. Simpson	Chairman, Health and Safety Commission
Prof. E.D. Acheson	Faculty of Medicine, Southampton University
A.C. Blyghton	Transport and General Workers Union
The Hon P. Bradbury	Confederation of British Industry
Dr. J.S. Gilson	formerly Director of the MRC Pneumoconiosis Unit
Mr. H.D.S. Hardie	Director, Turner & Newall
Mr. W. Lewis	Union of Construction, Allied Trades and Technicians
Mr. W.D. Lomas	National Union of Dyers, Bleachers and Textile Workers
Prof. A. Mair	Dundee University
Dr. M. Molyneux	Occupational Hygienist, Institute of Naval Medicine
R.C.J. Stairmand	Consultant Chemical Engineer and Physicist
Dr. J. Steel	Occupational Hygienist, University of Newcastle upon Tyne
Mr. F.C. Sugden	Chief Environmental Health Officer, Middlesbrough
Dr. Margaret Turner-Warwick	Professor of Thoracic Medicine, University of London
Mr. A.W. Ure	Director, Trollope and Colls Ltd
Mrs. R. Waterhouse	Consumer Association

Source: Greenberg 2004, p. 539.

[9] "We [the Ombudsman] provide a service to the public by undertaking independent investigations into complaints that government departments, a range of other public bodies in the UK, and the NHS in England, have not acted properly or fairly or have provided a poor service" (Parliamentary and Health Service Ombudsman 2007).

In 1977, public sessions were held during which evidence about the relative safety of asbestos was presented from all sides, with no cross-examination (Tweedale 2001, p. 244). In addition to the major industrial players, unions, public interest groups and activists who had lost family members to asbestos-related diseases (e.g., "Asbestos Action", the Society for the Prevention of Asbestosis and Dust Disease, the British Society for Social Responsibility in Science, the Consumers' Association, and the Royal Institution of Chartered Surveyors and Professor Julian Peto) were all heard (Greenberg 2004, p. 540).

The Simpson Report was published in 1979 (Simpson Committee 1979). Although damning some types of asbestos, it basically concluded that a doctrine of controlled use was possible for white (chrysotile) asbestos (Tweedale & McCulloch 2004, p. 247). The Simpson Committee only had the power to make recommendations and not to pass regulations or laws. It nevertheless recommended that blue (crocidolite) asbestos, as well as the spraying of all asbestos for insulation, be banned. However, the industry had already essentially abandoned these by this stage of the game. In addition to recommending that those responsible for removing asbestos be licensed, the Simpson Report suggested that the TLV for white asbestos be changed from 2 fibers/cm^3, as specified in the BOHS Asbestos Regulations of 1969, to 1 fiber/mL, with a target of 0.5 million fibers/m^3 (0.5 fibers/mL) for "brown" (amosite) asbestos (Gee & Greenberg 2001, pp. 56–57).[10]

The Other Players – The Rise of NGOs and the Role of the Media

The report of the Ombudsman in 1976 and the open consultation held by the ACA/Simpson Committee in 1977 spurred into action groups that had not previously participated in the debate over the management of asbestos risk. Not only did the outrage that occurred at the time of the Hebden Bridge episode drive people to action, but this was a time when there were more avenues for people to be legitimately heard. The HSC was – from its inception – dedicated to consulting as "widely as possible on all aspects of policy and regulation"[11] (Health and Safety Executive 2004, p. 10).

A number of non-governmental organizations (NGOs) rose out of these tumultuous times. In 1975, immediately following the Acre Mill episode, a group of concerned citizens who had been affected by the Hebden Bridge formed the Asbestos Action Group. Part of this group's success was its ability to unite politicians, doctors, lawyers, and former Cape employees to lobby the government and launch personal

[10]Do we want to include a rather technical critique of the exposure data that were used in the follow-up report commissioned by the ACA/Simpson Committee on the varying dangers associated with different kinds of asbestos?

[11]"This [commitment to consultation] led to the 1977 Safety Representatives and Safety Committees Regulations which established the system of safety representatives appointed by trade unions which has provided one of the cornerstones of health and safety consultation ever since" (Health and Safety Executive 2001, p. 10).

injury lawsuits (Tweedale 2001, p. 240). A similar vocal public interest group that joined the discourse around this time was the Society for the Prevention of Asbestosis and Industrial Diseases (SPAID), which was organized in 1978 by Nancy Tait. After her husband died from indirect asbestos exposure at the Post Office where he worked, Tait launched a lobbying campaign to ban all forms of asbestos. She also argued that white (chrysotile) asbestos was indeed carcinogenic (Tait 1983; Tweedale & McCulloch 2004, p. 247). SPAID, which later became the Occupational and Environmental Disease Association, was becoming vocal in public consultations on asbestos, and it forced reviews when the compensations provided by companies such as T&N were absurdly low (Tweedale 2001, p. 242).

Following the emergence of a number of public interest groups in the UK, international bodies began to get involved with asbestos research in the hope of influencing the regulations. In early 1970, the World Health Organization (WHO) launched the International Agency for Research on Cancer (IARC), which initiated research to explain the relationships between certain chemicals and cancer (Gee & Greenberg 2001, p. 57; Tweedale & McCulloch 2004, p. 248). Some of the IARC's early work involved the establishment of an Asbestos Working Group, which gave its first report in 1973 (Tweedale & McCulloch 2004, p. 248).

These early reports were not overly critical, but later in 1976 the Asbestos Working Group concluded "that all forms of asbestos caused lung cancer and mesothelioma and that it was impossible to designate a safe threshold" (Tweedale & McCulloch 2004, p. 248). It is important to note that the IARC and its Asbestos Working Group were nothing more than a research-based offshoot of the WHO. None of the groups' findings on the risks of asbestos led to mandatory legislative or regulatory action. Even though their results were "findings with no teeth," it is important to note that this was the time when international agencies began to take interest in the effects of asbestos on exposed populations and that the discourse about asbestos risk was no longer confined to Britain. These facts became much more evident when the European Commission and the EU ultimately forced the UK's hand on asbestos in the late 1990s.[12]

Another key transformation of this period that affected the management of asbestos risk was the increasing role of the media in raising awareness and broadening the discourse. During earlier wars and post-war periods, asbestos had a relatively positive image, which was managed by the industry. Much of the information concerning asbestos was mediated by the ARC. As cases of asbestosis, lung cancer, and mesothelioma continued to emerge, and reports began to surface that contradicted the earlier industrial pronouncements of the safety of asbestos, the media began to pick up on these cases and report them.

There were a number of media "furors" in the mid-1970s. Many of these concerned the Ombudsman report and the Simpson Committee, but none of them had as

[12] In this section on the "other players" we could also include a brief discussion of the fact that the British Occupational Health Society (BOHS) kept meeting until 1982 to analyze data and discuss asbestos TLV. That being said, the scope of their influence in terms of setting or suggesting asbestos thresholds was severely limited by the establishment of the Health and Safety at Work Act as well as the HSC and HSE. Such a discussion of their meetings until 1982 may not be warranted.

much impact as the 2-hour documentary, "Alice – A Fight for Life." Aired in 1982 during prime-time on Yorkshire TV, the documentary traced the life of a 47-year-old mother named Alice Jefferson as she struggled with mesothelioma, which she had contracted after working only a few months at the Cape Mill in Hebden Bridge (Tweedale 2001, p. 251). It painted a broad picture of the mining, production, and administration of asbestos with the general goal of raising awareness of the associated health risks. Overall, it attacked the industry for denying the problem. For example, it showed officials of T&N testifying that there was no asbestosis in the waving shed at its largest factory, TBA. The documentary also criticized the government for its inefficiency in dealing with the hazard, including its delay in implementing the recommendations of the Simpson Committee, as well as the deceptive statements of some medical professionals in defense of the industry (Tweedale 2001, p. 251). The documentary was so provocative and the accusations so hard-hitting that the directors and producers were summoned to appear before the House of Commons Employment Committee in 1983 to present evidence for the claims made in the documentary. The evidence was reviewed, but the makers of the documentary remained committed to their position and did not retract any of the evidence. The "Alice" documentary outraged the public, and T&N's stock price plummeted (Tweedale 2001, pp. 252–253).

Clearly the winds had shifted regarding the provision of information about asbestos and the concurrent discourse about its risks. The industry and the ARC no longer were a monopoly in this regard, and no longer did they have sole lobbying access to the regulators who were to set the ever-crucial fiber concentration limits. New players were now involved in the asbestos debate, and information about the potential risks of asbestos was being communicated through new channels.

The Regulations – Hyper-Regulatory Innovation

The regulations, limits, and laws governing the management of asbestos proceeded at a snail's pace throughout the 1930s, 1940s, 1950s, and 1960s. It is important to recall that there were only two regulations to speak of during this 40-year period: the 1931 regulations and the 1969 regulations. Conversely, the 1980s and 1990s can be characterized as having a relatively large amount of regulatory hyperactivity around the issue of asbestos risk, which resulted in five new asbestos regulations in the 11-year period from 1983 to 1994 alone. Arguably, this regulatory vigor can be traced to the establishment of a specific institution (the HSC) responsible for health and safety at the workplace. There was increased pressure from the NGOs, public interest groups, and the media, and it became necessary for the UK to play a catch-up game to extract itself from the dreadful state of affairs resulting from the prior decades of uninhibited asbestos use.

With the HSC in place, and with the legislative authority and responsibility for passing relevant regulations for asbestos management clearly and solely in its grasp, the 1980s and 1990s saw a boom in regulations that steadily reduce the acceptable asbestos concentration levels. We will discuss the individual regulations only briefly, because the institutional processes that gave rise to them are generally part of the operations of the HSC (the structure of which is described in detail in Part 3) and because of the drastic decline in the industry's impact on said regulations.

Implementation of the Simpson Committee Recommendations (1982/3)

Although the HSE denied that the implementation of the Simpson Committee's recommendations had anything to do with the public pressure resulting from the "Alice" documentary, enforcement of the recommendations of the Advisory Committee on Asbestos began within days of the TV broadcast. As a result, at the end of 1982, new limits on fiber concentrations took effect: 1 f/cc for chrysotile (white asbestos), 0.5 f/cc for amosite (brown asbestos), and 0.2 f/cc for crocidolite (blue asbestos). The latter limit for crocidolite was the same as the previous one (Tweedale 2001, p. 255).

Asbestos (Licensing) Regulations of 1984

In 1983, almost immediately after implementing the Simpson Committee's recommendations, the HSC requested a review of the existing limits and regulations for asbestos. A new set of regulations was subsequently issued that tightened the exposure limits even further. This tightening resulted in the 1984 Asbestos (Licensing) Regulations of 1984, which changed the fiber concentrations that had been mandated by the Simpson Committee from 1 f/cc to 0.5 f/cc for white (chrysotile) asbestos, and from 0.5 f/cc to 0.2 f/cc for brown (amosite) asbestos; the concentration of 0.2 f/cc for blue (crocidolite) asbestos remaining unchanged (Tweedale 2001, p. 256). Crucially though, given that these were licensing regulations, the import and manufacture of brown and blue asbestos were banned, but exposure limits were still needed for the other newly banned substances that were already present in schools, ships, and other public and private buildings (Tweedale 2001, p. 256).

(Control of) Asbestos (at Work) Regulations of 1987

"The limit of 1 fibre/mL for chrysotile, averaged over 8 h, and the overriding requirement to reduce exposure as far as is reasonably practicable were incorporated in the 1983 European Directive" (Council of the European Communities 1983). This was implemented in Britain through the 1987 Asbestos Regulations,[13] which incorporated a stricter limit of 0.5 fibers/mL over 4 hours. Currently, the 4 hour chrysotile limit in Britain is 0.3 fibers/mL, and the use of chrysotile is prohibited (with minor and temporary

[13] We are currently having difficulties accessing this Directive in order to unpack it, as well as its implications for UK asbestos regulation.

exceptions) – a measure that the ACA did not feel able to recommend" (Ogden 2003, pp. 5–6). It is important to note that this is the first time the European Community directly influenced British regulatory procedures vis-à-vis asbestos. As the UK continued throughout the 1990s to harmonize its health and safety policies with the European standards, the EC/EU came to play an increasingly important role in the management of asbestos risks at the local level.

The Asbestos (Prohibitions) Regulations of 1992

The (Prohibitions) Regulations, as amended, banned the importation, supply, and use of raw asbestos and asbestos containing materials (Health and Safety Executive 2005, p. 4).

Asbestos Exposure Limits of 1994

As in Britain, stricter measures on the manufacture, import, and processing of asbestos and products containing asbestos followed in the 1980s and 1990s. A permitted exposure limit of 0.1 ff/mL (fibres per millilitre) was introduced in 1994 (Bartrip 2004, pp. 74–75).

The Successes and the Failures of Asbestos Risk Management From 1974 to 1999

There is little doubt that the establishment of a clear institutional framework for assessing asbestos risks (i.e., the HSE) and the setting of concurrent regulations (i.e., the HSC) was a massive step forward in the UK. This centralized structure, which was outlined in the Health and Safety at Work Act of 1974, meant that clear lines of responsibility could be drawn from regulator to industry, lines that may have been blurred in the earlier period.

One of the major criticisms of this institutional development was that it came too late. By the time the regulations – which were not those of the asbestos industry – started being enforced, the UK had been subjected for over 60 years to dangerous exposure levels of the mineral. Unfortunately, this establishment of an institutionalized risk framework for asbestos in the UK could not undo the damage that had been caused from the 1920s up to the 1970s. However, the establishment of the HSC and HSE did set in motion a number of important developments in the communication and management of risk.

First, the processes by which the standards and concurrent regulations were determined were subjected to a negotiation process involving a wide variety of actors, including NGOs, labor representatives, and public interest groups. This is a clear departure from the previous era, which saw industry and professional associations (which were often connected) play the major role.

The more stringent regulation of fiber concentrations for blue (crocidolite) and, subsequently, brown (amosite) asbestos in effect meant the abandonment of these minerals, because in practice the standards could not be met. It is disconcerting that these bans did not become official until 1984. Such reluctance to place an all-out ban on blue, brown, and even white asbestos highlights one of the major criticisms of the this period, which has been characterized as "an era of controlled use" (Tweedale & McCulloch 2004, p. 247). Despite mounting evidence of asbestos' dangers, and mounting death tolls from asbestos-associated diseases, regulators seemed to have gotten stuck with "doing regulation." The 1980s and 1990s saw a series of revised regulations that increased the stringency of fiber concentration limits. Although at face value a relative increase in the stringency of asbestos regulations may seem to be a sign of an institutional risk management approach at work, this should not be confused with a risk management process that is really working.

What was needed during this period was what we currently have in the UK – a total ban on asbestos, all forms of it. The fact of the matter is that by the time the HSWA had passed and the HSC and HSE had been established, nearly half of all the asbestos imported to the UK was being used for cement floor tiling. Substitutes were available for other asbestos-related processes, and if any of these substitutes provided the same structural support as asbestos, then the use of asbestos for flooring could be abandoned (Tweedale 2001, p. 247). Regulators either did not feel that such a ban was justified based on the scientific evidence at hand, or they believed that the economic impact of a ban would eliminate the asbestos industry. Either way they were wrong, so they tried to hedge their bets by issuing ever more stringent regulations.

Perhaps one of the brightest spots of this period of asbestos risk management was the integration of the European Directives. International bodies, such as the International Agency for Research on Cancer (IARC), and in particular their Asbestos Working Group, had long been advancing the claim that any kind of regulation other than a ban would be counter-productive, as it was impossible to set a safe threshold for any kind of asbestos. The IARC was constrained by being a research body, and thus it had no "teeth" to impose its views. This restriction, however, did not apply to the EC/EU, and as they added risk management to their discourse on asbestos, things really began to change in the UK.

Period 3 (1999 to 2007): The "Europeanization" of Asbestos Risk and the Current Management Framework for Asbestos Risk

Characteristics of Asbestos Risk Management in This Period

The past 5 to 10 years have seen the European Commission (EC) and European Union (EU) take an increasingly larger role in matters concerning risk regulation in general. This applies to asbestos, as well as BSE. In light of the expanding role of supra-governmental bodies, domestic risk management has become, for individual European nations, more a matter of implementation. For the UK, this meant the development of regulations that meshed with the existing European framework. In fact, as we shall see, after decades of a rather cavalier approach to asbestos management, the UK even went so far as to pre-empt some of the European legislation and pass their own regulations, or they adopted the European regulations well before their implementation deadline.

The recent era of asbestos risk management can therefore be characterized not only by the increasing role of the EC/EU, but also by the continued employment of institutional measures to regulate asbestos at the domestic level. The HSC and the HSE continue to play a central role in asbestos risk management, with new policies passed as recently as November 2006, with enforcement beginning in April 2007. Similarly, the domestic regulatory focus shifted very much toward safety in the maintenance and removal of asbestos, as its production had more or less stopped.

The recent era of asbestos management in the UK can therefore be characterized as demonstrating a clarification and solidification of institutional and regulatory mechanisms to deal with the asbestos problem. However, this may not be entirely true for the communication of the risks. To be sure, asbestos has not "fallen off the radar screen" – the risk issue per se is still regularly brought up in newspaper articles describing in detail the deaths caused by asbestos-related cancers (Pye 2007), fire fighters still receive treatment after asbestos exposure (Forster 2007), and the asbestos risks associated with ship dismantling are still reported (Churchard 2007). The terrible legacy of the early periods of asbestos management – or lack thereof – has etched itself into the minds of the British public, with the negative side-effects of exposure now being common knowledge. That being said, it is still difficult to characterize these recent approaches to risk communication approaches because (1) the risks are now well known, and (2) there seems to be a separation between the HSE's provision of information about asbestos and its associated approach to risk-management regulation on the one hand, and the mere reporting of asbestos-related problems by the media on the other.

The following – and final – section on the strategic management and communication of the risks related to asbestos consequently focuses on the institutional mechanisms now in place for dealing with these risks in the UK, and it provides the basis for drawing conclusions from this case study.

Relevant Actors in the Use and Management of Asbestos

As mentioned in the introductory section, the period from 1999 to the present has continued to be one in which the HSE and HSC have played a central role in the strategic management of asbestos. For example, in 1999 the EC/EU introduced a ban on chrysotile asbestos that started to be implemented in 2005.

A new actor in this recent period of the strategic management of asbestos in the UK has been the insurance industry. In the context of the current broad acceptance of a clear association between asbestos exposure and respiratory diseases such as mesothelioma, and the history of negligence on the part of the asbestos industry, the insurance companies have come to play a central role in negotiating risk relationships and compensation. Although the evolution of compensation levels for asbestos exposure are briefly covered below, it is important to note that between 2001 and 2004 two large insurance companies "went bust" in the UK, "partly due to the escalating number of claims for asbestos-related diseases" (Budgen 2004, p. S78).

The Institutional Framework for the Management of Asbestos Risk

Clearly, the British courts and the insurance industry are parts of the institutional framework responsible for dealing with the management of asbestos risks. These institutions are responsible for responding to the failures of risk management strategies, and are thus reactive in nature rather than policy and regulatory institutions that would hopefully be more proactive, preventative, and precautionary.

In the UK, there are currently three courses that can be followed for settling a compensation claim for asbestos exposure: "The first is state social security through a prescribed industrial disease system called 'Industrial Injuries Disablement Benefit' (IIDB). IIDB covers those suffering from asbestosis, mesothelioma, lung cancer (with pleural thickening, or asbestosis) and bilateral diffuse pleural thickening. The second route is through the state no fault compensation scheme: under the Pneumoconiosis etc. (Workers' Compensation) Act 1979, payments can be obtained if no relevant employer is still in operation (and provided disablement benefit has also been paid to the applicant in respect of the prescribed disease). The condition is that the applicant has not already brought any action, or compromised any claim, for damages in respect of the injury complained of. Thirdly, a civil claim can be made for damages. These can be pursued for all of the diseases referred to previously, plus symptom-free pleural plaques" (Budgen 2004, p. S78).

The more proactive institutions involved in the risk management of asbestos are the EC and the domestic HSC and HSE. It should be noted that the WHO has also been involved with risk management related to asbestos,[14] but since the EC adopted its ban on the further use of asbestos in production processes, its role has been that of arbitrator (Greenberg 2004, p. 540).

Given that the HSC and HSE are the main institutions responsible for the strategic management – as well as, to a certain extent, the communication – of many risks,

[14] An issue that will surely come up in the French case study of asbestos.

including asbestos, it is perhaps worthwhile to describe in detail its organizational structure and current mode of operation.

The HSC is responsible for health and safety regulations in Great Britain. The HSE and local governments are the enforcement mechanisms that support the Commission. The HSC is lodged within the Department of Work and Pensions and reports to the Parliamentary Under Secretary for Work and Pensions.

The HSE and HSC try to base their operations on five principles, which then govern decision-making. The principles are (1) *targeting of action*: a focus on the most serious risks or where the need for greater control of hazards is greatest; (2) *consistency*: the adoption of similar approaches to similar circumstances to achieve similar ends; (3) *proportionality*: the requirement for actions that are commensurate with the risks; (4) *transparency*: openness to how decisions were arrived at and their implications; and (5) *accountability*: making clear who is accountable when things go wrong (Health and Safety Executive 2001, p. 19).

The literature published by the HSC and the HSE states – in keeping with the above mentioned principles – that these British risk-management institutions have followed six procedural stages: (1) deciding whether the issue is primarily one for HSC/HSE to handle, (2) defining and characterizing the issue, (3) examining the options available for addressing the issue, as well as their merits, (4) adopting a course of action for addressing the issue efficiently and in good time, informed by the findings of the second and third stages and with the expectation that as far as possible the actions will be supported by the stakeholders, (5) implementing the decisions, and (6) evaluating the effectiveness of the actions taken and revisiting the decisions and their implementation if necessary (Health and Safety Executive 2001, p. 21).

These stages do not apply universally to all risk-management decisions undertaken by the HSE and HSC. Because the system was built over time, because many of the regulations emanate from the EC/EU, and because some situations call for the circumvention of one or more of the steps, the six stages should be considered more as a sociologically "ideal type" of risk-management decision-making stages (Health and Safety Executive 2001, pp. 21–22).

In the document entitled "Reducing risks, protecting people: HSE's decision-making process," the HSE states a number of conventions that they follow when undertaking risk assessments. To be sure, these are some of the same conventions used in the HSE's risk assessment decision-making process, but they are nevertheless listed as follows:

Actual and Hypothetical Persons This convention refers to the fact that when conducting a risk assessment exercise, real people need not be involved; rather, the assessment can be undertaken by a "hypothetical person [...] as an individual who is in some fixed relation to the hazard." This use of the hypothetical person is beneficial in the sense that the authorities do not need to wait for a hazard to take place and to affect people before they conduct a risk assessment; it allows the authorities to assess the risk for people in general rather for variations in individuals' "physical make up, abilities, age, and the circumstances giving rise to their exposure" (we should note that a number of hypothetical persons are normally incorporated in the risk assessment to represent a particular at-risk population). It thus avoids the difficulty of

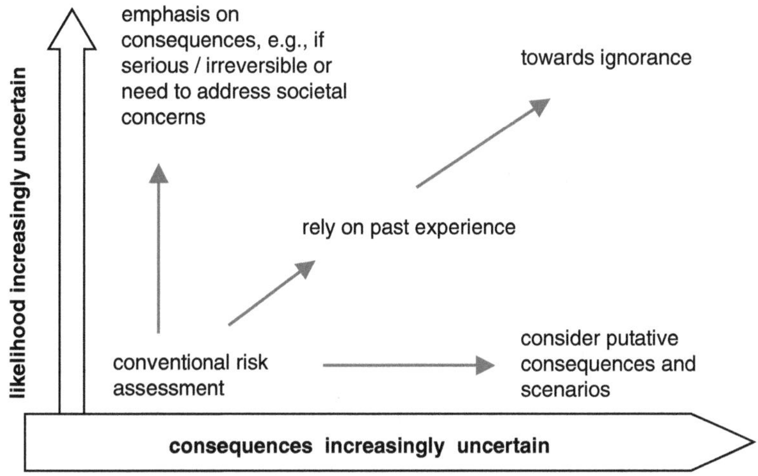

Fig. 3.3 Procedures for tackling uncertainty when assessing risks

having to "extract and distil useful information from all the individual assessments" (Health and Safety Executive 2001, p. 53).

Standards The results of assessments done in relation to hypothetical persons are also used for the adoption of standards. Standards can be regarded as generic control measures that must be applied to eliminate or reduce the risks for a particular hazard. The scope of the standard is set by specifying the circumstances in which the hazards give rise to the risk. One feature of using standards is that once adopted they may be regarded as applying to the hazard rather than to the risk in the sense that they are applied to control risks whatever the circumstances, for example, however short the actual exposure to the hazard (Health and Safety Executive 2001, pp. 55–56).

Procedure for Handling Uncertainty This is perhaps the most interesting of the three conventions and is illustrated in Fig. 3.3, which is reconstructed from the HSE document (Health and Safety Executive 2001).

In this figure, the vertical axis corresponds to the relative certainty that a risky event will actually take place, whereas the horizontal axis corresponds to the relative certainty about the outcome of that risky event. In the case of asbestos, a conventional risk assessment can be reliably applied because it is well known that the handling and maintenance of asbestos is bound to take place, as the mineral is present in our walls, floors, and ships. A conventional risk assessment for asbestos is also feasible, because the consequences of asbestos exposure are now well known: death by mesothelioma or asbestosis. For the HSE, uncertainty is handled at the bottom of the vertical axis – where likelihood of the risky event is increasingly unknown – by assuming "that the event will be realized by focusing solely on the consequences" whereas the far right of the horizontal axis – when the outcome of a risky event is increasingly unknown – "putative consequences are deliberately assigned to the hazard" (Health and Safety Executive 2001, p. 56).

Policies, Regulations, and/or Legislative Acts Based on the Institutional Frameworks

Given that the EU and EC, as well as the HSC and HSE, are now the main institutions responsible for the strategic management of asbestos risks, it is worthwhile to briefly outline some of their primary policy contributions to resolving the asbestos problem.

In July 1999, the EC announced an EU ban on the use of all remaining chrysotile by January 1, 2005. This ban does not mean that all chrysotile should be removed and disposed of, but rather that it should no longer be used in any production processes. An important rider was attached to the ban, stipulating that exemptions could be applied for if no substitute substance could be identified and there would be no health or environmental damage. The UK implemented the ban in October of 1999, 3 months later – and 5 years ahead of schedule. Bartrip notes that other EU countries beat the deadline and/or introduced their own domestic bans (Bartrip 2004, pp. 74–75).

The most recent piece of domestic legislation to be introduced in the UK was the Control of Asbestos Regulations that was issued on November 13, 2006, and came into force in April 2007. This 2006 regulation brought together and replaced three existing pieces of legislation: the Control of Asbestos at Work Regulations of 2002; the Asbestos (Licensing) Regulations of 1983, as amended; and the Asbestos (Prohibitions) Regulations of 1992 (Prohibitions Regulations), as amended.

"The Regulations prohibit the importation, supply and use of all forms of asbestos. They continue the ban introduced for blue and brown asbestos in 1985 and for white asbestos in 1999. They also continue the ban on the second-hand use of asbestos products such as asbestos cement sheets and asbestos boards and tiles; including panels which have been covered with paint or textured plaster containing asbestos" (Health and Safety Executive 2009). Again, these regulations apply solely to future (new) use of asbestos. Existing asbestos materials that are maintained in good condition can be left as they are, as long as they are monitored and they are not tampered with or removed without protection (Health and Safety Executive 2009).

Successes and Failures of Those Frameworks and Resultant Policies

It is, for a number of reasons, difficult to appraise the above-mentioned risk-management strategies regarding asbestos because (1) they are relatively recent, and (2) the damage has already been done.

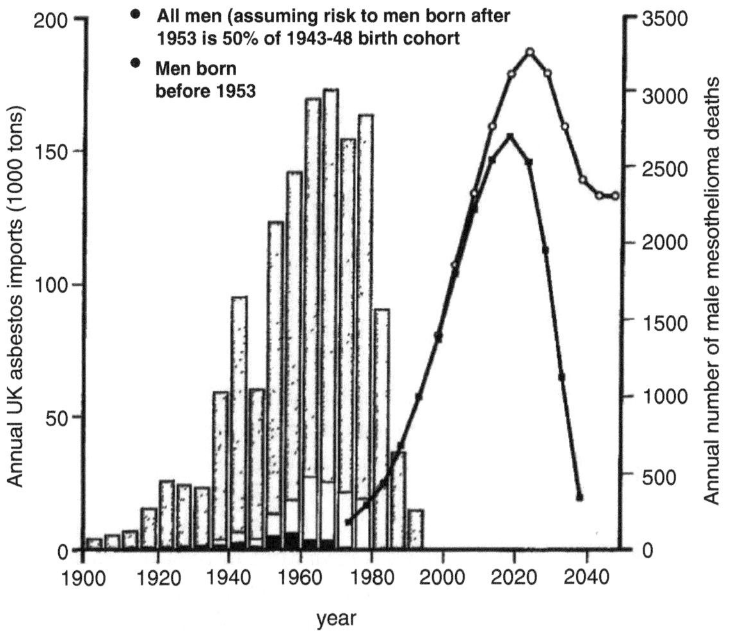

Fig. 3.4 Predicted mesothelioma death in British men and UK asbestos imports (adopted with revisions from Peto et al., 1995)

Because most of the "risky" asbestos exposure took place in earlier eras, the UK will soon enter a stage in which it can truly assess the significance and magnitude of the risk-management policy failures of these earlier periods. In 1995 the well-known Professor Julian Peto, who holds the Cancer Research UK Chair of Epidemiology, predicted that male deaths from mesothelioma in Britain will peak at between 2,700 and 3,300 per year around the year 2020, as illustrated in Fig. 3.4 (Peto et al. 1995, p. 537).

Bartrip cites De Vos (1995) and Webb (1995) in predicting more startling figures, as high as 10,000 per year for British males by 2020 (Bartrip 2004, p. 75). As a result of these late-onset side-effects from earlier exposure, the current risk-management strategies adopted during this most recent era may be inadequate. Further, because exceptions are allowed under both European and UK domestic bans only **if** alternative sources are absent **and** the health and environment damage can be contained, there will no longer be serious question about whether we have really "seen the last" of the asbestos issue. One remains hopeful that the legislation and regulations in place to deal with current and future problems related to asbestos maintenance and removal will protect the workers in question; however, because it takes so long for asbestos-related diseases to develop and kill their victims, by the time such a judgement is made it may be too late. This is the lesson that history has taught us in this case.

Improvements and Reforms of Possible Policy Failures

The recent era of risk management and communication with regard to asbestos has borne witness to a certain crystallization of the institutional mechanisms responsible for the strategic handling of the now well-known hazard, which has meant clear roles and responsibilities for both the HSC/HSE and the EC/EU. We have already remarked that shifting asbestos risk-management from the private industrial sector and institutionalizing it within the state apparatus represents a major step forward. Similarly, successes and failures of management and communication in this 1999–2007 era are difficult to assess at this juncture because the mechanisms are relatively new and the UK is just beginning to feel the brunt of the failures of previous eras.

With that said, additional remarks can be made about a number of significant characteristics of this era of risk communication and the management of asbestos. First, in Table 3.4 we compare the major areas of policy that the Health and Safety Commission has identified since its inception in 1974/6 to the more recent priorities it identified in 2003/4.

An examination of Table 3.4 shows a clear shift in priorities of the main risk-management agency in the UK. This shift in priorities arguably signals a movement away from law enforcement and/or the regulation of dangerous products such as asbestos, towards securing and enforcing the safety of all work environments.

Does this change signal an "end" to the asbestos threat? Does it mean that British risk managers are "getting soft" on hazardous materials such as asbestos? Although the HSE readily admits that "the mid-1990s saw some decline in enforcement activity while the focus shifted to addressing some of the organizational and management deficiencies underlying poor health and safety" and that "there has been a move back towards enforcement in recent years ... in particular ... enforcement powers to target high-risk areas" (Health and Safety Executive 2004, p. 9), a broader – and perhaps more historical – explanation can aid our understanding of this transition. The time for regulation of hazardous products and heavy chemicals such as lead, vinyl chloride, and asbestos has passed. Either there were no UK regulators as such

Table 3.4 Priority areas targeted by the Health and Safety Commission

Major areas of policy work listed in first annual report 1974–1976	Priority programmes listed in 2003/2004 annual report
Vinyl chloride code of practice	Falls from height
Lead code of practice	Workplace transport
Dust	Musculoskeletal disorders
Asbestos	Work-related stress
Fire precautions	Agriculture
Tanker marker scheme	Construction
Safeguarding of machinery	Health services
Flixborough report	Slips and trips
Major hazards branch	Government 'setting an example'

Source: Health and Safety Executive 2004.

when regulation was needed or they simply missed the boat. The entire nation has paid for this tardy ineptitude.

Further, the climate of the risk world has changed. It would appear that the days have passed in which the mining and handling of hazardous chemicals and compounds presents serious uncertainty for regulators. Dealing with asbestos is a dangerous activity, and the certainty of health damages following asbestos exposure has been well known for a long time. However, this fact has been denied or debated for political and economic reasons:

> Those unfamiliar with the history of asbestos may wonder why over 40 years have been spent in such intensive research on a mineral that already had a disastrous record by the 1960s. It is explicable only in terms of the actions of asbestos mining and manufacturing interests, which in the mid 1960s set out to prove that chrysotile did not cause mesothelioma, thereby turning this cancer into a problem of fibre type. (Tweedale & McCulloch 2004, pp. 257–258).

Although asbestos has certainly not left public consciousness or entirely disappeared from the HSE and HSC agenda, there is little doubt that attention is now focusing on areas in which there is high uncertainty about the probability of a risky event taking place. As the HSE itself explains:

> It is also worth noting that though more information frequently leads to a decrease in uncertainty, it does not necessarily change the probability of an event. For example, though frequent inspections of a critical component may reduce the uncertainty regarding the probability of the component failing within a period of time, the inspections do not reduce the probability of the component failing unless action is taken to remedy the situation (Health and Safety Executive 2001, p. 56).

Consequently, the regulatory environment in the UK appears to be concentrating on developing an understanding of the large and complex systems that can potentially account for new risks, such as biohazards and infectious diseases. Given the speed at which these risks can spread as biological agents, and the opaque network of actors involved with, for example, the agriculture and farming industries, new approaches to the strategic management and communication of risk must be developed in Britain.

Final Observations

Of course, asbestos continues to be used despite the WHO/WTO reports. This fact demonstrates that the argument is now essentially a political and economic one. In a sense, that has always been the case because – as this report shows – asbestos has been and still is too profitable to be abandoned without a fight by commercial interests. The influence of these interests largely explains why the so-called "precautionary principle" was never able to operate freely in regard to asbestos. Similarly, it explains why there has been far more debate and research about asbestos than about the merits of substitute materials, which since at least the 1970s have been much safer alternatives. It also explains why the current debate over asbestos has subtly shifted to the question of relative risk – thus opening the way for continued asbestos manufacture under "controlled" conditions.

References

Alleman, J. E., & Mossman, B. T. (1997). Asbestos revisited. *Scientific American, 272*(1), 70–75.

Agency for Toxic Substances and Disease Registry, United States Center for Disease Control and Prevention. (2001). *Asbestos exposure and your health*. Retrieved (last access), March 12, 2009, from http://www.atsdr.cdc.gov/asbestos/.

Asbestos Research Council. (1966). *1/275-280. The Asbestosis Research Council: Achievements over the First Eight Years*. Asbestos Research Council.

Asbestos Research Council. (1973). *304/2382. 56th meeting of ARC Research Committee, 10 July 1973*. Asbestos Research Council.

Asbestos Research Council. (1974). *304/2367. 59th meeting of ARC Research Committee, 23 April 1974*. Asbestos Research Council.

Asbestos Research Council. (1987). *361/1174-1180. The Asbestosis Research Council: The First Thirty Years*. Asbestos Research Council.

Bartrip, P. W. J. (2004). History of asbestos related disease. *Postgraduate Medical Journal, 80*(940), 72–76.

Berry, G. (1978). Contribution to discussion. In H. W. Glenn (Ed.), *Proceedings of the asbestos symposium, Johannesburg, 3–7 October 1977* (p. 56). Randburg: National Institute of Metallurgy.

Borron, S. W., Forman, S. A., Lockey, J. E., Lesasters, G. K., & Yee, L. M. (1999). An early study of pulmonary asbestosis among manufacturing workers: Original data and reconstruction of the 1932 cohort. *American Journal of Industrial Medicine, 31*(3), 324–334.

Budgen, A. (2004). Asbestos: A clear and present danger – a UK perspective. *Lung Cancer, 45*(Suppl. 1), S77–S79.

Churchard, C. (2007, August 13). Napoli wreck recycling to begin in Belfast. *Recycling & Waste Management News & Information*. Retrieved July 17, 2009, from http://www.mrw.co.uk.

Council of the European Communities. (1983). *Council Directive of 19 September 1983 on the protection of workers from the risks related to exposure to asbestos at work (83/447/EEC)*. Retrieved July 17, 2009, from http://eur-lex.europa.eu.

De Vos, I. H. (1995). Mesothelioma. *The Lancet, 345*(8959), 1233.

Deane, L. (1898). Report on the health of workers in asbestos and other dusty trades. In HM Chief Inspector of Factories and Workshops, 1899, *Annual Report for 1898* (pp. 171–172). London: Her Majesty's Stationary Office.

Department of Employment and Productivity. (1970). Standards for asbestos dust concentration for use with the Asbestos Regulations 1969. Technical Data Note 13. London: Her Majesty's Stationary Office.

Doll, R. (1955). Mortality from lung cancer in asbestos workers. *British Journal of Industrial Medicine, 12*(2), 81–86.

Egilman, D. S., & Reinhart, A. A. (1995). The origin and development of the asbestos threshold limit value: Scientific indifference and corporate influence. *International Journal of Health Services, 25*, 667–696.

Forster, J. (2007, August 8). Fire closes city centre store. *Sunderland Echo*. Retrieved July 17, 2009, from http://www.highbeam.com/doc/1G1-184235712.html.

Gaze, R. (1967). *81/57 Letter to Holmes, 12 December 1967*.

Gee, D., & Greenberg, M. (2001). Asbestos: From 'magic' to malevolent mineral. In P. Harremoës et al. (Eds.), *Late lessons from early warnings: The precautionary principle 1896–2000* (Environmental Issue Report No. 22, pp. 52–61). Copenhagen, Denmark: European Environmental Agency. Retrieved February 18, 2009, from http://www.eea.europa.eu/publications/environmental_issue_report_2001_22.

Greenberg, M. (1997a). The 1968 British Occupational Hygiene Society chrysotile asbestos hygiene standard. In G. A. Peters & B. J. Peters (Eds.), Sourcebook on asbestos diseases, 14: Asbestos disease and asbestos control (pp. 219–255). Dayton, OH: Lexis Law Publishing.

Greenberg, M. (1997b). Correspondence: Mesothelioma in a community in the north of England. *Occupational and Environmental Medicine, 54*(1), 67.

Greenberg, M. (1999). A study of lung cancer mortality in asbestos workers: Doll, 1955. *American Journal of Industrial Medicine, 35*(3), 331–347.

Greenberg, M. (2004). The British approach to asbestos standard setting: 1898–2000. *American Journal of Industrial Medicine, 46*(5), 534–541.

Health and Safety Commission. (2002). *The health and safety system in Great Britain* (3rd ed.). Norwich, UK: HSE Books. Retrieved February 18, 2009, from http://www.hse.gov.uk/pubns/ohsingb.pdf.

Health and Safety Executive. (2001). *Reducing risk, protecting people: HSE's decision-making process*. Norwich, UK: HSE Books. Retrieved February 18, 2009, from http://www.hse.gov.uk/risk/theory/r2p2.pdf.

Health and Safety Executive. (2004). *Thirty years on and looking forward: The development and future of the health and safety system in Great Britain.* Retrieved February 23, 2007, from http://www.hse.gov.uk/aboutus/reports/30years.pdf.

Health and Safety Executive. (2005). *Proposals for revised Asbestos Regulations and an Approved Code of Practice.* Norwich, UK: HSE Books. Retrieved February 18, 2009, from http://www.hse.gov.uk/consult/condocs/cd205.pdf.

Health and Safety Executive. (2009). *Asbestos Regulations*. Retrieved February 25, 2009, from http://www.hse.gov.uk/asbestos/.

Jeremy, D. (1995). Corporate responses to the emergent recognition of a health hazard in the UK asbestos industry: The case of Turner & Newall, 1920–1960. *Business and Economic History, 24*(1), 254–265.

Johnston, R., & McIvor, A. (2000). *Lethal work: A history of the asbestos tragedy in Scotland.* Edinburgh, UK: Tuckwell Press, Birlinn Limited.

Leydesdorff, L., & Etzkowitz, H. (1998). The triple helix as a model for innovation studies. *Science and Public Policy, 25*(3), 195–203.

London Hazards Centre. (1995). *The asbestos hazards handbook.* Retrieved February16, 2007, from http://www.lhc.org.uk/members/pubs/books/asbestos/asb10.htm.

MeasuringWorth. (2007). *Five ways to compute the relative value of a UK pound amount, 1830 to present.* Retrieved February 15, 2007, from http://www.measuringworth.com/calculators/ukcompare/.

Merewether E. R. A., & Price, C. W. (1930). *Report on effects of asbestos dust on the lungs and dust suppression in the asbestos industry.* London: Her Majesty's Stationary Office.

Ministry of Labour and Factory Inspectorate. (1949). Annual report of the Chief Inspector of Factories … for 1947. London: Her Majesty's Stationary Office.

Monopolies Commission. (1973). *Asbestos and certain asbestos products: A report on the supply of asbestos and certain asbestos products.* London: Her Majesty's Stationary Office. Retrieved February 18, 2009, from http://www.competition-commission.org.uk/rep_pub/reports/1970_1975/066asbestos.htm.

Mossman, B. T., Bignon, J., Corn, M., Seaton, A., & Gee, J. B. L. (1990). Asbestos: Scientific developments and implications for public policy. *Science, 247*, 294–301.

Newhouse, M. L., & Thompson, H. (1965). Mesothelioma of pleura and peritoneum following exposure to asbestos in the London area. *British Journal of Industrial Medicine, 22*, 261–269.

Ogden, T. L. (2003). Commentary: The 1968 BOHS Chrysotile Asbestos Standard. *The Annals of Occupational Hygiene, 47*(1), 3–6.

Parliamentary and Health Service Ombudsman. (2007). *Role, vision and values.* Retrieved March 6, 2007, from http://www.ombudsman.org.uk/about_us/role_purpose.

Peto, J., Hodgson, J. T., Matthews, F. E., & Jones, J. R. (1995). Continuing increase in mesothelioma mortality in Britain. *The Lancet, 345*(8949), 535–539.

Pye, C. (2007, August 13). Widow grieves stolen companion. *Lancashire Telegraph.* Retrieved July 17, 2009, from http://archive.thisislancashire.co.uk/2007/8/13/1016649.html.

Robens Committee. (1972). Robens report: Safety and health at work. *House of Lord Debates, 19 July 1972.* HL Deb 19 July 1972, volume 333, cc. 785–791. Retrieved February 25, 2009, from http://hansard.millbanksystems.com/lords/1972/jul/19/robens-report-safety-and-health-at-work.

Selikoff, I., & Greenberg, M. (1991). A landmark case in asbestosis. *JAMA, 265*(7), 898–901.

Selikoff, I. J., Hammond, E. C., & Churg, J. (1964). Asbestos exposure and neoplasia. *JAMA, 188*, 22–26.

Simpson, R.C. (1973). Safety and health at work: Report of the Robens Committee 1970-72. *Modern Law Review, 36*(2), 192–198. Retrieved February 25, 2009, from http://hansard.millbanksystems.com/written_answers/1979/nov/14/asbestos-simpson-report.

Simpson Committee. (1979). *Asbestos: Simpson report. House of Commons Debates, 14 November 1979.* HC Deb 14 November 1979, volume 973, c. 616w.

Tait, N. (1983). The role of SPAID … in the prevention of disease and the welfare of sufferers. In S. S. Chissick & R. Derricott (Eds.), *Asbestos* (pp. 9–50). Bath: John Wiley & Sons.

Tweedale, G. (1999). Management strategies for health: J.W. Roberts and the Armley asbestos tragedy, 1920–1958. *Journal of Industrial History, 2*, 72–95.

Tweedale, G. (2000). Science or public relations? The inside story of the Asbestosis Research Council, 1957–1990. *American Journal of Industrial Medicine, 38*(6), 723–734.

Tweedale, G. (2001). *Magic mineral to killer dust: Turner & Newall and the asbestos hazard.* Oxford, UK: Oxford University Press.

Tweedale, G. & Jeremy, D. J. (1999). Compensating the workers: Industrial injury and compensation in the British asbestos industry, 1930s–1960s. *Business History, 41*(April), 102–120.

Tweedale, G., & McCulloch, J. (2004). Chrysophiles versus chrysophobes: the white asbestos controversy, 1950s–2004. *Isis, 95*(2), 239–259.

Wagner, J. C., Sleggs, C. A., & Marchand, P. (1960). Diffuse pleural mesothelioma and asbestos exposure in the North Western Cape province. *British Journal of Industrial Medicine, 17*(4), 260–271.

Warren, R. C. (1997). The enforcement of social accountability – Turner and Newall and the asbestos crisis. *Corporate Governance, 5*(2), 52–69.

Webb, J. (1995). Tragic asbestos error will kill thousands. *New Scientist, 145*(1968), 4.

Chapter 4
Development of Asbestos Regulation in France: Policy Making Under Uncertainty and Precautionary Principle

Bernard Reber and Hajime Sato

Introduction

Asbestos is probably one of the most feared contaminants on earth, and in practice it is the most expensive pollutant to regulate and remove. Between 1965 and 1995, 35,000 people have died in France from asbestos-related illnesses, and another 50,000 to 100,000 deaths are expected before 2025. The cost to indemnify these claims is estimated to be between 26.8 and 37.2 billion Euros over the next 20 years. Few commercial commodities have generated such intense scientific, legal, and political scrutiny as asbestos (Hncharek 1993). Yet, as a component of many industrial and construction materials, asbestos offers numerous advantages.

A question remains as to why the French government decided to ban asbestos around 1996.[1] This choice reflects not only changes in the power equilibrium, public mobilization, media coverage of the crisis, epidemic outbreaks, and the emergence of scientific evidence; it also reflects a change in decision rationality. Although all the elements of the above factors are important, they do not adequately explain why and how the public authorities, who had been aware of the potential dangers of asbestos since 1945,[2] changed their mind on several key points and proceeded with the asbestos ban.

B. Reber
Research Center in Meaning, Ethics, Society, National Center for Scientific Research, University of Paris V (The Paris Descartes University), France

H. Sato (✉)
Department of Public Health, Graduate School of Medicine, The University of Tokyo, Tokyo, Japan
e-mails: hsato-tky@umin.net; hsato@post.harvard.edu

[1]We would like to thank M-C. Peter, an occupational health expert, physician and co-author of *Les risques du travail: Pour ne pas perdre sa vie à la gagner* (Cassou 1985, La Decouverte), for her comments regarding this article.

[2]"How can we claim 'we didn't know' when asbestos has been listed as an infection source on the table of occupational disease since 1945? To this question, the mission has not received any convincing answer" (Dériot & Godefroy 2005, p. 52).

H. Sato (ed.), *Management of Health Risks from Environment and Food*, Alliance for Global Sustainability Bookseries 16,
DOI 10.1007/978-90-481-3028-3_4, © Springer Science+Business Media B.V. 2010

There are a vast number of criteria for good risk management decisions under conditions of uncertainty (Golstein 1981; Lave 1981; Schulze & Kneese 1981). A fairly recent one is the precautionary principle, which has attracted substantial attention from policymakers, scientists, environmentalists, and the public. Some analysts see uncertainty and the need for action as the crux of the precautionary principle. This conclusion recognizes the complexity of the systems under investigation and the difficulty of easily predicting the outcomes of policy implementation. The precautionary principle implies the need for measures to protect health and the environmental, as implied by the proposition that precautionary action should be taken *despite* existing and/or lingering uncertainty (Bannister & Barrett 2006). It is sometimes noted that these actions are all the more important because of possible socio-political and cultural biases: "Aspects of ecological knowledge are removed from their defining cultural context and published in academic literature or exploited for commercial gain, thereby exaggerating the uncertainties and potential for harm" (Bannister & Barrett 2006, p. 230).

This article will first identify the most important time periods and describe the key historical events regarding asbestos in France during the past century. The events include interactions involving various actors (e.g., scientists, industries, citizens' groups, stakeholders, the French State, the courts of justice, foreign countries, and the media). In particular, recent events surrounding *Le Clemenceau, a* French warship, will be reviewed. Then, several factors that have hindered and/or facilitated the regulatory policies concerning asbestos will be identified, and the uncertainty surrounding these policies as well as the attendant political processes will be described. In the discussion section, several different typologies for policy decisions under uncertainty will be described and the rationales of the decisions to regulate or ban asbestos in France will be considered.

Method: Conceptual Lenses and Research Questions

To examine the process of making policy for the regulation of asbestos, we conducted a comprehensive search for historical documents. After a thorough search of archives of public records, both printed and on-line, we collected relevant documents from libraries, government archives, non-governmental organizations, and international organizations. Sources include academic databases such as the Social Science Citation Index, the Science Citation Index, and MEDLINE. We also gathered other types of documents and books, including theses, essays, recollections, and information from websites.

We will examine the history of French policies regarding asbestos regulation from the viewpoint of political science, especially in light of changes in rationale, or how decision makers decide to weigh the hierarchy of scientific uncertainties, the presumptions of major damage, and the need to act proportionally for the public good. The chapter is thus intended to elucidate the major factors that have hindered and promoted the regulation or banning of asbestos use in France.

The concepts underlying the precautionary principle will be succinctly presented, and the its role in the formulation of regulatory policy will be discussed.

The History of Asbestos in France

Asbestos is a family of silicate minerals that contain oxygen and silicon. The term asbestos, which has a broad commercial connotation, labels a group of naturally occurring hydrated silicates that crystallize into fibrous structures. However, this definition is controversial (Skinner et al. 1988). Asbestos minerals can be subdivided into serpentine and amphibole fibers. Chrysotile is the most common fibrous serpentine (over 90% of the world's production). Amphiboles include the minerals fibrous crocidolite, amosite, anthophyllite asbestos, actinolite asbestos, and tremolite asbestos (Mossman et al. 1990).

Asbestos has been recognized since ancient times. Theophrastus, one of Aristotle's students, spoke of it around 300 B.C., referring to it as resembling rotten wood and noting that when doused with oil it burns without being damaged. In 50 A.D., the Greek physician Dioscorides described in *De Materia Medica* reusable asbestos handkerchiefs that became white when burned. He gave the items the name *amiantus* (undefiled) to reflect their resistance to fire and to identify the quarry on Mount Olympus in Cyprus from which the asbestos was obtained.

It was not until the first half of the nineteenth century that dozens of other products incorporating asbestos were introduced. These products included blends of plastic, telephone buttons, fireproof ships, and insulation – all of which were lighter in weight and more thermally resistant than their predecessors. Until the end of World War II engineers placed a high value on the strength, fireproof nature, and durability of asbestos cement, especially for use in high-rise buildings. Asbestos was also used as a filter to purify fruit juice, sugar, and wine. The commercial world of asbestos grew rapidly, and by 1973 there were thousands of different products. In the same year, 1 million tons of asbestos were being used in the US. During the whole twentieth century, approximately 174 million tons of asbestos were consumed, and 3,000 different asbestos-containing products were created. The types of asbestos fiber differ in chemical composition and durability, and asbestos is noted for its fibrous structure. Seemingly blessed with useful attributes such as resistance to heat, flexibility (high tensile strength), softness, and a lower cost compared to man-made materials, asbestos is often spoken of as if it were silk from a magical kingdom (Alleman & Mossman 1997).

Asbestos use in France dates back to the 1930s, when the Turner & Newal Corporation moved its asbestos textile factory to Normandy. Up to that time, asbestos from mines never represented more than 20% of the total asbestos consumption in France. The last one, the Corsica Cape Mine in Corsica, was shut down in 1962. From 1962 to 1997, all asbestos in France was imported, with 80% of it was used as asbestos cement that was distributed to various industrial sectors for ovens, steel, boiler construction, thermal and nuclear power, foundry industries, metallurgy, and shipbuilding.

Although some of the damaging effects of asbestos had been known since the end of the nineteenth century (Castelman 1996), various publications describing the history of asbestosis in France increased French awareness of these health issues (Chateauraynaud & Torny 1999; Got 1998;[3] *Institut National de Recherche et de Sécurité* 2007; Thébaud-Mony 1990). These reviews suggest that there were eight distinct periods in the French history of asbestos: 1900–1964, 1964–1974, 1974–1977, 1977–1994, 1994–1996, 1996–1998, 1998–2002, and 2002–2007. The major events of each period will be presented in the following sections.

Period 1 (1900–1964): Quasi-Confidentiality

For many years, the clinical symptoms of respiratory disease in workers exposed to asbestos were attributed to tuberculosis (Rosner & Markowitz 1991; Thébaud-Mony 1990). In 1906, Denis Auribault, Inspector of Work in Condé-sur-Noireau, near Caen (Calvados), wrote *Note sur l'hygiène et la sécurité des ouvriers dans les filatures et tissages d'amiante,* in which he noted 50 deaths in a single mill over a 5-year period. He denounced the working conditions and warned the authorities, but the warnings had little effect (Corn & Starr 2007). It was common during this period for physicians to attribute the responsibility for such illnesses to the workers themselves, arguing that "along with some good workers, one finds alcoholics and other men idle because of their weaknesses. They say that there has been an understanding of the effects of dust on these weak natures, and that the lungs of formerly sick workers are attacked by the silica crystals leading to the onset of tuberculosis in workers predisposed to illness" (Teissonnière & Topaloff 2002). Information about the link between cancer and asbestos filtered through the narrow circles of French occupational medicine without leading to any action on the part of the public health authorities or any other governmental institution. It was not until 1945 that France finally put asbestosis on its official list of compensable occupational diseases (*Institut National de Recherche et de Sécurité* 2007a).

The danger of asbestos has been recognized since the beginning of the century. In 1945, it was listed in the table of occupational diseases (3.08), along with silica. In 1950, asbestosis and its complications were recognized as specific diseases. The particular activities that were addressed included the carding, spinning, and weaving of asbestos. When an asbestos-related illness is diagnosed by a doctor, the victims receive an inclusive payment (pension) proportional to their salary. If they are dead, the spouse has the right to a limited life annuity (a certain percentage of the victim's annual wage).

[3] This report is very complete, with more than 1,000 pages already uploaded to the website.

Period 2 (1964–1974): Heavy Consumption of Asbestos and Evidence of Health Risks

The consumption of asbestos in France peaked in the mid-1970s, with around 150,000 tons being used in 1975. Through their participation in the Asbestos International Association (AIA), French asbestos industry executives later learned of the link between asbestos and cancer (or at least its plausibility). However, the relationship between asbestos and cancer became a public concern only in the 1960s. One of the primary research papers at that time was the famous Selikoff et al.'s (1964) study in the United States entitled "Asbestos exposure and neoplasia." These researchers proved the link between asbestos exposure and cancer (Fig. 4.1).

Period 3 (1974–1977): The Coalition Between Workers and the University of Paris 6–7, and the Strategy of Controlled Use

One of the first public debates on asbestos in France concerned the textile manufacturer Amisol in Clermont-Ferrand. Thousands of workers, mostly women, had been exposed to asbestos for decades, but company physicians were unwilling to attribute their symptoms to dust exposure in the factory. Once the factory closed, the workers decided to strike against this closure and occupied the plant for 31 months. The main issue in the strike was economic security for the 250 families. It is only after this closure that the question of asbestos poisoning arose in the public domain.

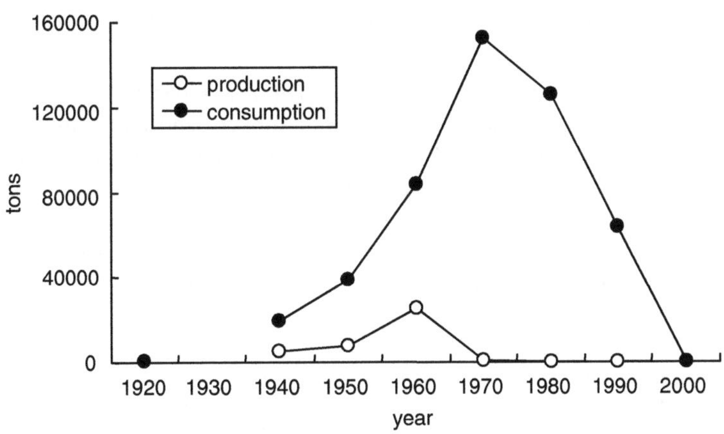

Fig. 4.1 Trend of asbestos production and consumption in France

At the same time, another crisis emerged and widely discussed. It took place on the Jussieu campus of the University of Paris 6 and 7, and the Institute of Global Physics. It was subsequently seen as "the first scandal of asbestos." Beginning in 1974, university physics researchers discovered the presence of unidentified dust during their regular experiments. Ultimately, the dust proved to be the largest quantity of sprayed asbestos in Europe. Henri Pézerat, a researcher in toxicology, led the protest. The group to which he belonged, the *Collectif Amiante*, carried out the protest with another group, the *Collectif Intersyndical de Sécurité des Universités-Jussieu*. These groups tried to alert the public authorities about the problem, claiming that the asbestos dust would only increase in the next 30 years. Ultimately, the authorities decided to cover over the asbestos on the ground floor, where numerous labs and offices were located, but not on the other floors. Experts conducted four studies during this period, culminating in a report by the *Cherchar* of the *Centre d'Etudes et de Recherches des Charbonnages de France* (Research Center of the French Coal Board), which concluded that "...the quantity found in Jussieu area should be considered unacceptable."[5] The trade unionists at Parisian University joined the Amisol workers in 1976. Their goal was no longer simply to reopen the factory; but also to obtain salary compensation, reemployment at another workplace, and free medical examinations plus follow-up medical care for those affected. Some trade union members of the *Confédération Française Démocratique du Travail* (CFDT), *Confédération Générale du Travail* (CGT), *Fédération de l'Education Nationale* (FEN), and the *Collectif Intersyndical de Sécurité des Universités-Jussieu* went to Condé-sur-Noireau, a factory of the Ferodo group that was involved in a case similar to Amisol. They wrote in *Danger! Amiante* (*Collectif Intersyndical de Sécurité des Universités-Jussieu, Confédération Française Démocratique du Travail, Confédération Générale du Travail, Fédération de l'Education Nationale* 1977):

> "As for asbestos, we cannot consider it a scourge, because it is our first material at work, and, if we put it into question, the employer will take advantage to justify its investment abroad to reduce employment (...) We have not said that we have to stop the use of asbestos immediately (...) but the less that is possible [the less we have] to envision its substitution in the long run" (*Collectif Intersyndical de Sécurité des Universités-Jussieu, Confédération Française Démocratique du Travail, Confédération Générale du Travail, Fédération de l'Education Nationale 1977*, p. 6).

This period was marked by protests, strikes, demonstrations, and media coverage. The movement spread to the RATP (regional subway and bus company in the Ile-de-France region) and the SNCF (French National Railway). Claims also were brought regarding asbestos found in drinks drawn from filters. Examples include Guérin Article (1978), "Amiante: Enquête sur des Assassinats Au-dessus de Tout Soupçon," published in *Les Temps modernes*, several reviews in consumer protection magazines such as *Que choisir?* (1976), and a piece in the lay science journal *Science & Vie*.

---

[5] This was an important step in banning asbestos there, a ban that did occur in 1978, according to the *Decree no 78-394, relatif à l'emploi des fibres d'amiante pour le flocage des bâtiments*, Ministry of Health, *Journal officiel*, March 20, 1978. The withdrawal of asbestos and the rebuilding of the university should have been completed in 1999 but has been delayed until 2012–2017 (Dépriot & Godefroy 2005).

Around the same time – most notably, December 20, 1976 – the *Chambre Syndicale de l'Amiante* and the *Amiante-ciment* trade union sent a letter to Prime Minister Raymond Barre, accusing Pr. Jean Bignon of being too alarmist when he held a conference at the International Agency for Research on Cancer (IARC, in Lyon, December 14–17, 1976). Along with other international experts, Bignon emphasized the health dangers of asbestos. On April 5, 1977, he sent a letter to the Prime Minister, alerting him to the asbestos problem and asking for effective prevention. He concluded by saying, "We have to admit that asbestos is a physical carcinogen [for] which the extent of damage to humans is now well known." He mentioned that the conclusions of his colleagues differed from those of the *Livre Blanc de l'Amiante*, which had been elaborated on by the industry. Finally, on August 17, 1977, the first French regulations concerning asbestos risk were published.[6] However, this warning still failed to provoke a strong reaction in French society.

The legislation was not implemented in an appropriate manner (*Senat Rapport*, October 26, 2005). No preventive measures at all were implemented by the health services at workplaces. The workers could sometimes not comprehend that symptoms could appear 40 years after asbestos exposure. The trade unions were more focused on silicosis than on asbestos.

Period 4 (1977–1994): Years of Silence due to the Innovative Permanent Asbestos Committee (CPA)

Some analysts of the history of the asbestos controversy in France see this as a quiet period (Chateauraynaud & Torny 1999). In truth, an innovative dialogue structure emerged at this time, during which the primary stakeholder was the *Comité Permanent Amiante* (Permanent Asbestos Committee), or CPA. This informal organization dominated the scene with regard to the prevention of and compensation for asbestos-related disease. Created in 1982 by a former General Director of the *Institut National de Recherche et de Sécurité* (INRS), it included representatives of the asbestos producers, physicians, trade unionists, and members of the ministries concerned.

The policy of the CPA was to promote the "controlled use" of asbestos rather than to ban it outright. The trade unionists on the committee claimed that there were no adverse health effects of asbestos, because there were no data showing otherwise; indeed, the union movement in France has a long tradition of protecting wages and employee rights. There were numerous scientific assessments suggesting the health risks of asbestos (Institut National de Recherche et de Sécurité 2007b). The French Government

[6]The *Décret no 77–949 du* (August 17, 1977) *relatif aux mesures particulières d'hygiène applicables dans les établissements où le personnel est exposé à l'action des poussières d'amiante* puts the Threshold Limit Value at 2 fiber/cm^3 in the air for one day of work, and only for fibers longer than 1 µm and wider than 3 µm wide, and with a report on both up to 3 µm in each dimension (Art. 2). If this level is exceeded, the work should be stopped until the problem is remedied (6 V).

tried to defend its asbestos control policy against the European Economical Community (EEC), where some of the member countries had filed pleas to ban it. The EEC published directives for the protection of workers (83/477/CEE), the prevention and reduction of environmental pollution (87/217/CEE), and limits on the marketing of asbestos products and on the employments of workers in the industries manufacturing those products (91/659/CEE).

Adverse events for the asbestos industry also appeared during this silent period. These included, for example, a warning letter sent by the CPA to Prime Minister M. Rocard in 1989 and various local lawsuits involving, for example, the City Hall of Montpellier (1980–1981), the Postal Sorting Center at the St-Lazare Railway Station in Paris (1985), the Anglade (Ariège) Mine (1985), the Berlaymont Center of the EEC in Brussels (1987), a hasty evacuation of the IARC in Lyon (1989), the Beaulieu tour of Nantes (1989–1991), the Prison of Fleury-Mérogis (1990), the School of Marcouville (1991), and the Thermic Central EDF (national electric company) in Chevire (1992). None of these cases, however, had any notable impact on national policy.

During this period, hundreds of asbestos victims approached *La Commission d'Indemnisation des Victimes d'Infraction Criminelles* (Commission of Indemnity for the Victims of Criminal Acts), or CIVI. This commission was established to implement the French law of July 8, 1983, which had been enacted to compensate victims of terrorism or other criminal acts.

On the international scene, April 1993 witnessed the birth of the Ban Asbestos Network (BAN), a private international organization based in France, which initiated action against asbestos use and its hazards: It published the manifesto *Le livre noir de l'amiante* (The Black Book on Asbestos) (Ban Asbestos Network 1993).

Period 5 (1994–1996): Trials, the Collective Expertise of the INSERM, and a Decision to Ban Asbestos

An important scientific paper written by Julian Peto was published in *The Lancet* and entitled "Continuing increase in mesothelioma mortality in Britain" (Peto et al. 1995). This paper brought new attention to the asbestos issue. The important word here is *increase*, with the careful application of uncertainty providing the upper and lower boundaries of the changes in mortality rates.

This paper coincided in time with four other major events in the French legal system. The first occurred in 1994 during the trial of Gérardmer (Vosges), which concerned the deaths of six teachers in a professional technical school in 1991. Their widows and other stakeholders asked the court to place an administrator and a politician in charge of the case. These individuals had been exempted from their legal, economic, and ethical responsibilities for the decisions they made. The media coverage of this event was remarkable and national. Victims groups had created the *Association Nationale de Défense des Victimes de l'Amiante* (Defense of the Asbestos Victims National Association), or ANDEVA, which had three objectives: (1) a complete ban on asbestos manufacturing in France, (2) effective public action to assess and remove asbestos hazards from the community so as to prevent future cases of cancer, and (3) recognition in the criminal and civil courts of victims' rights to compensation and industry's liability for the adverse effects of asbestos.

The second event was connected with the previous one, although it did not concern asbestos per se. Some politicians had been condemned at a trial for distributing HIV (AIDS virus)-contaminated blood (1995). The trial was known as "*L'affaire du sang contaminé* (blood contamination case)." These two crises together underscored both the mismanagement and the responsibilities of public decision makers.

The third event was the adoption on February 7 1996, of a decree requiring the declaration of asbestos levels in buildings. In the same year, Jussieu's *Amiante Comité* complained about an article of the new penal code on "empoisoning" and "criminal abstention" (Article 63) and used it to draw an explicit link between the code and one interpretation of the precautionary principle, namely, that "in the case of risk management, a strong (verisimilar) hypothesis should be taken as valid." This rule indicated the impossibility and undesirability of using scientific uncertainty as an excuse to avoid making decisions to prevent the potential risk of irreversible damage, whether it be to the public or an individual.

Finally, the Government requested the *Institut National de la Santé et de la Recherche Médicale* (INSERM) to provide a detailed report on the health effects of asbestos. This group of experts drew on 1,200 international scientific publications. In reviewing these documents, the experts found no French research on asbestos and health.[7] Nonetheless, the INSERM concluded that in 1996 alone asbestos could be considered responsible for over 2,000 cancer cases in France (Institut National de la Santé et de la Recherche Médicale 1997). Its report further noted major epidemics of pulmonary fibrosis and lung cancer and a pandemic of mesothelioma, insisting that scientific research had already clearly established that all asbestos fibers were carcinogenic.

During the same year, an Academy of Medicine report asserted that asbestos was not a major danger; the academy further stated that it did not favor a ban on asbestos. However, days after the INSERM report was released (i.e., December 24, 1996), a political decision was made that resulted in Decree 96–1133 to ban asbestos. The ban was to take effect just a week later, on January 1, 1997.[8] The decree was intended to protect workers and consumers by forbidding the manufacture, marketing, import, export, or detention for shipping of any product containing any type of asbestos fiber. Compliance was required within 4 years. The law included exceptions for chrysotile (which was not substituted for until 2002) and asbestos products used for vehicles made before 2003. As a consequence, France did not stop importing asbestos until 2002.

[7] The testimony of one of these experts before the commission, for the *Rapport d'information au nom de la mission commune d'information sur le bilan et les conséquences de la contamination de l'amiante* (Dériot & Godefroy 2005).

[8] The *Décret no 96–1133 du 24 décembre 1996 relatif à l'interdiction de l'amiante, pris en application du code du travail et du code de la consommation* bans the manufacture, modification, sale, import, and introduction to the national market of all varieties of asbestos fiber, whether or not they are incorporated in materials, products, or devices. The decree applies to both workers and consumers (Art. 1). For a limited time, an exception was made for chrysotile, provided there is no substitute.

Period 6 (1996–1998): Creation of the Public Health Agency for Prevention and Precaution

Prior to July 1, 1998, when the health sanitation law came into effect, no entity had ever been charged with the mission to alert public authorities about the risks of toxic products. This authority changed with the creation of the *Institut de Veille Sanitaire* (the French Institute for Public Health Surveillance), or InVS,[9] the mission of which includes continually monitoring public health, issuing health alerts, and contributing to the management of health crises. In 1998, the InVS developed a national screening program for mesothelioma, thus creating the first database for the incidence of this disease in France (Fig. 4.2).[10]

Period 7 (1998–2002): New Version of the Law Regarding Employers' Inexcusable Fault

After the 1997 asbestos ban was implemented in France, the Canadian Government and other asbestos-exporting countries adopted a defensive posture, arguing before the WTO's dispute-resolution panel in 1998 that the French decision was a violation

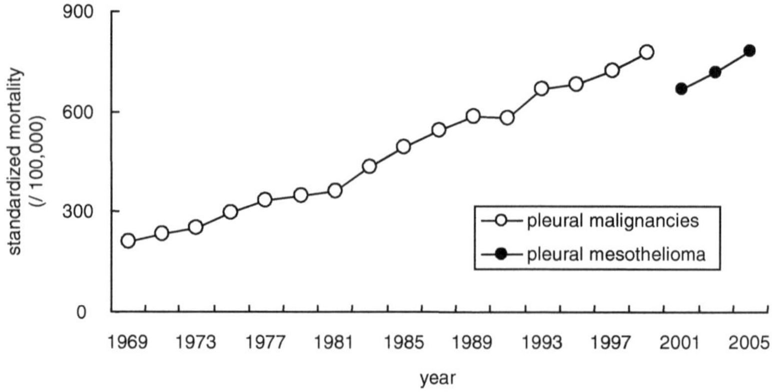

Fig. 4.2 Mortalities due to Pleural Tumors in France

[9] See: http://www.invs.sante.fr/presentations/default_en.htm (retrieved November 11, 2007).

[10] In 2007, the In VS had a budget of 59.8 million Euros and a staff of 387, mainly epidemiologists from various health disciplines and information systems.

of free trade.[11] The Canadian claim was definitively rejected in March 2001 by the WTO panel (Castelman 2002).

One of the most important changes in the management of asbestos hazards during this period in France occurred on February 28, 2002, a historic date for victims of asbestos exposure. The French Court of Cassation reversed the established rules regarding an employer's inexcusable faults. Prior to 2002, making improvements in the workplace (i.e., stopping asbestos use) was voluntary, even if the employer knew that the employees were in danger. Based on the understanding that it was very difficult – if not impossible – for victims to prove the danger of asbestos, in 2002 the assignment of (fault) liability was shifted. The law now stated that the employer knew or should have known that there was a risk and simply failed to take the necessary protective measures. It should be noted, however, that the law is not specific to asbestos. Nonetheless, the Court of Cessation's decision applied to the 29 claims of negligence by companies that manufactured or handled asbestos.

Because of the claims before the CIVI and the inexcusable negligence by employers, in 2000 Parliament created a fund called the *Fonds d'Indemnisation des Victimes de l'Amiante* (FIVA, Indemnity Fund for Asbestos Victims) to indemnify the victims of asbestos. The fund was authorized under the Social Security Law and drew on its indemnification fund. The FIVA was preceded by the funds for automobiles and those for victims of HIV-contaminated blood. The insurance mechanism does not consider the "cause" of a disease or death, just its "occurrence."

Period 8 (2002–2007): Double Standards and Substitutes for Asbestos

Some ministers still do not understand the issue. They consider the publication of the Institut National de la Santé et de la Recherche Médicale (1997) report to be a "shock and a surprise."[12] On March 3, 2004, the *Conseil d'Etat* (French Council of State) acknowledged the state's responsibility and its failure to take proper steps to prevent the risks associated with occupational exposure to asbestos dust. This failure is particularly notorious for its lack of research and inadequate regulations. It is a sad illustration of the problem created by having a vast amount of information that is rendered indigestible because of poor organization and too little cross-referencing (Lewis 1978; Rasmussenn 1975). However, decision makers continued to defend themselves. "If the INSERM report would have been published ten or twenty years before, we [would have] banned [asbestos, but] we were not always [aware] … [of the] causal links, before [the presence of a] presumption that we had never had

[11] Some analysts noticed that victims associations, and the INGO that represented them, were not allowed to offer their points of view in this debate.

[12] Such as the Minister of Social Affaires, Jacques Barrot.

previously. (…) only in 1994 and 1997 (…) appears not absolute certainty, but [only] the presumption that, even below these levels … it could be risky."

In 2006, a new and famous event revealed a double standard in French government behavior. The warship *Le Clemenceau* was supposed to have its asbestos removed while in India, but following a long international discussion with various interlocutors the ship was forced to return to Brest (in Northeastern France). The interlocutors included the Egyptian government (for permission to cross the Suez Canal), the Supreme Court of India (regarding whether such asbestos removal could take place in India), the French *Conseil d'Etat* (which attacked the decision to send the ship to India), and various French ministries (e.g., industry, defense) as well as the President. Greenpeace played an important role in the media coverage, choosing to highlight the case as a symbol of asbestos problems to come and an illustration of unethical environmental practices. The United Kingdom and the United States have fared no better in their dealings with old warships.

The second point concerns the difficulties in establishing and maintaining safety levels for the removal of asbestos from such warships, as well as from buildings. In 2006, approximately 100 million square meters of buildings in France still contained asbestos (Dériot & Godefroy 2005). More important is the third point, which concerns acceptable alternatives to asbestos. The 1998 INSERM report asked for more research on this question.

Currently, only 40 countries (15 of them in the European Union) have banned asbestos. It may be true that after 1996 France played an important role in the WTO's addressing of the regulation and banning of asbestos. However, three issues remain regarding France's apparent health-protection position in the broader social context: double standards, the removal of asbestos, and asbestos substitutes.

First, a strategy of double standards has been pursued by the asbestos industry and the French government. The strategy reveals a contradiction between the high level of security in France itself and the low level of security in Third World or emergent countries with which France has relationships. When asbestos was banned in France, the French multinational firm Saint-Gobin, for example, continued to mine and manufacture it in Brazil, with the full cooperation of the Brazilian government.

This habit of double standards – an asbestos ban domestically and loose safety regulations overseas – was vigorously defended in France. The government argued that no asbestos-related diseases had been reported in these developing countries, ignoring the fact that there had been no epidemiological studies in these countries that reliably established either the presence or absence of adverse health effects from mining, processing, or using asbestos. Because of this lack of research and information, individuals and institutions in these underprivileged countries cannot make a persuasive case for legislative reforms that protect workers and their environment from the dangers of asbestos (Harris & Kahwa 2003). Another application of the double standard can be seen in the defense of the social utility of asbestos for the construction of cheap housing in these countries (Thébaud-Mony 1990). This argument reflects a double standard because it attaches a different (and greater) value to convenience and economic benefits than to public health.

Discussion

Because of the many positive attributes of asbestos, especially its fire-resistant capacity, it was difficult to argue that its risks outweigh its benefits. Above all, asbestos was very cheap. Consequently, it is everywhere – in buildings products, ships, etc. – which means that it would be very costly to remove it all. Given the high cost of indemnification, can anything be cited that would lead us to conclude that the risks do indeed outweigh the benefits? In the historical part of this chapter, we mentioned several such factors: scientific uncertainty, special-interest politics, national and local politics, litigation, the AIDS scandal, and institutional change (creation of governmental agencies).

As stated at the beginning, the purpose of this chapter is to continue the analysis of the scientific and public controversy surrounding asbestos (Reber 2006) and to understand the rationality of the decision-making process for regulating or banning asbestos, with reference to several[13] key elements of the precautionary principle as defined by the Conference of Rio (1992). In other words, the purpose of the chapter is to elucidate the key factors affecting the application of the precautionary principle to policy making, such as when a lack of scientific certainty is or is not put forth as a reason to postpone proportionate regulations. We first addressed various aspects of this scientific uncertainty, how various institutions have confronted it, and the uncertainties that remain, particularly concerning the use of asbestos substitutes. Then we considered the factors that have contributed to progress in asbestos regulation, with the ultimate goal of totally banning its use.

Lack of Full Scientific Certainty Regarding Health Hazards

In retrospect, we can admit that the health danger of asbestos has been known since 1964. Nonetheless, according to Claude Got (Dériot & Godefroy 2005), one of the major French advocates of asbestos regulation, "France took the first measures of precaution" 46 years later than the United Kingdom, 31 years later than the United States,[14] and 13 years later than the International Conference of New York. The proceedings of this conference can be accessed at the medical faculty library of Paris University. Surprisingly, the INRS[15] published several articles in its review *Travail et Sécurité* (1954), admitting the dangers of asbestos and advocating preventative technical steps to overcome these dangers. However, none of these conclusions were propagated or utilized to inform appropriate political decisions.

The INRS was indeed playing a devious game. Between 1950 and 2004, it published 362 papers on asbestos in *Travail et Sécurité*. The INSERM's 1997 claim that

[13] In this chapter, it is not possible to make the same assertion for all these characteristics.

[14] In 1918, American insurance companies refused to issue life insurance policies for employees who worked in asbestos-related industries.

[15] See: http://en.inrs.fr/ (retrieved November 11, 2007).

there were no French articles on asbestos before 1996 is patently false. The INRS had long established that such documents existed. Prior to the creation of the InVS, the INRS was the only institution that could have alerted the public authorities about the asbestos problem, yet they failed to do so. Created in 1947, the INRS is composed of an equal number of employers and employees, thereby guaranteeing both an absence of research independence and a lack of expertise on public health. In addition, it was the former General Director of INRS, Dominique Moyen, who spear-headed the creation of the CPA in 1982. Even though the INRS did not play a role in alerting the public authorities, "it has taken the occasion to prevent certain risks."[16]

The "lack of full scientific certainty" was always present in the controversies surrounding asbestos. There are different types of asbestos fiber and the term asbestos applies to all these types collectively. Biological effects should be considered for each type individually. The identification of specific types in air samples requires sophisticated technologies, such as electron microscopy, x-ray diffraction, or energy dispersive x-ray spectroscopy. Even with such equipment, however, the chemical composition of each fiber type is complex (Mossman et al. 1990).

In their now famous article, Mossmann et al. recapitulated these difficulties, summarizing the medical and scientific controversies concerning asbestos-related diseases. In the following year, Bignon (1992) presented a set of four unresolved scientific questions about asbestos-related diseases: (1) Is the development of asbestos-related lung cancer always preceded by pulmonary fibrosis? (this issue was still unsolved in 1992); (2) How are the causality of mesothelioma by different fiber types to be understood? (controversial interpretations appeared in 1992); (3) In what percentage of mesothelioma cases is there no exposure to asbestos? (most epidemiological studies indicate approximately 20% to 30%);[17] (4) What are the health effects of exposure to low amounts of asbestos. Asbestos substitutes have also come into question. It seems that nothing is completely free of danger.

Efforts of the Asbestos Industry Against Stricter Regulations

Why was it possible for the government to have these scientific articles available but yet take no coherent policy action? The production of asbestos is globalized. After 1964, the French – and more importantly, the international – industry developed two strategies. The first strategy was to promote "controlled use," through which asbestos manufacturing purportedly could be made safe. This option allowed the asbestos industry to gain the support of most trade unions,[18] which defended it as an acceptable compromise to protect jobs. The second strategy was to provide research grants to scientists and physicians. However, their role in promoting

[16] Recognized by P. Huré, who was in charge of the Department of Chemical and Biological Risks of the INRS (Dériot & Godefroy 2005).

[17] "This question will probably take more than 20 years to be resolved" (Bignon 1992).

[18] The trade union *Force ouvrière* asked for a ban on asbestos and did not attend the Permanent Asbestos Committee meeting in 1982.

science is now suspected. Some analysts now speak of the dubious nature of the research contracts with the asbestos companies. These contracts came with the stipulation that the asbestos industry/companies would have permanent control over the publication of the results (Lenglet 1996; Thébaud-Mony 2003), which means that the industry could control the generation and dissemination of scientific knowledge to serve their own interests. It seems that this strategy allowed the industry to dominate for many years the international scientific controversy regarding the different toxicity levels of various asbestos fibers.

Several interesting issues arose during the CPA period (1977–1994). First, some of the scientists involved were not informed of asbestos substitutes being used by the industry, despite the fact that an INRS article in *Travail et Santé* recommended the use of such substitutes. The industry defended its position by attributing the possible health risks of asbestos to the factors other than their products themselves, stressing the benefits of the latter. Thus, they used the press in an attempt to trivialize the hazardous nature of asbestos products. For example, one article states that "asbestos is everywhere … the dangers are essentially caused by the inadequacy in implementation … Be quiet. If your house is in good condition, you have on the contrary more security (with some construction materials using asbestos)" (*Chambre syndicale de l'amiante* 1976).

Indeed, among public policymakers, industries, unions, and CPA members, the lack of scientific certainty regarding asbestos hazards and the differences in hazardousness properties between asbestos and its substitutes was used to postpone radical decisions such as banning asbestos. Controlled use was presented as a workable compromise.

From the historical perspective, one can all too easily say, as many analysts have, that policymakers have made only "instrumental" use of the sciences, treating them as a black box and preferring to follow short-term prevention strategies with the "controlled use" of asbestos. However, to make their case, these analysts selected a particular experiment to sell as "the right one, before the others," forgetting about the complexity of the controversy as a whole and thus increasing scientific uncertainty. In a sense, this use of science can be viewed as both instrumental and anachronistic. A traditional concept "holds that the science and engineering community should lay out the 'facts,' making it clear what is known and what is not known" (Morgan & Henrion 1990, p. 17). Their actual retrospective position is not that of the former decision-makers. One could hypothesize, of course, that the decision makers were not cynical, just that they were accepting (false) economical calculations and waiting for complete or at least greater scientific certainty, while pretending that they could justify all this to the public and themselves later (Dériot & Godefroy 2005).

Institutional Change in the French System of Government

Bignon once pleaded, "[W]e must admit our limitations and express the wish that scientists, industrialists, politicians, and consumers will endeavor to work together in order to reach a consensus on acceptable risk" (Bignon 1992, p. 445). This comment represents one way to take account of the precautionary principle concept without actually mentioning it. In the late 1990s, such a lack of adequate communication among the different spheres came to be regarded as lethal in France.

Around 1998, new agencies in the French government were created to address the issues discussed above. In 2000, Kourilsy and Viney reported to the Prime Minister on the interpretation of the precautionary principle. In establishing a link to these agencies, they mentioned the *Agence Française de la Sécurité Sanitaire des Aliments* (AFSSA, The French Food Safety Agency), [19] an agency that was to be concerned with communication problems during crisis periods. Also added was the *Agence Française de Sécurité Sanitaire des Produits de Santé* (AFSSAPS, the French Health Products Safety Agency).[20] Of more direct relevance to asbestos in the workplace, the *Agence Française de Sécurité Sanitaire de l'Environnement et du Travail* (AFSSET, The French Agency for Environmental and Occupational Health Safety)[21] was created to coordinate expertise in health and the environment. These specialized agencies were expected to provide transparent and multi-faceted expertise in more areas than just the dominant one. What is new about this approach to risk assessment and management is that these agencies are independent, formal, and well-structured; this was not the case with the CPA.

In reality, the state was not well enough organized to tackle the scientific uncertainty about asbestos before the INSERM started to provide collective expertise in 1996, and before the sanitation agencies were created in 1998. More transparency was required for this assessment, and it was necessary to take into account the relationship between science and decision making within the framework of the precautionary principle. The efforts also needed to be cooperative – and of a scale that is only possible in specialized agencies.

During the period of the CPA the health effects of asbestos exposure were for the most part simply kept invisible (Pezerat 1995). There were no assessments of asbestos exposure at workplaces or in public and private buildings, and no systematic survey was conducted of asbestos-related diseases. There was no risk assessment by public health authorities or labor inspectors (Thébaud-Mony 2003)[22] before 1996. If risk management was deficient, so was risk communication. The French Government's position between 1977 and 1996 can be summarized as "don't alarm the public, but take inventory of the public buildings concerned; do the best you can to control asbestos." This was not a true policy of vigilance, but it was perhaps a policy to reassure and quiet the public.

We also have to raise the question of how the state delegated its authority to the CPA, which was presented as an innovative space to be used for stakeholder dialogue. Some scientists were regularly invited to participate, to give their audience an

[19] Created on April 1, 1999. See: http://www.afssa.fr/ (retrieved November 11, 2007).

[20] Created in March 1999. See: http://afssaps.sante.fr/ang/indang.htm (retrieved November 11, 2007).

[21] The former *Agence de sécurité sanitaire de l'environnement* (AFSEE) was transformed in 2005. See: http://www.afsset.fr/index.php?pageid=779 (retrieved November 11, 2007).

[22] For investigations completed during this period of time, see Cordier et al. (1987).

impression of state-of-the-art research on asbestos. The primary asbestos companies were also present. As in England and the United States, the asbestos company executives in France were the first to know about the serious health effects of asbestos exposure among their workers (Lilienfield 1991). Yet, the companies insisted that the prevention measures be applied first to the workers, through their controlled use of asbestos, and only later to the public. The industry reaction to these warnings allowed company executives to avoid for many decades, and throughout the world, the adoption of measures that would restrict the manufacture and use of asbestos (Thébaud-Mony 2003).

In July 2004, the AFSSET received a solicited request within the framework of the French National Health and Environment Plan (PNSE) to assess the risks of exposure to man-made mineral fibers to both asbestos workers and the general population. In its conclusions, the PNSE issued a series of recommendations for improving the traceability of asbestos products, increasing knowledge about their potential hazards, and reinforcing preventative measures; further, it argued for the continuation of research into other types of siliceous man-made mineral fibers, including mineral wools (glass wool, rock wool, and slag wool) and continuous glass filaments. On February 7, 2005, the same agency received a solicited request from its funding ministry, *Direction Générale de la Santé, Ministère de la Santé* (the French Ministry of Health, Environment, and Employment) to conduct an assessment of the risks associated with asbestos fibers shorter than 5 μm.

On January 9, 2006, the expert group reported that the data gathered up to that time demonstrated no asbestos risk to humans. However, the group did not believe the data were sufficient to rule out a "potential" risk, including in metrological terms; nor were they able to provide a satisfactory solution for managing the risk. The group, therefore, recommended that the further investigations planned by AFSSET be carried out.

After the introduction of asbestos substitutes, it became even more difficult for epidemiologists and clinicians to determine the types of fiber that were responsible for the disease, especially in workers who had been exposed to more than one type during their lifetime (i.e., asbestos, natural fibers, or synthetic fibers).

Public Policy Making Under Conditions of Uncertainty

In 1990, Morgan and Henrion (1990) presented decision criteria under uncertainty: (1) *utility-based* criteria, involving decisions based on the value of outcomes; these values in turn are based on a cost-benefit analysis that is either (1.1) *deterministic* (the conventional choice; Mishan 1973) or (1.2) *probabilistic*, which applies when there is uncertainty. A (1.3) *cost-effectiveness* criterion may be used, sometimes on non-economic grounds, to help determine the alternative that can achieve the objective at the lowest cost. A (1.4) *bounded cost* criterion is used to maximize the amount of risk reduction within budgetary constraints. To determine if this criterion is met, one uses the *maximize multi-attribute utility*, which is based on multi-attribute utility theory (Keeney & Raiffa 1976), by specifying a utility function that evaluates outcomes in terms of all the important attributes, including risks and uncertainties. Morgan and Henrion add a last option, which has close proximity to the elements of the precautionary principle. It can be roughly described as (1.5) trying to *minimize the chance of the worst possible outcome ... maximize the chance of the best possible outcome.*

Rather than utility-based criteria, there proposed (2) *rights-based criteria.* These criteria are not concerned primarily with outcomes, but rather with the process of making choices. They are usually the choice for assessing "new" human risks. The four specific criteria are the following: (2.1) *zero risk*, for which benefits and costs cannot be considered or introduced; (2.2) a *bound* or *constrained risk*, when a *specific level* of a risk, such as of a nuisance or reckless endangerment, is entered; (2.3) *approval-compensation* (Howard 1980), which allows risks to be imposed on people who voluntarily give consent and receive compensation for inconvenience or a possible loss; and (2.4) an *approved process*, for which all relevant parties must act in accordance with a set of procedural rules, such as those defining due process. In addition, there are two other types of criteria: (3) *technology-based criteria*, which involve selecting the best available technology for environmental regulation and risk reduction in terms of cost-effectiveness; and (4) *hybrid criteria*, which combine utility and rights-based criteria (Table 4.1).

Above all, these decision-making models are based on the different sets of factors that must be considered in policy making: economics, legal rights, risk of damage, and technology. Second, one must take account of how these factors can be combined for purposes of calculation, compensation, or other procedures. Third, the calculation of the utilities of the different choices can be deterministic or probabilistic, or, if neither of these is possible, the problem can be shifted to procedures (e.g., due processes) or rights. In the case of highly controversial technologies, such as those involved with genetically modified organisms (GMOs) or asbestos regulation at the beginning of a crisis, the gaps between the different hypotheses make it impossible for decision makers and experts to calculate effective probabilities, reliable costs and benefits, and therefore, all the social consequences of a problem in the distant future.

Table 4.1 Decision criteria applicable to policy analysis for risk management

1. Utility-based criteria	Deterministic benefit-cost
	Probabilistic benefit-cost
	Cost-effectiveness
	Bounded cost
	Maximize multi-attribute utility
	Minimize chance of worst possible outcome … maximize chance of best possible outcome
2. Rights-based criteria	Zero risk
	Bounded or constrained risk
	Approval and compensation
	Approved process
3. Technology-based criteria	Best available technology

In other words, there remain: (1) uncertainties about technical, scientific, economic, and political variables; (2) uncertainty about the appropriate functions of technical, scientific, economic, and political models; and (3) disagreements among experts about the functional form of the models or the variables entered into them (Morgan & Henrion 1990, p. 39). In reality, however, these decision-making criteria are now commonly used for risk management strategies, helping those analysts who cannot avoid addressing a broader philosophical framework. It is not difficult to recognize some forms of consequentialism in the utility-based models and deontologism in the rights-based models, because these two categories of criteria are derived from normative moral theories.

Criterion (1.5), *minimize the chance of the worst possible outcome ... maximize the chance of the best possible outcome*, although not fully developed in this chapter, does offer an introduction to the precautionary principle. Theoretically, the application of *rights-based criteria* is not restricted to the regulation of new technologies and/or new products; it can also apply to existing ones, such as asbestos, which is a very old natural material. What makes the precautionary principle important as an analytical framework is the undesirability of public authorities (and producers) using scientific uncertainty as an excuse for inaction, as well as the need for "rapid action to limit or eliminate a risk whose effects will not surface until ten or twenty years later" (Commission of the European Communities on the Precautionary Principle 2000, pp. 15–17). The precautionary principle transcends some of these difficulties in cases where there is a strong presumption of major damage.

The Precautionary Principle as an Ideal Policy Choice

A simple definition of the precautionary principle has been proposed[23] and widely adopted in regulations regarding marine pollution, climate change and biodiversity loss, dangerous chemicals, and GMOs. After the World Charter for Nature was adopted by the United Nations General Assembly in 1982 and the precautionary principle was first mentioned, an archetypal and globally influential formulation of the concept appeared as Article 15 of the 1992 *Rio Declaration on Environment and Development*. The Article states that "in order to protect the environment, the precautionary approach shall be widely applied by States according to their capabilities. Where there are threats of serious or irreversible damage, lack of full scientific certainty shall not be used as a reason for postponing cost-effective measures to prevent environmental degradation" (United Nations Environment Programme 2007).

A less extreme version of this argument can be found in the *Communication from the Commission on the Precautionary Principle*, one of the main documents defining the Precautionary Principle for European Policy (Commission of the European Communities on the Precautionary Principle 2000). The *Communication* states,

---

[23] This concept found its first coherent form in the *Vorsorgeprinzip*, and it was first introduced into German environmental policy in the early 1980s. Strictly speaking, this German word focuses more on anticipation than on responsibility or attention.

very strongly, that "An assessment of the potential consequences of inaction should be considered and may be used as a trigger by decision-makers. The decision to wait or not to wait for new scientific data before considering possible measures should be taken by decision-makers with a maximum of transparency. The absence of scientific proof of the existence of a cause–effect relationship, a quantifiable dose-response relationship or a quantitative evaluation of the probability of emergence of adverse effects following exposure should not be used to justify inaction, even if scientific advice is supported only by a minority fraction of the scientific community" (Commission of the European Communities on the Precautionary Principle 2000, p. 16). Those who are skeptical about the precautionary principle prefer its weaker definition, because "its requirement [says] that bounds be put on the uncertainty surrounding scientific knowledge (...) when there is a very great uncertainty regarding the likely impact of technology" (Morris 2000, pp. 14–15).

Compared with the traditional decision criteria for uncertainty, the application of the precautionary principle to practical policy issues requires that the various considerations (i.e., economics, legal rights, risks of damages, and technology) are explicitly matched. A novel feature of precautionary principle-based criteria is their characterization of this meta-principle as "serious and irreversible damage," a "lack of full scientific certainty," and the impossibility of postponing action to address these conditions. This last point is stated more clearly in Article 10 of the *Cartagena Protocol on Biosafety*, ratified in June 2003.[24] The Decision Procedure therein states: "10.6. Lack of scientific certainty due to insufficient relevant scientific information and knowledge regarding the extent of potential adverse effects of a living modified organism (...) shall not prevent that Party from taking a decision (...) in order to avoid or minimize such potential adverse effects." The key terms here seem to be "potential" and "adverse," with "potential" being the more important.

Indeed, the precautionary principle proposes a new relationship between scientific uncertainty and policy decisions. An alternate sense of scientific uncertainty, which refers to uncertainty about what should be assessed and is often used as an excuse by governments to avoid making beneficial but radical decisions, is also possible. The precautionary principle has sometimes been used in real decision-making processes and actual court trials (Foucher 2002), and it has been debated and interpreted as part of participatory technology assessments (Reber 2007, 2008). Furthermore, deliberations by ordinary citizens has led them to conclude that application of the precautionary principle is also desirable for them (Dryzek et al. 2008). Because of its possible arbitrary application, however, some political and legal philosophers prefer the classical risk-benefit assessment to the precautionary principle (Sunstein 2005), and others are outright hostile to the precautionary principle (Morris 2000).

The precautionary principle came into existence as part of French law in 1995 (Kourilsky & Viney 2000). It was explicitly introduced to the French public around 1997,[25] when it was decided to ban asbestos in the country. The precautionary

---

[24] See, for example, Meyers and Raffensperger (2006, p. 323).

[25] The precautionary principle was mentioned in the first paragraph of the mission letter signed by the Ministers of Labor and Health (M. Aubry and B. Kouchner) for the Got Report (Got 1998).

rationale at least partly explains the choice of the French administration to change its strategy regarding the controlled use of asbestos and the creation of public health agencies.

French Asbestos Regulations from the Precautionary Principle Perspective

The health risks from asbestos have been on the agenda for a long time. However, as noted above, the long latency period before the illness was identified has contributed to the failure in utilizing that recognition for effective preventive measures. The scientific uncertainty regarding the danger of minimal exposure to asbestos clearly led to a delay in the implementation of official measures to prevent the health hazards of asbestos. Nonetheless, the INSERM report of 1997 states that some uncertainty has remained about the risks of lung cancer and mesothelioma associated with exposure to 1 fiber/mL of asbestos. The uncertainties include the exact nature of the risk of exposure less than or equal to 1 fiber/mL, and the general risk of asbestos exposure that has continued to exist in the French population.

Several socio-political factors have further hindered progress in the development of health-oriented regulatory policy. The involvement of researchers from Jussieu (Paris University 6 and 7), coupled with the mobilization of workers affected by asbestos, has raised awareness of the health dangers of asbestos. However, the workers' activities were crippled at this time, because they did not dare risk their employment for a danger that in the eyes of many was remote. Consequently, even as an occupational hazard, asbestos has not had a prominent place on the agenda.

Although only semi-official, the *Comité Permanent Amiante* (CPA) was established to settle the asbestos issue after the focusing event. It was neither proactive nor effective in regulating asbestos, although it did help to soothe the public outcry about the problem. However, in the final analysis, this delegation of regulatory authority to the CPA proved to be nothing more than a authorization for lobbying by the asbestos industry, "a 'model' of lobbying, communication and manipulation … [which was] known to exploit, in the absence of the State, pseudo-scientific uncertainties which were even withdrawn, for the most part, in the serious Anglo-Saxon literature" (Dériot & Godefroy 2005). The role of the INRS as a source of research and information in policy making remained ambiguous: Its research outputs were not employed for the advocacy of policy to any great extent. In retrospect, the French asbestos case during this period might be one in which scientific uncertainty was used as a tactic to fend off more stringent official regulations.

In the mid- to late-1990s, the numerous lawsuits filed by asbestos victims, court decisions, and the AIDS scandals provided a favorable environment for the French government to become more responsibly involved in asbestos-related health issues. Certainly, the increasingly noticeable asbestos-related health hazards were present in the background.

Through institutional changes, especially the creation of the InVS, the varying degrees of lingering scientific uncertainty no longer posed a major obstacle to introducing the regulations and, ultimately, the total ban of asbestos. Finally, the French Government waited until 2005 to take measures to reinforce its own expertise and its global monitoring of occupational risks. The most important of these measures was the French Occupational Health Plan (PST) 2005–2009. In a sense, the French policy changes at this stage were induced by factors outside the bureaucracy.

The scientific analysis of risks is unavoidably and inextricably intertwined with subjective framings, assumptions of values, as well as with trade-offs and expectations of surprise (Stirling 1999; Weill & Hughes 1986; Bates 1991). The appraisal of technological risks should, therefore, be conducted in an open and pluralistic fashion, fully allowing for incisive critical discourse as an essential part not only of the regulatory process, but also of appraisals of the technological options themselves. For instance, most of the articles in *Precautionary Tools for Reshaping Environmental Policy* (Myers & Raffensperger 2006) make this link between the precautionary principle and collective assessments.

A growing recognition of the limitations of science has led to calls for new styles of governance. Analysts have argued that the effective regulation of potentially hazardous technologies indeed requires that the so-called "scientific" (natural and engineering sciences) point of view be broadened to include increased public involvement in policy making (Fischer 2000; Harremoës et al. 2001; Wynne 2001).

A search for past rationales for the regulation of asbestos in France revealed that no explicit calculations of costs and benefits were ever included in formal policy discussions. An examination of the course of events prior to the mid-1990s suggests that the benefits of asbestos use collectively outweighed their costs for French society, at least in the short term. The problems of distributing the costs among various social actors, policy myopia, and power imbalances, all of which had the potential to substantially affect public policy decisions, were not openly discussed. Even in the periods when regulation progressed toward a total ban, no explicit calculations of any kind were performed.

Only in the final stages of the asbestos ban, when the advocacy for stringent regulation was gaining momentum, was the precautionary principle employed to justify policy change. As a presumption, it is safe to say that the State cannot always wait for scientific certainty; it must proactively take appropriate measures to reduce the asbestos risk for the population and attempt to answer the scientific questions later. The key question of presumption, then, at the core level is one of precaution; there has been no clear discussion of this point, either quantitative or qualitative. The stated goal, then, was "the elimination of possible hazards," toward a zero risk. This goal is consistent with that suggested by the rights-based model, namely minimizing hazards at any cost. Such goals were sometimes put forward during discussions of litigation, which tended to focus on individual cases of asbestos poisoning rather than overall costs and benefits.

Conclusions

Although politics is not always cynical (i.e., an exercise in economic power after reassuring the public), accusations of such have sometimes been leveled retrospectively at the French government: Rather than regulating exposure, policymakers chose instead to increase asbestos production and keep the health concerns about asbestos off the agenda (Scheberle 1994, p. 81). In making policy decisions under conditions of uncertainty, several key forms of rhetoric are equally vulnerable to manipulation and can even be exploited by various parties to achieve political or commercial objectives. The precautionary principle-fits well with the needs for flexibility and the learning (trials and errors) nature in the process of developing such regulations. These factors are also important for scientists and decision makers. It should be noted, however, that the precautionary principle might be just another tool in symbolic politics.

The management of technological risk is necessarily and inevitably an incremental and context-specific undertaking. Different technological risks warrant greater or lesser degrees of precaution at different times, and different regulatory instruments are considered appropriate in different contexts.[26] First, each risk of damage has its own implications for practice – for instance, different conventions for determining the burden of proof. Attention, then, should be paid to the different crucial (and often qualitative) characteristics displayed by different types of risk (Klinke et al. 2006). Second, policy making must be based on the available scientific information, which, as we have seen for France, is ever-changing, never complete, and occasionally insufficient. This point is crucial for understanding the conduct of heath safety agencies, whether they are in Europe or elsewhere.

Very often when there are multiple explanations of a event and one must choose to cite one piece of research among many possibilities, the uncertainties and controversies surrounding science and risk assessment are ignored in favor of a "complex combination of motives – economic, electoral, and symbolic" (Tanguay 1985). We thus consider it imperative to look beyond "what is on the agenda" and "what are the equilibriums of power." Instead, we should ask, "what could be, … in different possible worlds," always keeping in mind the available scientific knowledge, the possible damage, and the limits of our ability to determine an appropriate proportional response. This tendency of uncertainty in the context of decision making and science seems to be more evident when comparing possible scenarios (possible social changes with different policy choices) than when examining a single issue or agenda item to the exclusion of all others.

---

[26] This point takes into account Sunstein's (2005) attack against the precautionary principle for pretending that some laws adopted "in the name of the precautionary principle" could be worse than the risks they were meant to avoid. In most cases, such as a particular asbestos crisis in the US, suspect measures do not invalidate the precautionary principle. Indeed, in that case, the removal of asbestos in schools was accomplished too quickly and in an unsafe manner. To the contrary, Sunstein's claimed association between asbestos, GMOs, and the Iraq War is unfair.

References

Alleman, J. E., & Mossman, B. T. (1997). Asbestos revisited. *Scientific American, 272*(1), 70–75.

Ban Asbestos Network. (1993). *Le livre noir de l'amiante* [The Black Book on Asbestos]. Paris: Ban Asbestos Network.

Bannister, K., & Barrett, K. (2006). Harm and alternatives: Cultures under siege. In N. J. Myers & C. Raffensperger (Eds.), *Precautionary tools for reshaping environmental policy* (pp. 215–239). Cambridge, MA: MIT.

Bates, D. V. (1991). Asbestos: The turbulent interface between science and policy. *Canadian Medical Association Journal, 144*(5), 554–556.

Bignon, J. (1992). How are we going to change our attitudes and opinions regarding asbestos and cancer in the next 20 Years? *American Journal of Industries Medicine, 22*(3), 443–446.

Cassou, B. (1985). *Les Risques du Travail (Reliure inconnue)*. Paris: Editions La Découverte.

Castelman, B. I. (1996). *Asbestos: Medical and legal aspects*. Englewood Cliffs, NJ: Aspen Law & Business.

Castelman, B. (2002). WTO Confidential: The case of asbestos. World Trade Organization. *International Journal of Health Services, 32*(3), 489–501.

Chambre Syndicale de l'Amiante. (1976). *Rigueur Scientifique ... Ou manipulation des chiffres*? Retrieved, July 17, 2009, from http://www.sante-publique.org/amiantte/communication/chsyndamiante77.htm

Chateauraynaud, F., & Torny, D. (1999). *Les sombres précurseurs: Une sociologie pragmatique de l'alerte et du risque*. Paris: Editions de l'Ecole des Hautes Etudes en Sciences Sociales.

Collectif Intersyndical de Sécurité des Universités-Jussieu, Confédération Française Démocratique du Travail, Confédération Générale du Travail, & Fédération de l'Education Nationale, (1977). *Danger! Amiante*. Paris: Maspero.

Commission of the European Communities. (2000). Communication from the commission on the precautionary principle. Brussels, Belgium: Author. Retrieved February 28, 2009, from http://europa.eu.int/comm/dgs/health_consumer/library/pub/pub07_en.pdf

Cordier, S., Lazar, P., Brochard, P., Bignon, J., Ameille, J., & Proteau, J. (1987). Epidemiologic investigation of respiratory effects related to environmental exposure to asbestos inside insulated buildings. *Archives of Environmental Health, 42*(5), 303–309.

Corn, J. K., & Starr, J. (2007). Historical perspective on asbestos: Policies and protective measures in world war II shipbuilding. *American Journal of Industrial Medicine, 11*(3), 359–373.

Dériot, G, & Godefroy, J.-P. (2005). *Rapport d'information fait au nom de la mission commune d'information sur le bilan et les conséquences de la contamination par l'amiante* (Les Rapports du Sénat, n° 37). Paris: Sénat.

Dryzek, J. S., Goodin, R. E., Tucker, A., & Reber, B. (2008). Promethean elites encounter precautionary publics: The case of GM foods. Science, *Technology & Human Values, 0*, 0162243907310297v1.

Fischer, F. (2000). *Citizens, experts, and the environment: The politics of local knowledge*. Durham, NC: Duke University Press.

Foucher, K. (2002). *Principe de précaution et risque sanitaire: Recherche sur l'encadrement juridique de l'incertitude scientifique*. Paris: L'Harmattan.

Golstein, S. N. (1981). *Uncertainty in life cycle demand and the preference between flexible and dedicated mass-production systems*. Doctoral dissertation, Department of Engineering and Public Policy, Carnegie Mellon University, Pittsburgh, PA.

Got, C. (1998). *Rapport sur la gestion du risque et des problèmes de santé publique posés par l'amiante en France*. Paris: Ministère du Travail et de la Santé.

Guérin, A. (1978). Amiante: Enquête sur des assassinats au-dessus de tout soupçon. *Les Temps Modernes, 387*, 438–496.

Harremoës, P., Gee, D., MacGarvin, M., Stirling, A., Keys, J., Wynne, B., et al. (Eds.). (2001). *Late lessons from early warnings: The precautionary principle 1896–2000* (Environmental Issue Report No. 22). Copenhagen, Denmark: European Environmental Agency.

Harris, L. V., & Kahwa, I. A. (2003). Asbestos: Old foe in 21st century developing countries. *The Science of the Total Environment, 307*(1–3), 1–9.
Hncharek, M. (1993). Exporting asbestos: Disease and policy in the developing world. *Journal of Public Health Policy, 14*(1), 51–65.
Howard, R. A. (1980). On making life and death decisions. In R. C. Schwing & W. A. Albers (Eds.), *Societal risk assessment: How safe is safe enough?* New York: Plenum Publishing.
Institut National de la Santé et de la Recherche Médicale. (1997). *Effets sur la santé des principaux types d'exposition à l'amiante.* Paris: Author.
Institut National de Recherche et de Sécurité. (2007a). *Amiante: l'Essentie* [Asbestos: The essentials]. Retrieved September 9, 2007, from http://www.inrs.fr.
Institut National de Recherche et de Sécurité. (2007b). *Travail et Sécurité* [Work and security]. Retrieved November 11, 2007, from http://www.inrs.fr.
Keeney, R. L., & Raiffa, H. (1976). *Decisions with multiple objectives: Preferences and value tradeoffs*. New York: Wiley.
Klinke, A., Dreyer, M., Renn, O., Stirling, A., & Van Zwanenberg, P. (2006). Precautionary risk regulation in European governance. *Journal of Risk Research, 9*(4), 373–392.
Kourilsky P, Viney G. Le principe de précaution: Rapport au Premier ministre. Paris: Odile Jacob, La Documentation française; 2000.
Lave, L. B. (1981). *The strategy of social regulation: Decision frameworks for policy.* Washington, DC: Brookings Institution.
Lenglet, R. (1996). *L'affaire de l'amiante.* Paris: Découverte Enquêtes.
Lewis, R. W. (1978). Risk Assessment Review Group Report (NUREG/CR-0400). Washington, DC: US Nuclear Regulatory Commission.
Lilienfield, D. (1991). The silence: The asbestos industry and early occupational cancer research-a case study. *American Journal of Public Health, 81*(6), 791–800.
Meyers, N. J., & Raffensperger, C. (Eds.). (2006). *Precautionary tools for reshaping environmental policy.* Cambridge, MA: MIT Press.
Mishan, E. J. (1973). *Economics for social decisions: Elements of cost-benefit analysis.* New York: Praeger.
Morgan, M. G., & Henrion, M. (Eds.). (1990). *Uncertainty. A guide to dealing with uncertainty in quantitative risk and policy analysis.* Cambridge, UK: Cambridge University Press.
Morris, J. (Ed). (2000). *Rethinking risk and the precautionary principle.* Oxford, UK: Butterworth-Heinemann.
Mossman, B. T., Bignon, J., Corn, M., Seaton, A., & Gee, J. B. L. (1990). Asbestos: Scientific developments and implications for public policy. *Science, 247*(4940), 294–301.
Peto, J., Hodgson, J. T., Matthews, F. E., & Jones, J. R. (1995). Continuing increase in mesothelioma mortality in Britain. *The Lancet, 345*(8949), 535–539.
Pezerat, H. (1995). Evaluer et réduire les risques dans les immeubles floqués à l'amiante. *Archives Maladies Professionnelles, 56*(5), 374–384.
Rasmussenn, N. C. (1975). Reactor safety study: Assessment of accident risks in U.S. commercial nuclear power plants, WASH-1400 (NUREG-75/014). Washington, DC: US Nuclear Regulatory Commission.
Reber, B. (2006). Les controverses scientifiques. In *Encyclopædia universalis, supplement: La science au présent* (pp. 156–159). Paris: Encyclopædia Britannica.
Reber, B. (2007). Technology assessment as policy analysis: From expert advice to participatory approaches. In F. Fischer, G. J. Miller, & M. S. Sidney (Eds.), Handbook of public policy analysis: Theory, politics and methods (pp. 493–512). New York: CRC Press.
Reber, B. (2008). La dissémination des interprétations du principe de précaution: Le cas des essais OGM au champ. In E. Brun-Rovet, S. Groyer, S. Plaud, & A. C. Zielinska (Eds.), *Les institutions saisies par le principe de précaution.* Paris: Centre National de la Recherche Scientifique.
Rosner, D., & Markowitz, G. (1991). *Deadly dust: Silicosis and the politics of occupational disease in twentieth-century America.* Princeton, NJ: Princeton University Press.
Scheberle, D. (1994). Radon and asbestos: A study of agenda setting and causal stories. *Policy Studies Journal, 22*(1), 74–86.

Schulze, W. D., & Kneese, A. V. (1981). Risk in benefit-cost analysis. *Risk Analysis, 1*(1), 81–88.
Selikoff, I. J., Churg, J., & Hammond, E. C. (1964). Asbestos exposure and neoplasia. *JAMA, 188*, 22–26.
Skinner, H. C. W., Ross, M., & Frondel, C. (Eds.). (1988). *Asbestos and other fibrous materials: Mineralogy, crystal chemistry and health effects.* Oxford, UK: Oxford University Press.
Stirling, A. (1999). *On science and precaution in the management of technological risk: Volume I – a synthesis report of case studies* (EUR 19056 EN). Sevilla, Spain: Institute for Prospective Technological Studies.
Sunstein, C. R. (2005). *Laws of fear: Beyond the precautionary principle.* Cambridge, UK: Cambridge University Press.
Tanguay, A. B. (1985). Quebec's asbestos policy: A preliminary assessment. *Canadian Public Policy, 11*(2), 227–240.
Teissonnière, J.-P., & Topaloff, S. (2002). *L'affaire de l'amiante. Semaine Sociale Lamy, 1082*(Suppl.).
Thébaud-Mony, A. (1990). *L'Envers des sociétés industrielles: Approche comparative franco-brésilienne.* Paris: L'Harmattan.
Thébaud-Mony, A. (2003). Justice for asbestos victims and the politics of compensation: The French experience. *International Journal of Occupational Environmental Health, 9*(3), 280–286.
United Nations Environment Programme. (2007). *Rio declaration on environment and development.* Retrieved October 12, 2007, from http://www.unep.org/Documents.Multilingual/Default.asp?DocumentID=78&ArticleID=1163&l=en.
Weill, H., & Hughes, J.M. (1986). Asbestos as a public health risk: Disease and policy. *Annual Review of Public Health, 7*, 171–192.
Wynne, B. (2001). Expert discourses of risks and ethics on genetically manipulated organisms: The weaving of public alienation. *Notizie di Politeia, 17*(62), 51–76.

Chapter 5
Asbestos in the United States

Rose Campbell, James S. Webber, and Hajime Sato

Introduction

Asbestos, once extolled as the miracle mineral, has been a focus of national debate for many years due to delayed and uncertain responses by government and resistance by corporate entities to protect people from its known toxicity. Asbestos represents an important case study in public health; government, medical professionals, and industry managers first recognized the harmful properties of asbestos almost a century ago in America, and yet decades passed before regulatory agencies began acting on this information.

Background

Asbestos was not originally a product of modern industrial society, however. Historians believe that asbestos may have been used in pottery making and home building 5,000 years ago. Even in antiquity, accounts of asbestos harm were recorded. For example, in ancient Rome, it was noted that slaves who wove asbestos for clothing became infirmed and had difficulty breathing with even normal exertion (Castleman *1996)*. In addition to clothing, asbestos was also used for creating wicks and other products by Greeks and Romans 2,000 years ago (Barbalace *2004)*.

R. Campbell
Eugene S. Pulliam School of Journalism, Butler University, USA

J.S. Webber
Wadsworth Center, New York Department of Health, and Department of Environmental Health Sciences, School of Public Health, State University of New York at Albany, USA

H. Sato (✉)
Department of Public Health, Graduate School of Medicine, The University of Tokyo, Tokyo, Japan
e-mail: hsato-tky@umin.net; hsato@post.harvard.edu

H. Sato (ed.) *Management of Health Risks from Environment and Food*,
Alliance for Global Sustainability Bookseries 16,
DOI 10.1007/978-90-481-3028-3_5, © Springer Science+Business Media B.V. 2010

While the use of asbestos declined over the centuries, its appearance spiked in the eighteenth century and again in the late nineteenth and early twentieth centuries during the industrial revolution. Because of its effectiveness at insulation and resistance to heat, asbestos was needed for steam pipes, boilers, building materials, and many other products that had to withstand extreme temperatures. Its value increased as industry grew, and as jobs shifted from rural to urban centers, more people were exposed to asbestos. Today, asbestos is found in hundreds of products, ranging from brakes and insulation to building materials and household products (Bowker 2003).

Asbestos is a term that is generally applied to a group of hydrated mineral silicates that formed naturally as fibers during metamorphosis. For this report, asbestos includes "chrysotile, crocidolite, amosite (cummingtonite-grunerite asbestos), tremolite asbestos, actinolite asbestos, anthophyllite asbestos, and any of these minerals that have been chemically treated and/or altered" (Occupational Safety and Health Administration [OSHA] *2007)*. This is the official or legal description that US government agencies consider for policy discussions. Chrysotile is the only asbestos from the serpentine mineral group, with the last five from the amphibole mineral group. Similar minerals, such as fibrous talc or other fibrous amphiboles, also are included in our asbestos discussions as they can share asbestiform properties.

Exposure to airborne asbestos can lead to serious health problems, from minor respiratory ailments to asbestosis, lung cancer and mesothelioma, a rare cancer affecting the membranes surrounding the lungs and other organs. The most perplexing problem is that threshold levels of asbestos that cause these diseases have not been adequately established.

Asbestos is not always harmful to product users, however. Friable (easily crumbled or pulverized) materials that contain asbestos present a danger to those exposed to it through manufacturing (workplace) or product use (home or office). Non-friable materials (typically cement, vinyl, asphalt) that contain asbestos generally do not liberate asbestos fibers during normal handling. However, any action that degrades or abrades these materials (fire, sawing, shattering, sanding) may liberate airborne asbestos.

Although friable asbestos is regulated now, many products, such as building or insulation materials, are commonly found in homes and workplaces today; their construction predated asbestos policy. Additionally, removing existing materials that contain friable asbestos can sometimes do more harm than good, because asbestos particles can be released when disturbed. Because airborne asbestos is odorless and invisible, persons inhale it without knowing it. Thus, removing asbestos materials requires professional handling. While some manufacturers voluntarily recalled products that contained friable asbestos, many banned products remain in circulation or can be found in existing structures because they predate regulations.

The asbestos case represents a complex issue that draws international attention due to its continued use in society in spite of its documented health risks. This case also illustrates how government regulation develops and is influenced in the process, making asbestos an important lesson in government action. Our report reviews key events that have led to contemporary views on asbestos, including: the medical discovery of its health effects; US mining, manufacturing, and use of asbestos; development of policy to protect workers and consumers; the agents driving the policy process; and the information environment that influences public opinion. Although comprehensive books and reports have already been written on many different asbestos topics, this specific report humbly attempts to summarize critical information, with a goal to provide insight regarding what events and agents drive US asbestos policy.

Medical Evidence of Asbestos Harm in the US

Medical cases involving asbestos have been documented in the United States by both scientists and industry sources since the early part of the twentieth century. Not until 1970, however, were regulations passed in the US to create protections for American consumers and workers. The delayed industry and policy actions are the focus of millions of asbestos lawsuits in the US today. Scores of articles and books have been published accusing the US asbestos industry of suppressing this medical information over decades to avoid the expense and liability associated with protecting workers and subsequent medical expenses (Schepers *1992, 1995*; Bowker 2003; Brodeur *1985)*.

Europe predated the US in evidencing asbestos harm to shape public health policy. Observations of asbestos-related illnesses were recorded in Vienna in 1897 by a physician who reported that "emaciation and pulmonary problems in asbestos weavers and their families left no doubt that dust inhalation was the cause" (Castleman *1996*, p. 2). In 1898, factory inspectors in Great Britain reported damage to asbestos workers' bronchial tubes and lungs. In formal reports each year, the inspectors continued to describe asbestos' effects on workers' health and urged studies be initiated to track the health condition of former workers (Castleman *1996*, p. 3). Several deaths attributed to asbestos inhalation were reported in Italy, France, and Great Britain in the early 1900s, and a German pathologist in 1914 is believed to be the first scientist to report scarring in lung tissue of a deceased asbestos worker (Castleman *1996)*.

Asbestos-related illnesses and deaths also were noted in the US at about the same time, but these reports did not lead to policy action. Epidemiological evidence from 1900–1910 revealed an inordinate number of asbestos workers contracted lung disease compared with other worker populations (Bowker 2003). In the US insurance companies were among the first to note the connection. According to Bowker (2003) and the Environmental Working Group,[1] in 1918 some insurance companies began charging higher rates for asbestos workers or denying them coverage completely. The US Bureau of Labor Statistics published a report conducted by an insurance company that noted a statistically unexpected number of early deaths for asbestos workers (Bowker 2003; Hoffman *1918)*. In spite of growing evidence regarding the deleterious health effects of asbestos exposure, both in Great Britain and the United States, many owners of asbestos mines, mills and processing plants had no procedures in place to protect workers and consumers from exposure.

[1] See: Environmental Working Group website: http://www.ewg.org/

Great Britain was prompted to enact worker protection laws after independent scientific studies were published. In 1924, the first published medical report of a death due to asbestos exposure appeared in *BMJ (British Medical Journal).* The author, Cooke *(1924)*, described the case study of a female employee who had worked in the spinning room of an asbestos factory. Many asbestos-focused medical papers followed, including one by British occupational health specialist, Thomas Oliver *(1925)*, who first coined the term "asbestosis" shortly following Cooke's *BMJ* report. In 1927, Cooke *(1927)* published an additional paper based on his original case study, and he also used the term asbestosis; thus the term became part of the permanent medical nomenclature. Several other reports followed and were published in esteemed British scientific journals. In 1931, responding to mounting medical evidence, Great Britain adopted safety regulations for asbestos workers (Asbestos Resource Center *2008*; Bowker 2003).

The US lagged behind Great Britain in the science and medicine of asbestos. Italy, France and Canada also were reporting asbestos-related illnesses and were well ahead of the US Asbestos exposure was, however, scientifically linked to lung disease as early as 1917, at the University of Pennsylvania School of Medicine. Dr. Henry Pancoast found lung scarring present in fifteen asbestos workers. Ten years later the first asbestos-related disability claim for workmen's compensation was supported. One of the most comprehensive early reports was authored by an employee of the Prudential Insurance Company and published by the US Department of Labor and Statistics in 1918 (Castleman *1996).* The report of the "dusty trades" showed that asbestos workers experienced a significantly higher premature death rate than other workers. The comprehensive monograph illustrated the insurance company's ability to monitor worker hazards and subsequent health effects. Except for insurance companies' proprietary research, no US-based medical studies were published until 1930, when Millls *(1930)* described a single case of asbestosis diagnosed in a man who formerly worked in an asbestos mine in South America. The report was published in the journal *Minnesota Medicine* and later referenced in *JAMA (Journal of the American Medical Association)* (Brodeur *1985*, p. 14; Castleman *1996*, p. 17).

Through the 1920s and 1930s, the British factory inspectors continued to issue asbestos reports and publish studies on both sides of the Atlantic, providing critical information about the disease anatomy and progression. Two inspectors in particular, Merewether and Price, argued for policy to protect asbestos workers and raised the issue of future medical needs. They acknowledged the principal problem that delayed policy decisions – asbestosis could be mistaken for other chronic diseases, such as bronchitis, pulmonary tuberculosis, and broncho-pneumonia (Castleman *1996*, p. 12). Several US journals published international asbestos studies, such as the *Journal of Industrial Hygiene* and *JAMA*. The *JAMA* publication on asbestosis likely had the greatest impact on American physicians, as it was mailed directly to about 80% of all licensed physicians (Castleman *1996*, p. 17). Asbestosis studies in prominent UK journals such as *The Lancet* and *BMJ* also were widely read by American physicians. In 1930, asbestosis was the focus of a paper presentation at the annual meeting of the Radiological Society of North America. According to

Castleman *(1996*, p. 19), the presenter, British scientist Dr. J. V. Sparks, expressed surprise at the dearth of US-based studies on asbestosis. Because Canada was the major supplier of asbestos for the Great Britain, Sparks imagined this was likely true for the US, as well. He expected US asbestosis confirmation to parallel Great Britain's medical discoveries.

Following the alarming number of international reports of serious asbestosis cases, the 1930s represents a surge in US-based research and publication related to asbestos exposure. Independent scientific articles also were published in US medical journals soon after the *Minnesota Medicine* study was published. The studies were both internal and external to the industry (for a review, see Castleman *1996*; Brodeur *1985)*. For example, a major insurance company, Aetna, published a definitive report in 1934 stating there was no known cure for asbestosis and it usually resulted in death (Bowker 2003, p. 18).

Castleman *(1996)* describes a large-scale study sponsored by the asbestos industry that resulted in less serious conclusions than Aetna's. The Metropolitan Insurance Company was approached by asbestos industry officials and asked to conduct studies of workers' health. Although the study uncovered alarming details, such as 53% of workers had asbestosis, the results only reflected active workers and not those who had left work due to illness. The disease rate likely was even higher. In spite of the overwhelming evidence, the study's authors minimized the data and did not recommend regulation of the industry. They argued that no serious disabilities were noted among workers. Further, the study made no mention of British factory inspectors' reports and the British workplace regulations that resulted, nor were published medical studies used to support their conclusions (Castleman *1996)*. The report was completed in 1931, but was not published until 1935. The investigators did recommend that asbestos-related industries more regularly monitor workers' health, address asbestos dust in the workplace, and pre-screen job candidates to eliminate those whose current health conditions put them at greater risk for asbestos-related illnesses, and continue the study of workers' health (Castleman *1996*, p. 34).

American insurance companies and scientists continued to conduct studies of health records and workers in the thirties and forties. This was aided by the science of radiology, which allowed more definitive diagnoses of lung diseases (Castleman *1996)*. Scientists could determine the point at which symptoms would begin to occur after exposure and when lung-scarring appeared. Findings continued to indicate elevated levels of asbestos-related illnesses and deaths, but still no government regulations were in place.

The only concrete outcome of the insurance companies' research findings was the continued practice of charging higher insurance premiums for asbestos workers than for other employee groups (Bowker 2003). Although no formal policies were in place at the time, some mines had voluntarily created worker rules to reduce direct exposure to asbestos. The guidelines were not enforced, however; few workers were inclined to follow rules when their inconvenience seemed to outweigh the risks.

Asbestos miners received the most attention in medical studies, both scientific and proprietary, but later research examined workers exposed to the mineral through the product manufacturing environment. For example, a study of a Massachusetts cigarette filter manufacturing plant found that workers who were employed in the factory in 1953 had unusually high rates of asbestos-related disease and mortality compared with cohort groups (Talcott et al. *1989*; Dodson et al. *2002)*. The researchers concluded that the unusually high morbidity and mortality rates in the cigarette factories were caused by "intense exposure to crocidolite asbestos fibers" (Talcott et al. *1989*, p. 1220). Additionally, a case study of two deceased cigarette filter factory workers, one who worked a short time in the cigarette filter factory during the 1950s and one who worked for several decades beginning in the fifties, suggested that even short exposure led to high levels of crocidolite asbestos accumulation (Dodson et al. *2002)*. Both died from asbestos-related diseases, confirmed by lung tissue analysis. Similar postmortem studies conducted on asbestos workers in mines and other types of manufacturing environments found the more common form of chrysotile asbestos.

Trade journals, such as one published for the International Association of Heat and Frost Insulators and Asbestos Workers, also issued reports on asbestosis in miners, but without describing the implications to those in related industries (Castleman *1996*, p. 19). It is unlikely that workers exposed to asbestos outside the dusty mines considered themselves to be at risk, without the line being drawn for them.

Asbestos Industry Initial Response to Mounting Medical Evidence

Contrary to the incontrovertible evidence linking asbestos exposure to serious illness and premature death, the US asbestos industry expanded during the depression and beyond. It did so in part by developing an international cartel and strategically eliminating competition that had developed alternative products, by using the weapons of buy-outs and price-fixing. While the Federal Trade Commission ruled that some degree of corporate malfeasance was evident, no penalties were issued to any companies involved. The industry also successfully prevented trade unions from gaining increased compensations for those who worked around asbestos. Concern regarding shared liability may have prevented the unions from pushing harder for special compensations for asbestos workers (for a comprehensive review of the asbestos industry's successful attempt to delay regulations, see Castleman *1996*; McCulloch & Tweedale *2008)*.

As this review of medical discovery shows, hundreds of industry reports and scientific studies proved the serious health problems that resulted from direct or indirect exposure to asbestos. The first reports recognizing asbestos' effects on workers emerged in the late 1800s in Europe, then these medical studies accelerated in the 1920s and became prevalent in the US in the 1930s and beyond. Yet these reports seemed to gain little traction with government. For example, the Department of Natural Resources (DNR) states on its official website for Wisconsin that the medical problems associated with asbestos exposure did not occur until the late 1960s:

> In the United States, asbestos became popular in the early 1900s and its use peaked during WWII into the 1970s. While use of asbestos is not banned by legislation, it is not commonly used by American manufacturers anymore due to health concerns and liability issues. However, there is a strong international market, so imported materials may contain asbestos.
>
> During the late 1960s, evidence emerged indicating that asbestos fibers were a dangerous health risk and by the 1970's, the federal government began to take action. During the 1980's, the concern regarding asbestos resulted in the new industry of asbestos abatement (Wisconsin Department of Natural Resources *2006, 2007)*.

Esteemed US medical journals such as *JAMA* and the *New England Journal of Medicine* published studies detailing disease rates among asbestos workers beginning in the 1930s (Lerman *1992)*. A recent search for articles on MEDLINE and other health databases revealed thousands of published reports using only the two key term "asbestos" and "mesothelioma," which limited the search significantly. It is likely that thousands of other asbestos-related health articles were published in the past 70 years. Clearly there is a time lag between medical evidence of asbestos-related health risk and government acknowledgement of it. The above statement from the DNR suggests that the regulations passed in the 1970s represented a timely reaction to new findings.

Opportunities for Asbestos Exposure in the US

Scientific and asbestos industry-sponsored health studies were conducted while workers and consumers were routinely exposed to asbestos and suffering the consequences (Brodeur *1985*; Lilienfeld *1991)*. Many workers were unknowingly exposed to asbestos, in spite of mounting evidence of health risks related to naturally occurring and mined asbestos, as well as products containing asbestos. There were multiple ways in which people were exposed to asbestos.

Mining Operations

Those at greatest risk of asbestos-related health problems are those who are exposed to it daily in the workplace. Asbestos miners represent one of the high-risk groups for serious health effects. Asbestos was first commercially mined in Canada in 1879 and then Russia began asbestos mining shortly thereafter. Chrysotile accounts for more than 90% of the world's production of asbestos, but asbestos has not been commercially mined in the United States since 2002, when the last chrysotile mine was closed in California. Most of the asbestos mines in the US, however, were located in the eastern states of Georgia, Maryland, Massachusetts, North Carolina, Pennsylvania, and Vermont. Reports indicate as many as 60 asbestos mines previously operated on the east coast, with the first major anthophyllite asbestos mine founded in 1894 in Sall Mountain, Georgia. The last operational chrysotile asbestos mine in eastern United States, the Lowell quarry in Vermont, closed in 1993 (Van Gosen *2005)*.

Some miners of fibrous talc are also exposed to asbestos or asbestos-like dust, leading to lung ailments similar to those seen in asbestos miners. Officials from companies that produce products from talc argued that the mineral fibers found in talc have different properties and thus should not be labeled as asbestos. The first fibrous talc mine in the US was opened in 1878 in upstate New York and by the end of World War II, several talc mines and mills were in operation in that area. Industry records indicated that the mine and mill owners were aware that talc contained asbestos, yet withheld the information from workers. Miners were not provided information about health risks associated with talc mining or asbestos mining more directly (Schneider *2001b*; Schneider & McCumber *2004)*. Recently the fibrous talc industry has been the subject of litigation regarding workers' health or product toxicity (e.g., putty for wall patching). As recently as May 2007, however, industry officials denied that the mined talc contained asbestos, saying the tremolite mineral in the talc is nonasbestiform. This has been disputed, however, by environmental scientists (Webber et al. *2006*; Webber et al. *2004)*.

Amphibole asbestos has also been found in some vermiculite mines (Meeker et al. *2003)*. Vermiculite is an ore similar to mica used in products such as housing insulation, fertilizers, and cement mix. Major commercial mines for vermiculite are in Australia, Brazil, China, Kenya, South Africa, USA and Zimbabwe. In the US, Libby, Montana, was the site of the world's largest vermiculite mine; it shipped vermiculite for processing to over 60 plants in the US and Canada, and the mine supplied over 80% of the world's demand during the middle of the twentieth century (Vollers & Barnett *2000*; Schneider & McCumber *2004)*. The vermiculite at this particular site was laced with deadly amphibole fibers. The Libby mine, which was in operation from 1924 through 1990, now is designated as an Environmental Protection Agency (EPA) Superfund site. Estimates ranging from 18% to 25% of Libby residents showed signs of lung scarring related to asbestos exposure. This is in comparison to 50% of Libby's mineworkers who reportedly had lung scarring or more advanced asbestosis. Once environmental contamination was identified, Libby qualified for the site designation. Libby currently is on the final national priorities list in the "active" stage of cleanup (Environmental Protection Agency *2008)*.

Similarly, Virginia Vermiculite operated a mine in Louisa, Virginia. The Mine Safety and Health Administration tested the site and found both tremolite and actinolite asbestos fibers. As a result, the company offered free health screenings for workers to test for asbestos-related diseases (Schneider *2000)*. The mine is still in operation but has safety standards for asbestos handling in place.

It is surprising that in the latter part of the twentieth century, that mines such as W.R. Grace & Co. could operate without effective risk management practices, given the history of asbestosis disease noted above. Mining operations of talc and vermiculite continue to exist in the US and elsewhere, although generally with stronger worker safety guidelines and better site management to prevent environmental contamination and protect employees (more thorough discussion of the US government's response to asbestos is in a later section of this report).

Asbestos in Non-mining Workplaces

Workers who process asbestos from mines, work directly with products containing asbestos, or manufacture products made from asbestos are at risk for serious health problems similar to those experienced by asbestos miners. Each year, over a million Americans are employed either in shipyards, construction (renovation, asbestos abatement, construction), general manufacturing (textiles, friction and insulation products), or the automotive industry (brake or clutch repair), and are consequently at elevated risk for asbestos exposure (White *2004)*. These are the employment categories currently protected by the Occupational Safety and Health Administration (OSHA, under the US Department of Labor) and the EPA in regard to asbestos.[2] Firefighters, electricians, and plumbers also are at risk of asbestos exposure. Additionally, small appliance repair workers who accept older models of coffee pots, popcorn poppers, hair dryers, crock-pots, and portable heaters, for example, also are at risk of asbestos exposure. Factory workers who produced cigarette filters using crocidolite form of asbestos in various plants in the US, including Kentucky and Massachusetts, were exposed to high levels of asbestos, as well (Talcott et al. *1989)*. Overall, an estimated 27 to 100 million Americans have been exposed to asbestos (White *2004)*.

Because vermiculite was not previously considered asbestiform, vermiculite mines and manufacturers continued to process it without regulations. Some vermiculite mines and processing plants in the US, however, had conducted their own testing, which revealed high concentrations of amphibole fibers. Plant officials were aware of asbestos' deleterious effects on health, well documented since the 1930s, yet they did not warn employees about the asbestos test results related to vermiculite. Because of the lag in mines and manufacturers revealing information about unsafe levels of asbestos exposure, it is believed that millions of American workers may have been unknowingly exposed to asbestos through a variety of processes, e.g., factories where vermiculite was processed, gardens/lawns where vermiculite was applied, and attics where vermiculite was used for insulation. For these reasons, asbestos litigation has become the longest-running mass tort litigation in the US (White *2004)*. In 1978 the EPA and OSHA began to regulate asbestos, and strict guidelines were created to protect workers from inhaling asbestos fibers, yet vermiculite processors did not enact these protections, as they did not reveal their internal asbestos test results.

---

[2] See: Occupational Safety and Health Administration website: http://www.osha.gov/

In a 1991 report published in the *American Journal of Public Health*, Lilienfeld argued a more conspiratorial plan was in place: "This industry (asbestos), in concert with many of its insurers, systematically developed and then suppressed information on the carcinogenicity of asbestos. As a result, millions of workers were exposed to the carcinogen and hundreds of thousands died" (Lilienfeld *1991*, p. 791). Lilienfeld presents evidence showing that as early as 1933, physicians warned that American workers should be made aware of the serious health consequences of asbestos exposure. Industry officials believed that economics and production factors should be weighed equally with medical information, so workers were not warned (Lilienfeld *1991*, p. 792).

W.R. Grace & Company, which operated the Libby, Montana, vermiculite mine from 1963 until it closed in 1990, also had a processing mill in the same town and vermiculite processing plants throughout North America. Government records indicate that the Grace Company sent its vermiculite to over 200 processing and packaging plants throughout North America from Seattle to New Jersey. The types of products produced at these plants primarily were lawn and garden products and insulation. In 2005, the US government estimated that up to 35 million homes or businesses contained these products and the workers and residents could be at high risk for contracting asbestos-related illnesses (Bowker 2003). More were affected due to Grace & Company's community relations activities. The company encouraged employees to take mill waste home to use in their gardens. Some of the processed vermiculite or mill waste was donated to schools for construction use or for playground soil, for example (Bowker 2003, p.195). Because of the widespread distribution of Grace's products, it is not known how many people were exposed to asbestos-contaminated vermiculite.

The Libby, Montana, mine town serves as an exemplar case; studies indicate that even residents who were not directly exposed to asbestos through work or indirectly exposed through contact with workers, had unusually high rates of asbestos-related illnesses and disease compared to overall population rates. Asbestos fibers were found in Libby residents' homes, even if a mineworker did not live there (Bowker 2003).

Inadvertently Encountered Asbestos

As a naturally occurring mineral in many parts of the US, asbestos exposures can be created when veins are disrupted by man-made or natural activities (Van Gosen *2005*). Recently, construction projects that have disturbed such bedrock, e.g., northern California and suburban Virginia, have concerned public health officials regarding the safety of construction workers and residents.

An EPA study was conducted in 2004 in El Dorado County, California, in a densely populated area east of Sacramento. The EPA found that typical activities such as bike riding, playing baseball, and using playground equipment, disturbed the soil and caused high levels of asbestos fibers to be released in the air.

The *Sacramento Bee* published a series of articles beginning in October 2004 concerning the asbestos levels, and since then residents have expressed concern about the safety of their community.

Of even greater concern to residents was a recently published epidemiology report. A study by University of California-Davis researchers (Pan et al. *2005)* examined the California Cancer Registry and found unusually high rates of mesothelioma cancer, a rare form of cancer associated with asbestos exposure, among those who lived near ultramafic bedrock (which contained the asbestos), compared to normal population rates. The probability of having the disease dropped by over 6% for every 10 km people lived away from the asbestos source.

Industry professionals, however, have criticized such studies. Public health professionals have difficulty making a case for regulating construction on such sites or announcing risk management programs, due to four factors (Berg *2004)*: (a) difficulty in determining the inhalational exposure levels; (b) no universally agreed upon testing and sampling protocols for gauging this exposure; (c) no known threshold for exposure that leads to mesothelioma cancer; and (d) epidemiology reports today may not accurately predict asbestos-related disease, since the disease can take over 20 years to develop (there may be higher rates found later, or perhaps lower). For these reasons, scientists and public health professionals may be unable to make credible arguments that lead to regulation regarding activities around naturally occurring asbestos sites.

Asbestos in the Home and Businesses

No absolute ban of asbestos products has been established in the US; consequently, many products containing asbestos may be commonly found in homes today. Due to the well-documented association between asbestos and serious diseases such as mesothelioma and asbestosis, new uses of asbestos were banned in 1989. At one point, over 500 products containing asbestos were commonly sold for household use, from personal appliances such as hair dryers and coffee pots, to home construction materials like insulation and paint, to gardening supplies like potting soil and fertilizer. As mentioned above, only products containing less than one percent of friable asbestos can be produced and distributed in America today. Because this excludes existing products, however, Americans may have many products in their homes that put them at risk for asbestos exposure.

The 2005 ABC Nightline report, “A Killer in Town,” highlighted the Libby, Montana case and emphasized the magnitude of the asbestos problem in America. The news report presented alarming statistics of asbestos exposure in private homes, estimating that between 15 and 35 million homes in America likely contain asbestos insulation. With the average family size at four, the report predicted that “significant numbers of people (will be) coming down with asbestos disease for the next several decades” (ABC News Productions *2005)*. The two-part program stressed not only the health dangers to homeowners, but also the enormous expense of widespread abatement efforts to eradicate it from existing structures and the mounting health care costs associated with asbestos exposure.

Concerned consumers, including those frightened by the Nightline report, may turn to government agencies for guidance in regard to asbestos products. The US Consumer Product and Safety Commission (CPSC) provides consumer information regarding what kinds of products may contain asbestos and how people should respond when they detect asbestos in the home. According to its website, the CPSC protects the public "from unreasonable risks of serious injury or death from more than 15,000 types of consumer products under the agency's jurisdiction."[3] Asbestos-containing products commonly found today include roofing and siding shingles containing asbestos cement, insulation in homes built prior to 1950, paint and wall patching compounds applied before they were banned in 1977, artificial ashes used for aesthetics purposes for gas fireplaces, older home products such as pads for stoves or coatings for hot water and steam pipes, vinyl products such as floor tiles or sheet floorings and their corresponding adhesives, and protective structures built around wood stoves. Old small appliances that may contain asbestos and are still in circulation are not mentioned on the CPSC consumer information site.

The potential health threat from asbestos is also true for schools, businesses, and any older structures containing these products. It is expensive to renovate schools, as asbestos abatement is labor-intensive and costly. Because asbestos removal in buildings is highly regulated, necessary renovations often are ignored due to the expense, which disproportionately affects the poor. This is not dissimilar to the lead abatement situation: lead commonly found in paints was banned, as it was toxic when ingested. Small children who found paint chips were at greatest risk of poisoning. Lead paint still exists in older buildings and public housing, where dwellers can ill afford renovation.

Impact of Asbestos: US Mortality Rates

In 1979, the US began tracking deaths related to asbestos. From 1979–2001, 43,073 deaths were attributed to asbestos exposure. This is likely a very conservative estimate based on death records, because mesothelioma was not designated as an asbestos-related cause of death until 1999.[1] Previously it had been associated with other cancers. Asbestos mortality rates doubled from 1998 to 1999, once mesothelioma was counted. The Centers for Disease Control and Prevention lists asbestos death by occupation, and construction workers, insulation workers, and ship builders had the greatest number of deaths attributed to asbestos between 1985 and 1992 (Bartrip *2004)*. Based on previous statistics and accounting for the lag between asbestos exposure and disease presentation, an independent study estimated

[3] See: Consumer Product Safety Commission website: http://www.cpsc.gov/

131,200 US deaths from asbestos-related cancers from 1985–2009 (Lerman *1992*; Lilienfeld et al. *1988)*. These data indicate that asbestos exposure represents a serious problem that will have an impact on the healthcare industry – and the courts – for years to come.

The Regulatory Environment for Asbestos

The US government's response to asbestos-related illness and death centers on risk reduction and public education rather than eradication. While the US has not uniformly banned asbestos, there are safety regulations for the workplace and product-related regulations for manufacturers. Asbestos is also monitored and regulated as part of laws addressing air quality in general.

Several US agencies are involved in asbestos regulation, education, and monitoring. Federal agencies with regulation oversight or policy enforcement regarding asbestos products, wastes and emissions include: The Environmental Protection Agency (EPA), Occupational Safety and Health Administration (OSHA, within US Department of Labor), US Consumer Product Safety Commission (CPSC), Agency for Toxic Substances and Disease Registry (ATSDR), National Institute for Occupational Safety and Health (NIOSH), National Institute of Standards and Technology (NIST), and Mine Safety and Health Administration (MSHA). Several other government agencies provide information and public education about asbestos or are part of the enforcement structure for regulation. For example, in Wisconsin, the Department of Natural Resources (DNR) is responsible for enforcing EPA regulations in regard to building demolition and renovation, to prevent asbestos fibers from becoming airborne. Other states have their own agencies that serve the same function. Additionally, the Centers for Disease Control and Prevention (CDC) provides information on asbestos regulation, exposure prevention and health effects of exposure.

For the following report of asbestos policy development, the website of each agency were reviewed, as well as government and industry documents posted on the Environmental Working Group (EWG) website. The legislative tracking site at thomas.loc.gov from the Library of Congress also was consulted and the government action tracking database at govtrack.us provided details about status of recent legislative actions. Other relevant sources are cited within the text when referenced.

Timeline of Major Asbestos-Related Government Actions

The timeline of asbestos regulation does not parallel the timeline of medical discovery regarding asbestos-related illness and mortality. Although asbestos harms were documented in the early part of the twentieth century, asbestos regulations in the US were not passed until decades later: The earliest known independent asbestos medical studies were conducted in the US in 1917, but the first asbestos-related government regulation was passed in 1970.

Several policies were enacted beginning in 1970 with the Clean Air Act (CAA), a federal law that opened the door for asbestos regulation. The CAA "regulates air emissions from area, stationary, and mobile sources."[4] The CAA also empowered the EPA to create testable standards for air quality that protect the environment and people in the interest of public health: National Ambient Air Quality Standards (NAAQS). The EPA officially recognized asbestos as a hazardous pollutant in 1971. This led to the National Emission Standards for Hazardous Air Pollutants (NESHAP) regulation in April 1973, attached to the CAA. This was a significant act, as it made asbestos one of the first hazardous air pollutants under government regulation. For example, spray-on asbestos insulation was banned by the EPA as an air pollutant. NESHAP is designed to prevent asbestos fibers from becoming airborne when being handled in some form, whether friable or non-friable. It particularly specifies work practices to be followed during renovation, demolition or other abatement activities when friable asbestos is involved. Under the CAA, these national ambient air standards were to be met by every state by 1975. It was amended in 1977 to reset these goals, which were difficult for many states to achieve in the prescribed timeframe. The amendment also clarified the act in relation to issues such as ozone depletion and acid rain. The asbestos-related tenets of NESHAP, however, remained unchanged.

In response to the CAA's and the EPA's designation of asbestos as a hazardous pollutant, OSHA specified worker protections in 1972. These threshold standards for asbestos exposure were made even more stringent 2 years later. This was a significant year for asbestos regulation; the crux of current class action lawsuits is whether employers knowingly exposed workers to asbestos after this regulation was passed.

In 1976, the Toxic Substance Control Act (TSCA) was passed, authorizing the EPA to regulate asbestos in schools and public and commercial buildings. The TSCA also gave responsibility to the EPA for tracking industrial chemicals either produced or imported into the United States. The EPA has the discretion to test any of these chemicals and can ban those it deems a public health or environmental threat. Meanwhile, 1976 marked an all time high for asbestos production in the United States, with 1 million tons processed (Djerassi et al. *1979).*

In 1977, the US Consumer Product Safety Commission (CPSC) became involved in asbestos regulation. It approved a ban on two consumer products containing inhalable asbestos – consumer patching compounds and artificial fireplace ash used for aesthetics only – due to concerns that asbestos fibers would be released into the air while being used.

NESHAP next banned most spray-applied surfacing asbestos-containing material (ACM) in 1978, although this was revised in 1990. Under the revised 1990 rule, instead of an outright ban, spray-on applications were allowed if the asbestos material was encapsulated within a binder and the asbestos fibers were not released after the spray dried. This revised rule by NESHAP also prohibits spray-on application

[4] See: Environmental Protection Agency website: http://www.epa.gov/

of materials containing more than 1% asbestos to any structures, pipes, and conduits, unless the material is encapsulated with a bituminous or resinous binder during spraying and the materials are not friable after drying.

To alleviate costs associated with new asbestos regulations stemming from the CAA and EPA guidelines, the Asbestos School Hazard Abatement Act (ASHAA) was passed in 1984. This Act created funds for loans and grants to both public and private schools that otherwise could not afford to finance asbestos abatement during renovations. These funds were provided between 1985 and 1993, but no funds have been awarded since then. Complementary to the TSCA, the Asbestos Hazard Emergency Response Act (AHERA) was passed by Congress in 1986. Its main provisions were to set guidelines for asbestos inspection and abatement procedures for schools. AHERA also provided grants for schools not only for asbestos inspections but also to create management plans to address abatement needs. All of these regulations fall under the control of the EPA. The year 1986 is significant, because both the EPA and the International Agency for Research on Cancer, part of the World Health Organization, declared asbestos "a human carcinogen" (Landrigan *1998)*.

Also in 1986, OSHA created stricter guidelines limiting asbestos exposure in the workplace, and the EPA outlined new asbestos abatement guidelines. *The EPA took the stance that asbestos at any level is not safe.* Below is an excerpt from the EPA report, containing the conclusion and three of the most significant findings that support it:

> Therefore, EPA finds that unregulated removal, enclosure, or encapsulation of friable asbestos material presents an unreasonable risk to human health and proposes to require that certain measures be taken to reduce the risk faced by asbestos abatement workers and persons using and visiting buildings during and after asbestos abatement activities. This finding is based on the following points:
>
> 1. The health effects from asbestos are very serious. Asbestos is a demonstrated human carcinogen. The cancers caused by asbestos are usually fatal and cause much pain and suffering.
> 2. Available evidence supports the conclusion that there is no safe level of exposure to asbestos. This conclusion is consistent with present theory of cancer etiology and is further supported by the many documented cases where low or short-term exposure has been shown to cause asbestos-related disease.
> 3. Models developed to estimate the relative risk of developing cancer from exposure to asbestos show a linear dose-response relationship. Based on data from epidemiology studies, these models predict that humans exposed to even very low levels of asbestos incur some risk (Environmental Protection Agency *1986)*.

Two years after these guidelines were issued, the EPA banned new uses of asbestos. This later was empowered by the Toxic Substances Control Act of 1989, also known as "Asbestos Ban and Phase-out Rule" (this was appealed and never put into effect; see below). In 1988, considering vulnerable publics, the EPA created new requirements that schools must inspect for damaged asbestos and if identified, either remove it or encase it. It also clarified 1973 NESHAP guidelines for "acceptable use" of asbestos and defined risk levels of asbestos exposure (e.g., enforcement rules for "friable asbestos," materials containing more than 1% asbestos by weight). In 1990, the Asbestos School Hazard Abatement and Reauthorization Act (ASHARA) extended AHERA with minor revisions.

Other government efforts focused on informing relevant parties regarding asbestos exposure. For example, to address the increasing number of asbestos-related lawsuits, the Asbestos Information Act was passed in 1988, which required all companies that used asbestos in the manufacturing process to file reports revealing information on the types of products made, the years they were manufactured, and other information that would help identify their products. Companies had 90 days to file these reports once the Act was passed, and afterwards the EPA made the information available to the public. Having information readily available would also eliminate the time typically used by lawyers and courts to collect such evidential data.

Other regulatory actions reflected the need for accurate assessment tools to detect asbestos levels. In 1990, NESHAP announced that the transmission electron microscopy (TEM) method was the best tool for identifying asbestos in the air, compared to current methods. Later, this was restated. The EPA distributed pamphlets explaining the complex NESHAP regulations. The agency said there was no numeric emissions limit or standard for release of asbestos that may occur during building demolitions or renovations. The standard of "visibility" was described: If you can't physically see the asbestos in the air then it is within the law's guidelines. Any visible particles of asbestos would mean it is unsafe under these guidelines (Bowker 2003).

Other restatements were issued. The EPA announced in 1993 that new studies determined that some previously recommended dry vacuuming and wet cleaning methods recommended by the agency for reducing exposure to asbestos actually *increased* hazardous emissions of asbestos and also did not effectively remove asbestos fiber from carpets. The next year, guidelines for asbestos cleaning at demolition sites were issued, prescribing that after cleanup, the demolition site should match the average asbestos level found in the area surrounding it. An EPA advisory was issued again reinforcing use of TEM for testing levels of asbestos in air samples.

Clearly, the Clean Air Act of 1970 paved the way for future asbestos regulation, with the majority of asbestos-related policies being passed between 1970 and 1990. The CAA and the Toxic Substances Control Act are the two major federal laws under which asbestos policies are enacted and enforced. In the years following, however, many restatements were issued and some policies were over-ruled.

Asbestos in the Schools: Balancing the Science of Risk Assessment with Social Values

Although consumers can choose whether to worry about or ignore health dangers in their own homes, parents are less accepting of threats that potentially affect their children and are out of their direct control. Not surprisingly, the asbestos issue became salient to the public once children became the subject of risk assessment and government policy.

Jacqueline K. Corn *(1999)* addresses this issue in her book, *Environmental Public Health Policy for Asbestos in the Schools: Unintended Consequences.* Specifically, she describes the development of science-based risk assessment tools for government decision-making in the 1970s, in regard to environmental policy. She explains how these tools gained acceptance in the 1970s and early 1980s, and were questioned when risk estimates were applied to children and asbestos exposure, "where emotion overruled science" (Corn *1999*, p. 24).

The struggle emerged when policymakers applied objective tools to estimate risk (probability and severity of harm) and then weighed their results in consult with the social values of the affected community. Corn cites an influential book by William Lowrance (*Of Acceptable Risk*), whose philosophy, she argues, was highly influential in guiding federal policy decisions: "Safety is not measured. Only when those risks are weighed on the balance of social values can safety be judged: a thing is safe if its attendant risks are judged to be acceptable" (Lowrance *1976*, p. 8).

Although most studies conducted by the EPA and independent sources concluded minimal risk of children's death from exposure to asbestos in schools, Corn said that instead the EPA primarily relied on epidemiology studies of non-occupational exposure as the basis for its decisions. She said no effort was made to measure how much asbestos was in the ambient air in schools under study. Such studies would provide important data required for science-based decision making (Corn *1999*, p. 25). Thus, policy was determined and monies allocated without appropriate scientific testing. She charges the EPA with moving forward to regulate asbestos in the schools, including the expensive school asbestos-abatement policies, without adequate risk assessment to support its actions. In sum, she argues that AHERA and other policies were not based on scientific reasoning but on political pressures that had little to do with public health. Additionally, at the time, reports of environmentally caused cancer and asbestos workplace hazards were well publicized, and public health campaigns were launched to educate the public about toxic substances. In this way, asbestos in the schools became embroiled in a larger value issue rather than a risk assessment issue.

Corn's extensive analysis of policy actions in regard to the school asbestos issue represents an important case study of the perils of risk management in the public sphere. It is important to note, however, that Corn's book was funded in part by a grant from the W.R. Grace Company to the Johns Hopkins University. Other scholars, however, have reached similar conclusions without funding from the asbestos industry. For example, a report by Wilson et al. *(1994)* argued that New York City school administrators' actions following an asbestos emergency were not based on scientific assessments. The authors argued that the EPA's subjective reasoning was employed, and "(d)ata on concentrations of asbestos in the air, important for the calculation of risk to building occupants, were not required" (Wilson et al. *1994*, p. 161) for decision making. Comparable ambient air measures taken elsewhere indicated that exposure would bring no more risk than smoking 12 cigarettes in a lifetime, yet the school administration bowed to pressure by parent groups that perceived greater risk to their children.

The issue of asbestos in the schools illustrates how the science of risk management can get muddied by politics and public pressure. While effective assessment tools were available at the time to make sound science-based decisions, the judgment of social costs and values tipped the scale toward stronger precautionary actions.

Unsuccessful Asbestos Regulatory Efforts

While asbestos policy in regard to schools was very strict, given the minimal risk to children, not all government efforts to strengthen policy protecting consumers were successful. For example, the EPA's 1989 announcement to phase out all products in the US that contained asbestos (reported above), did not become federal law. It was overturned by the Fifth Circuit Court of Appeals in New Orleans in 1991. Following this ruling, the EPA ban covered only the following asbestos-containing products: flooring felt, rollboard, and corrugated, commercial, or specialty paper, as well as new uses of asbestos; i.e., products that historically had not been manufactured using asbestos (Environmental Protection Agency *1992)*. This legal action by the courts placed much of the responsibility for asbestos safety in the hands of consumers, who are urged to visit the EPA website to find out where asbestos is commonly found and how to minimize personal exposure to asbestos. Here, interested consumers will find a "commonly asked questions" section, as well.

Many attempts were made to create stronger protections for asbestos workers, as well. These efforts were bolstered by a series of government studies by the Asbestos Work Group, which was formed from two government agencies, OSHA and NIOSH. The group's studies indicated a strong link between asbestos exposure and cancer even in low exposure conditions. The strongly worded document described incontrovertible evidence that "there is no level of exposure below which clinical effects do not occur" (Asbestos Work Group *1980)*. This led to a new policy ordering reduced threshold levels for airborne asbestos, enacted in an emergency ruling to protect workers. In 1984, however, this standard was overturned by the United States Court of Appeals for the Fifth District. *The New York Times* reported:

> An intermittent 13-year struggle by the Occupational Safety and Health Administration to sharply reduce the exposure of industrial workers to asbestos, a known cause of lung disease, including cancer, has been dealt a setback by a Federal appeals court. A three-judge panel … has unanimously overturned an emergency regulation proposed by the agency last November to limit the asbestos content of workplace air to half a fiber per cubic centimeter. The agency had said the standard would have immediately begun saving lives, but the court disputed the agency's precise figures (Franklin *1984*, sect. 1, p. 22).

The New York Times also reported that OSHA attempted to pass nine emergency asbestos regulations and five of these were overturned in courts. This particular court ruling from the US Court of Appeals led to the resignation of OSHA's administrator (Franklin *1984*, sect. 1, p. 22).

The next year (1985), a controversy emerged when the White House Office of Management and Budget (OMB) attempted to prevent the EPA from enacting programs to protect the public from asbestos:

> A Congressional report issued today asserted that the White House Office of Management and Budget, in an "unlawful abuse of power," had for more than a year blocked measures to protect the public from asbestos. The report, prepared by the House Energy and Commerce Committee's Subcommittee on Oversight and Investigations, said that the OMB deliberately obstructed proposed rules that would ban some asbestos products completely and gradually eliminate most other uses over 10 years (Shabecoff *1985*, sect. A, p. 32).

This *New York Times* report described criticism of the Reagan administration for taking extraordinary measures to deregulate industry. The government attempted to control the EPA through the OMB; the EPA was one of the agencies the government attempted to control in this endeavor. "The report on the asbestos rules is one of the heaviest Congressional salvos fired to date at the Reagan Administration's efforts to reduce the burden on industry of regulations to protect public health and safety. The White House budget office has taken the leading role in carrying out President Reagan's program to cut back regulations" (Shabecoff *1985)*. It was unprecedented that the White House OMB would have oversight for established agencies such as the EPA. Interestingly, the head of Reagan's committee to reduce government regulations was J. Peter Grace, Chairman of the Board of the W.R. Grace Company, the Libby, Montana, asbestos mine that is the focus of a federal court case today.

More recently, the OMB was criticized for squelching an EPA public warning about W.R. Grace & Company attic insulation, branded "zonolite." The warning was to be issued in 2002 and is detailed in a formal report by Senator Patty Murray of Washington State, a key advocate for banning asbestos in America (Murray *2003)*. Internal documents showing e-mail conversations between OMB and EPA officials were leaked to Senator Murray, who formally launched a complaint with President George Bush and communicated it via a press release on her website (Murray *2003)*. The EPA report that never was released contained an estimate that several million homes in the US were insulated with zonolite. The EPA planned to warn homeowners and provide instructions for reducing exposure to it. Many people routinely remodeled their homes and were handling the asbestos-laced vermiculite insulation, unaware of its toxicity; zonolite was promoted as safe insulation.

It was no surprise to asbestos industry leaders, however, that the EPA warning on zonolite did not transpire. A former employee of the vermiculite mine in Louisa, Virginia, reported his bosses celebrating on Bush's inauguration day in 2001: "(T) he bosses at the plant were joyful and ordered all the red flags and orange cones removed … and the workers were told to excavate through the asbestos as they always had" (Schneider & McCumber *2004*, p. 280).

Other government agencies appear to have minimized the asbestos threat levels previously issued. For example, earlier consumer guidelines issued by the CPSC were more strongly cautionary regarding asbestos, quoting federal health department findings (e.g., "there is no level of exposure below which clinical effects do not occur") to punctuate its warnings. Current reports on the agency's website, however, are much more reassuring: "This booklet will help you understand asbestos: what it is, its health effects, where it is in your home, and what to do about it. Even if asbestos is in your home, this is usually NOT a serious problem. The mere presence of asbestos in a home or a building is not hazardous. The danger is that asbestos materials may become damaged over time. Damaged asbestos may release asbestos fibers and become a health hazard" (Consumer Product Safety Commission, Environmental Protection Agency and American Lung Association *2005)*.

Current Asbestos Regulation Efforts

More recent government actions related to asbestos policy have addressed the asbestos litigation issue. For example, the Fairness in Asbestos Injury Resolution Act of 2005 (FAIR Act of 2005) was designed to help facilitate the growing claims involving asbestos-related injuries and deaths. According to the legislative tracking site,[5] the FAIR Act of 2005 proposed establishing, within the Department of Labor, the Office of Asbestos Disease Compensation, a system to process all worker-related injury claims related to asbestos exposure and to distribute compensation to victims from the Asbestos Injury Claims Resolution Fund (also to be established in the Department of Labor). The FAIR Act of 2005 never became law. The next year it was presented again as the FAIR Act of 2006, only to die in committee.

The FAIR Act was not supported by most asbestos industry workers, due to what they believed were unfair provisions regarding proof of injury and caps on claims, among other issues (Creswell *2006)*. Indeed, many medical studies (reported above) have shown that injury from asbestos exposure may not be evident for decades after asbestos exposure. For example, a special ABC News television documentary profiled a young adult who had lived in Libby for a little more than 10 years when he was a child and later contracted asbestosis (ABC News Productions *2005)*. Although he never worked at Grace, he played on the ball field where high levels of asbestos were later discovered. He disagreed with the legislation because he was unlikely to be covered under its provisions. For example, there were debates regarding specific provisions, such as setting time limits for deliberating cases and compensating claimants, and treating some claimants differently than others (e.g., Libby residents would be in a different class of claimants).

The bill was criticized by asbestos workers, insurance companies, and asbestos industry officials alike. There was disagreement about the bill within each of these groups, as well. For example, smaller companies feared they would pay an inordinate amount to the Asbestos Claims Resolution Fund that would not reflect their fair share of the actual damage awards (Creswell *2006)*. Officials from larger companies, such as United States Gypsum, claimed they would have paid billions of dollars less than what would be paid for individual settlement claims if the FAIR Act had become law. It was predicted that smaller companies, such as

[5] See: Library of Congress Thomas website: http://thomas.loc.gov/

A. W. Chesterton, would go bankrupt, as the FAIR Act would require payments that reflected "more than double a year's profit" for the company (FAIR Act of 2005, Section 3). Although the bill was designed to end the lengthy and expensive legal proceedings and provide equitable distribution of compensation to those who truly were impaired from asbestos exposure, many affected groups were not supportive. Firms with the largest number of claims stood to benefit most from the bill, in comparison to their smaller industry counterparts who preferred gambling with the existing court system rather than allowing government to make preemptive decisions. Many companies argued that the bill was unconstitutional – it would take away their rights to a fair trial and force them to hand over their insurance assets to create a pool for all claimants, even those companies not affected by individual asbestos claims.

The flaws in this legislation delayed and ultimately blocked asbestos litigation for years. In a speech to citizens at Macomb Community College, President George W. Bush associated the asbestos claims with "junk lawsuits":

> A way to make America the best place to do business in the world, a way to make sure jobs continue to exist here is to tackle the tough issue of legal reform. We have too many junk lawsuits in our system, pure and simple, and frivolous and junk lawsuits cost our economy $240 billion a year. That's a problem...And why is (class action lawsuits) a national problem? Well, first of all we're spending $80 billion on asbestos litigation, and that could end up being $200 billion over time. Secondly, these asbestos suits have bankrupted a lot of companies, and that affects the workers here in Michigan and around the country. Thirdly, those with no major medical impairment make up the vast majority of claims, while those who are truly sick are denied their day in court (Bush *2005)*.

In his speech, Bush argued that most asbestos-related companies are bankrupt and cannot pay claims, so the burden lies on the smaller companies that "that aren't directly involved in the manufacturing of asbestos" (Bush *2005*, p. 24). He fails to mention that legislation enabled asbestos companies to file bankruptcy indefinitely, without firing employees or stopping any business as usual. The only difference is that they are no longer liable for any claims brought against them now, and most notably, *in the future*. This was an attractive position to be in; businesses could continue to make profits and be immune from punishment of criminal negligence of the past:

> While any form of bankruptcy is serious, it is clear that asbestos filings under Chapter 11 have not wreaked havoc on the economy. Between February 2000 and October 2001, the seven largest companies facing asbestos liability filed for bankruptcy protection under Chapter 11. These companies include Babcock & Wilcox, Owens Corning, Armstrong, W.R. Grace & Co., US Gypsum Co., Federal-Mogul and Building Materials Corporation of America. An analysis of 10 K filings for these companies for the years 1998 through 2002 concluded that: "The Chapter 11 companies have been able to continue operations successfully. Indeed, with few exceptions, they have prospered, increasing their sales. They have been able to maintain their assets and employment, meet their obligations to business creditors and employees, and make capital investments that will allow them to continue to prosper."[1]

In the 2007 Congress, two bill amendments were in progress that specifically addressed asbestos. Ban Asbestos in America Act of 2007 was proposed to amend the Toxic Substances Control Act to require NIOSH to (a) conduct scientific studies on asbestos and the health effects of exposure, (b) identify the mechanism for accurate measurement of asbestos, (c) create labeling practices for asbestos-containing materials, and (d) offer guidelines for future research on asbestos-related disease from exposure. The Act would have empowered the EPA to regulate importing, manufacturing, processing, and distributing asbestos-containing products with exceptions for the Department of Defense and National Aeronautics and Space Administration. The Act also would have amended the Public Health Service Act to lead and encourage research on asbestos and to serve as "a national clearinghouse for data and specimens relating to asbestos-related diseases" (Congressional Research Service *2007)*. Finally, the Act involved the Director of the National Institutes of Health (NIH) to create a network for asbestos research and particularly to support research on asbestos that directly affects military personnel. The bill passed committee on August 2, 2007, and was passed by consent in the Senate. However, the Senate never assigned the bill (S.742) to a House committee, and no action has been taken as of October 5, 2007. If a new bill like this is eventually passed, it would correct the action by the Fifth Court of Appeals in New Orleans, which in 1991 overturned the EPA's ban on all asbestos products. This bill from 2007 is one of several previously proposed by Senator Patty Murray of Washington.

Perhaps the only successful and non-controversial legislative action regarding asbestos was the simple resolution designating the first week of April 2008 as "National Asbestos Awareness Week," which passed unanimously in the Senate (the same resolution was passed the previous year).

Industry Risk-Management Response to Public Concerns: Self-Regulatory Actions

The 1970s were considered the "decade of the regulatory boom" (Kramer *1979)*, with over 20 federal regulatory agencies, including OSHA and MSHA, created during this period. In the 1970s, the public had a heightened interest in asbestos, as government agencies issued cautionary reports, asbestos lawsuits were widely publicized, and the media were saturated with stories about asbestos and other environmental harms. A search of the LexisNexis database yielded nearly 500 articles that were published by major US newspapers and newswires during this 10-year period, describing asbestos and corresponding health issues. For example, this headline from *the Washington Post*, "Office Pollution: On-the-Job Peril; Perils on the Job; Your Working Environment May Be Doing a Job on Your Health" (Von Eckardt *1979)*, is typical of the media reports of the era.

The decade of regulation corresponded with the decade of consumerism. Companies developed interesting rhetorical strategies designed to quell public fears, acknowledge health concerns, and yet discredit scientific evidence at the same time. Consequently, self-regulatory actions were taken by some manufacturers of asbestos-containing products (ACP). Although only a few of these ACP were banned outright (see above), consumers had a heightened awareness of the health risks associated with asbestos, due to the widespread media coverage of public health reports.

Some companies responded to consumer concerns although they were not required to do so. One of the more prominent examples involves hair dryer manufacturers. In 1979, manufacturers representing 90% of US domestic brands of electric hair dryers stopped using asbestos in their appliances. This was a voluntary agreement reached with the CPSC, which had issued reports on small household appliances including hand-held hair dryers. It was expected that the CPSC would issue a ban on asbestos in these consumer products; instead, several companies agreed to eliminate future use of asbestos but not admit that asbestos used in their products was dangerous (Consumer Product Safety Commission [CPSC] *1979).* They cited a lack of scientific agreement regarding the risks associated with asbestos exposure. Additionally, some companies that had not used asbestos in their dryers since the early 1970s, agreed to replace models through a massive recall of dryers that predated this. The following statements issued by manufacturers involved in the voluntary recall illustrate the rhetorical skill of avoiding admission of potential harm while announcing changes in their manufacturing practices. These statements were originally included in the CPSC report and now are posted on its website:

> While the Commission and GE are not in agreement as to whether such hair dryers present or could present a health hazard, and while GE does not believe that it is legally required to undertake this program, GE has urged its distributor customers to return hair dryers containing asbestos to GE and to advise consumers of the situation. GE will retrofit all units capable of being retrofitted with a non-asbestos insulator.
>
> Although Korvettes feels that it has not been determined whether or not asbestos in hair dryers is a health hazard, Korvettes in keeping with its policy of customer satisfaction is accepting returns for replacement or refund of purchase price on Korvettes-brand dryers with asbestos and other brands with asbestos sold in Korvettes' stores if the customer can furnish proof-of-purchase from Korvettes.
>
> Montgomery Ward said that although the health hazard controversy regarding asbestos-lined hair dryers has not been resolved, the company has been exchanging asbestos-lined hair dryers returned by consumers with some of its many other models which do not contain asbestos.
>
> Questions recently have been raised by CPSC and the news media as to whether there is a health hazard posed by the presence of asbestos in hand held hair dryers. Norelco feels (and has felt all along) that its discontinued models (which are no longer available for purchase) pose no health hazard to consumers. Nevertheless, the company was one of the first US marketers to cooperate with CPSC staff by supplying a listing of its dryers containing asbestos and their dates of manufacture (CPSC *1979).*

These businesses questioned the scientific findings regarding their products' potential health hazards, yet demonstrated a desire to address consumers' concerns through the voluntary recalls. In one case, media and scientists were blamed for causing the controversy in the first place, implying that the asbestos threats were exaggerated.

Recent Major Asbestos-Related Incidents and Litigation

The Libby, Montana Case

From 1970 to 1990, the EPA was an advocate for public health and took a strong stance on asbestos control. When the Libby, Montana, story broke about asbestos contamination in the vermiculite mine, however, the EPA was criticized for withholding information from residents about the level of contamination. Nearly 5 years after EPA studies determined the harmful effects of vermiculite mining in Libby, Montana, investigative stories published in the *Seattle Post-Intelligencer* (November, 1999), reported that exposure to tremolite asbestos from the vermiculite mine caused over 192 cancer deaths and 275 incidents of lung disease in this city. This reflected a much higher rate of cancer than expected in a community of this size. Two key stories in the *Seattle Post-Intelligencer*, "A Town Left to Die" (Schneider *1999a)* and "Miners' Search for Gold Led to Vermiculite" (Schneider *1999b)* generated international attention. Additional stories, including community reactions to the health findings, continued to be published for several weeks. In response to the news stories, the EPA retested the site and found that outdoor air samples indeed showed high levels of tremolite and tremolite-like asbestos fibers. Additionally, nearly half of all homes in Libby contained asbestos levels many times greater than the EPA cancer risk level threshold. The stories also brought attention to the former owner of the mine, W.R. Grace & Company, which operated the mine for three decades until 1990, when it was sold to a development company. The Libby controversy and W.R. Grace's refusal to acknowledge its responsibility in the crisis, was the subject of a two-part investigative documentary on ABC's Nightline (ABC News Productions *2005)*. The reporter who broke the news was interviewed, as were Libby physicians, asbestos abatement experts, and victims from Libby, Montana and other states where Grace processing plants existed.

The earlier news prompted attention by health professionals and other agencies. A comprehensive and rigorous environmental health study by the Agency for Toxic Studies and Disease Registry found that 30% of Libby residents had lung abnormalities, and only half of them had been employed by Grace & Company. This was a significantly higher rate than would be predicted in normal populations. All those tested had worked at W.R. Grace & Co. or were residents of Libby prior to the mine's closing in 1990 (Schneider *2001a*; Smith *2001)*. W.R. Grace officials denied knowing that the vermiculite was laced with asbestos.

After the discovery of contamination of Libby residents' homes, the EPA began removal of asbestos from many Libby, Montana, homes (now a Superfund site), and in some cases completely demolished homes that were highly contaminated. These were rebuilt after the site cleanup was completed.

In 2000, the EPA admitted knowing for nearly 5 years about the dangerous asbestos levels caused by the vermiculite mine in Libby. The EPA's damage control included a question-and-answer section on its website regarding asbestos safety in general and the Libby investigation (Environmental Protection Agency, Office of

Pollution Prevention and Toxics, 2007). Also in 2000, several studies of products using asbestos-contaminated vermiculite and fibrous talc/asbestos (such as crayons) were conducted, and the EPA concluded that there was only a minimal health threat for humans, yet many manufacturers of products previously using fibrous talc and other vermiculite had voluntarily stopped using the materials in their products. Additionally, the EPA issued guidelines for consumers, such as gardeners, to reduce the already "low" risk associated with vermiculite: use outdoors or in well-ventilated areas, change clothes that are contaminated with dust from vermiculites before going indoors, and use premixed soil. The strategy shifted from advocating for stronger asbestos regulation to encouraging people to take personal responsibility for reducing risks associated with asbestos exposure.[4]

By the time the Libby story broke, asbestos victim litigation was already well underway across the nation. The first successful American asbestos lawsuit was in 1969; an employee of Johns-Manville was awarded over $79,000, but he died before the trial ended. Until 1977, industry officials denied having any prior knowledge that asbestos exposure was harmful to health. In a trial that year involving Raybestos-Manhattan, a plaintiff's attorney uncovered indisputable evidence that contradicted these claims. The evidence proved industry officials knew for years about the dangers of asbestos exposure and yet did nothing to protect workers and consumers. Since then, the floodgates have opened, with over 900,000 asbestos-related lawsuits filed since then. Most of the earlier claims against W.R. Grace were settled out of court with a gag order attached. One mine worker refused to settle out of court (Bowker 2003). A former bagger who swept up and bagged anything at the mine that was "in the hopper," Les Skramstad was angry with the company for failing to warn workers of the health dangers. In 1996, suffering from asbestosis, he decided to sue the company and was eventually awarded over $600,000. W.R. Grace appealed and Skramstad waited over a year for a decision and compensation. Because Montana law does not place time limits on a judge's ruling, he was forced to settle out of court for much less than he was awarded, because he needed the money for his immediate health care needs. The judge who delayed the case, Michael Prezeau, previously worked as a staff lawyer for a firm in Missoula, Montana that represented Grace in previous worker lawsuits (Bowker 2003, p. 134).

As lawsuits against W.R. Grace grew, the company and over 60 affiliated companies filed for bankruptcy in 2001. In 2004, it attempted to reorganize the bankruptcy filing to limit payments to claimants. Asbestos now represents the longest-running mass tort litigation in US history. Asbestos lawsuits have been filed on behalf of nearly 900,000 citizens, who claim they were unknowingly exposed to health-threatening toxic levels of asbestos by those who knew it was dangerous. People most affected by asbestos are not necessarily the ones who have been compensated for their illnesses. Some reports indicate that millions of dollars have gone to those who have no medical evidence of impairment from asbestos. This is true because many people exposed to asbestos may not present symptoms for up to 40 years. Trying to sort through the victims' claims is difficult. The FAIR Act of 2005 (and 2006) was proposed to help settle cases efficiently but it was criticized as setting a precedent that is both dangerous and unconstitutional.

Lawsuits stemming from asbestos contamination originating at the Libby, Montana, vermiculite mine represent some of the most serious cases of asbestos harm. A federal indictment alleges that as early as the late 1970s, officers of W.R. Grace & Co. knew that the vermiculite mined was laced with toxic tremolite asbestos. Officials gathered evidence internally using a number of methods, all leading to the same conclusion that dangerous asbestos levels were present and posed a health threat to workers. Although required by the Toxic Substance Control Act of 1976 to disclose this information to the EPA, W.R. Grace & Co. officials continued to withhold information about the findings (Schneider & McCumber *2004)*. The company also is charged with denying access to NIOSH officials in the 1980s, when they attempted to test air quality and health conditions at the site. In addition to withholding health findings from employees and the EPA, Grace officials continued to distribute asbestos-laced vermiculite to processing plants and manufacturers all over the country and allowed it to be distributed throughout the Libby community for gardening and playground soil. For these reasons, it is impossible to estimate accurately the number of people affected by asbestos from Libby. The class action lawsuit received support when in February 2005, W.R. Grace and several top executives were indicted on federal charges for "knowingly exposing miners and residents in a small town to asbestos" (Associated Press *2005*; Johnson *2009)*.

Recently, progress was made toward resolution in both the federal trial and the criminal trial. In March of 2008, W.R. Grace agreed to pay $250 million to the federal government to cover the expenses related to the investigation and cleanup of Libby. This represents the highest sum in the history of the Superfund program (Associated Press 2005). The government has yet to receive the money, however; the bankruptcy court must evaluate the judgment, along with other debts and claims against the company. In April 2008, the company agreed to pay $3 billion to victims of the Libby, Montana plant (Scott *2008)*. The company successfully delayed the class action lawsuit for years, but a final appeal to the Supreme Court was unsuccessful, as the Court refused to hear the case.

W.R. Grace and other companies peddling toxic substances are awaiting the outcome of the latest government action, which may eliminate the number of allowable claims in the class action lawsuit. EPA's latest effort is to change the level of danger currently designated for the different types of asbestos. In order to enact the change, a panel of experts was appointed, which has created a controversy:

> In its final days in power, the Bush White House is rushing to have federal agencies water down the regulation of hazardous substances, lawmakers and public health experts say. A panel of scientific advisers this week denounced an Environmental Protection Agency plan to quickly alter the way it measures the cancer-causing risk of asbestos, but the thumbs-down doesn't prevent the agency from making the change anyway.
>
> The latest eleventh-hour toxic sparring match comes while members of Congress are asking why the Labor Department has sent plans for sweeping changes in how workers are protected from chemical hazards directly to the White House Office of Management and Budget.

> Many of the government scientists and physicians in the Labor Department and other agencies who are normally required to weigh in on these kinds of changes say they haven't had a peek at the proposal.
>
> Similar concern has been focused on the firing of John Howard, the popular director of the National Institute for Occupational Safety and Health.
>
> People at EPA headquarters say the rush to have them change the way asbestos hazards are calculated is caused in part to OMB's desire to appease the automotive, mining, construction and chemical industries being sued for harm done by asbestos-containing material they've used or sold.
>
> The 20 experts appointed to the Scientific Advisory Board's asbestos panel were to evaluate the validity of the EPA's plan to change how the toxicity of the six types of asbestos regulated by government differ in danger. The change centered on the EPA's desire to ignore decades worth of what are considered solid studies documenting the actual hazard of the most common type of asbestos – chrysotile.
>
> Instead, the EPA submitted other studies which, it said, showed chrysotile isn't dangerous and doesn't cause mesothelioma, an almost always fatal cancer, which often garners multimillion-dollar judgments in court cases brought by people sickened or killed by exposure to it (Schneider *2007, 2008).*

The Libby, Montana case has drawn much media attention, including a Nightline television special in 2005 and a book, *An air that kills: How the asbestos poisoning of Libby, Montana, uncovered a national scandal.* One of the book's authors is the reporter who first broke the story in the *Seattle Post-Intelligencer* in 2001 (Schneider & McCumber *2004)* and who continues to cover related stories (see above), although for a different newspaper.

September 11, 2001 Terrorist Attacks on the World Trade Center: Asbestos Concerns

Except for asbestos-related litigation, asbestos was not in the news much until the terrorist attacks on the World Trade Center on September 11, 2001. The destruction of the World Trade Center renewed concerns for asbestos; EPA scientists took samples of dust as they were escaping from their building, which later tested to have four percent asbestos composition, much above current regulatory levels. The city's mayor at that time, Rudolph Giuliani, reported on television shortly after the attacks that initial health department tests did not raise concerns; asbestos and other toxins were not apparent in the samples studied (James *2001).* A number of asbestos-related studies were conducted in the area, lasting until present. The dust, smoke and debris composition were assessed and found to have extremely high pH content and contained many toxic materials, including asbestos (Landrigan et al. *2004).*

Numerous missteps occurred during the months following the attack, due to the emergency communications break-down. For example, a few days after the terrorist attack, residents in an apartment near Ground Zero cleaned their building using guidelines issued by the New York City (NYC) Department of Health (Weissenstein & Ritter *2002)*. Private residents, not professional abatement workers, swept roofs and other areas both inside and outside of the structure. Seven months later the area was tested and found to have significantly higher levels of asbestos fibers than allowable regulatory levels. Many private residents, workers, firefighters, and others engaged in rescue, body retrieval, and cleanup of the area surrounding the former World Trade Center site were found to be exposed to elevated levels of asbestos. The EPA had assigned the New York City Department of Environmental Protection (DEP) to monitor cleanup to ensure EPA standards were met and buildings were safe before workers and residents reoccupied them. Within days of the attack, air quality monitors were placed all over the city and around the site to assess the air quality (Bowker 2003; Lyman *2003b)*.

In 2002, the city government of New York posted test results from a TEM test on air asbestos levels on the DEP website. There is a marked discrepancy between the data posted on the site and the results reported to the New York State Department of Environmental Conservation in November 2001. Specifically, results that originally had been declared as having "excessive asbestos levels" were either deleted from the website report or changed to the reassuring, "not detected" (Lyman *2003a)*. The EPA was investigated by its Office of the Inspector General, which released a report in 2003 finding that the EPA's response to September 11 (9/11) was mismanaged, citing at least ten areas of major concern, such as, "The agency did not have sufficient data to support its claim that air in Lower Manhattan following September 11 was 'safe to breathe.'" It also criticized the tests used and the agency's failure to test for all potential toxins one would expect to find with such a disaster. The report did not directly state the EPA knowingly deceived the public about safety of the city by concealing important test results, although some health advocates have disagreed (EPA *2003)*. Mayor Giuliani, now a failed presidential candidate for the 2008 primaries, was being criticized by firefighters and others who said they were not given proper guidelines to protect themselves from dangerous carcinogens, in spite of having access to data early on regarding the risk levels for workers at the site (Lyman *2003b)*.

There are scores of EPA documents relating to the 9/11 terrorist attacks that describe terms of risk and asbestos levels; nearly 6 years later the actions by the city and environmental agencies are still under scrutiny and lawsuits have been filed. For example, even the peer-reviewed study of the EPA's World Trade Center report (EPA *2003, 2004)* was criticized; one of the independent peer reviewers had served as an expert defense witness in asbestos lawsuits. This led to the EPA Office of Inspector General issuing a 58-page report entitled, "Review of Conflict of Interest Allegations Pertaining to the Peer Review of EPA's Draft Report, 'Exposure and Human Health Evaluation of Airborne Pollution from the World Trade Center Disaster'" (EPA *2004)*.

Lyman's 2003 report for the National Environmental Health Association, "Messages in the Dust," is a detailed analysis of the government's response to the aftermath of 9/11 and the resulting cover-up. Lyman wrote the monograph to provide a "lessons learned" account, so that mistakes would not be repeated.

International Focus

While North America, East Asia, and Western Europe currently have regulations and education programs in place to prevent unsafe asbestos exposure, there are no universal standards in regard to worker safety regulations, which is especially a concern in the developing world. McCulloch & Tweedale *(2004)* argue that major UK asbestos companies failed to apply their knowledge about asbestos-related diseases and worker safety procedures while operating asbestos mines in South Africa. Worker conditions were poor, and as a consequence, higher levels of asbestosis and mesothelioma are found among South African asbestos workers than their UK counterparts. The authors argued that company officials had three decades of medical and epidemiological evidence of asbestos effects yet continued to operate their mines in South Africa without protections in place. Although the day-to-day operations were managed by South Africans, UK managers visited regularly and had oversight of operations.

China has a serious problem with asbestos exposure in mines and product manufacturing environments. Best estimates suggest that 100,000 workers currently are exposed to asbestos, including 24,000 in the mining industry and 46,000 in more than 1,200 asbestos factories (Feng et al. *2002)*. These numbers are likely to increase due to preparations for the 2008 summer Olympics in Beijing, where a construction surge was prevalent. Many of the new buildings and renovations were done by migrant Chinese workers who lived in tent communities or small shelters at the base of the construction sites. This extended their exposure to toxic chemicals in the settling dust for longer periods than the normal workday. A quote from a Chinese worker is posted on the website of the largest US-based international non-profit advocacy group, Human Rights Watch: "Since I first arrived at the work site, every day I've worked, I don't have any money and I've never left [the work site] for relaxation."[6]

Many scientists have blamed globalization for the increase in asbestos-related morbidity and mortality in developing nations (Huncharek *1993)*. With demand for asbestos declining in North America and Europe and asbestos litigation hurting asbestos-related businesses, these companies are targeting the developing world for their products. Recent studies have revealed the deleterious effects of asbestos exportation on poorer countries. A study by Aguilar-Madrid et al. *(2003)* traced the processing, importing, and use of asbestos-containing products by Mexicans and compared trends with patterns of mortality cases of malignant mesothelioma from 1979 to 2000. The authors noted a significant increase in mesothelioma disease as asbestos use increased. "The commercial strategy of multinational asbestos industries has depended on the opening of markets within the global economy framework … and the establishment of free trade agreements. The prerogative of free trade has prevailed over human, environmental, and labor rights" (Aguilar-Madrid et al. *2003)*.

[6] See: Human Rights Watch website: http://www.hrw.org/

The authors argue that the appeal of developing nations to multinational companies is obvious: more relaxed labor laws, less stringent environmental codes, cheaper labor, ability to influence foreign governments due to the economic benefits claims, lack of professionals in occupational health who would advocate for worker safety, lack of government controls to monitor workplaces for compliance of existing laws.

In one case, the World Trade Organization (WTO) became involved in protecting countries' rights to block imports of asbestos, although the member nations of the WTO do not include nations of the developing world. Castleman *(2002)* describes the case involving Canada and France. Canada appealed to the WTO in 1999 when France banned asbestos imports due to public health concerns. Canada argued it was a violation of free trade rights. In the end, the WTO sided with France, finding that a member nation has a right to ban imports if it is vital to protecting public health. This was an unusual case, as Canada expected the WTO to support its claims under the banner of free trade. Castleman described the WTO as lacking medical, engineering, or scientific expertise, as most officials have a law or business background. In the end, however, the health and environmental consultants were able to influence the WTO's decision.

Because of the known harms associated with many asbestiform minerals, some scientists have called for an international ban on asbestos, primarily to protect workers in developing nations (Maltoni *2000)*. Scientists cite a lack of epidemiological data as the reason for delays in global regulations; however, it is imprudent to wait until such data exist to enact regulations as it will be too late to prevent new cases (Harris & Kahwa *2003)*. Harris and Kahwa also report that although exposure to asbestos through mining may be decreasing in developing nations, manufacturing and use of products containing asbestos has increased: "The regulatory mechanism for use, handling and disposal of asbestos and associated waste in developing countries are weak and information on asbestos-related diseases is scanty but emerging" (p. 230).

Asbestos-Related Actions

Asbestos Profiteers

It's estimated that the asbestos industry has caused serious health problems and kills over 10,000 people a year in the United States (Centers for Disease Control and Prevention *1999*–2001). Over a million people have been exposed to asbestos and are at risk for asbestos-related diseases (Environmental Working Group *2007)*. The discovery of asbestos' harmful effects and the resulting efforts to regulate and educate, however, have benefited some groups. Once asbestos abatement was introduced into policy, many new businesses emerged. "Asbestos Detection and Removal Services" and "Asbestos Consulting and Testing Services" are official categories in the business pages of phone books, with 34 businesses listed under the first category in Chicago, 17 each in New Orleans and New York City, 26 in Boston, and 11 in Indianapolis, for example.[7] Additionally, law firms specializing in asbestos litigation are ubiquitous. Four law firms that specifically have business profiles with "asbestos," "mesothelioma attorney," or "asbestos and chemical attorney" were listed in New York and three in Chicago. Many more emerge (334,000 results) on a Google search for "asbestos lawyer."

Asbestos Education and Public Health Advocacy

Media

Scores of scholarly articles on various asbestos topics, ranging from health effects to politics to litigation, have been published over the years, but the asbestos controversy also has led to a flood of information in the popular press. Over 5,000 books were published between 1970 (when the CAA was passed) and 2007 to educate the public about asbestos (e.g., *The Citizens' Guide to Geologic Hazards: A Guide to Understanding Geologic Hazards, Including Asbestos, Radon, Swelling Soils, Earthquakes, Volcanoes, Landslides, Subsidence, Floods, and Coastal Hazards* (Nuhfer et al. *1993)*), to explain how to prevent exposure (e.g., *Raising Children Toxic Free: How to Keep Your Child Safe From Lead, Asbestos, Pesticides, and Other Environmental Hazards* (Needleman & Landrigan *1994)*), to describe the medical effects of exposure (e.g., *Pathology of Asbestos-Associated Diseases* (Roggli et al. *2004)*; *Asbestos: Medical and Legal Aspects* (Castleman *1996)*), and To explain the legal proceedings that surround asbestos (e.g., *Asbestos in the Courts: The Challenge of Mass Toxic Torts* (Hensler et al. *1985)*). There are numerous critical books, as well, with titles such as *Outrageous Misconduct: The Asbestos Industry on Trial* (Brodeur *1985)*, *An Air That Kills: How the Asbestos Poisoning of Libby, Montana Uncovered a National Scandal* (Schneider & McCumber *2004)*, and *Fatal Deception: The Untold Story of Asbestos: Why It Is Still Legal and Still Killing Us* (Bowker 2003).

News coverage of asbestos has exponentially increased since the Clean Air Act put asbestos on the public's radar screen. *The New York Times*, which archives abstracts or articles online beginning in the year 1969 to present, published 34 articles between 1969 and 1979 that specifically relate to asbestos harm. This increased to 458 articles between 1980 and 1989, as asbestos victim lawsuits emerged and regulations continued to be debated. Over 300 articles were published in the newspaper between 1990–1999 and again between 2000–2007. *The Washington Post* archives revealed a similar jump in news stories from 34 (1969–1979) to 458 (1980–1989). Additionally, numerous editorials were published, with 191 published in *The New York Times* since 1969 through July 2007.

---

[7] See: Yellowpages.com website: http://www.yellowpages.com/

Advocacy Groups and Coalitions

Several legitimate asbestos victim support groups have been formed in the US and abroad, such as the Asbestos Awareness Disease Organization, but many support groups appear to be sponsored by law firms trying to capture some business, as mentioned earlier. A Google search for "US asbestos support group" yields numerous law firm websites, some of which list the names of major asbestos support groups as a public service. Once on their websites, however, people with a connection to asbestos illness can find legal help, as well.[8] When one clicks on another site Google captured with "US asbestos support group," a pop-up ad featuring a smiling female telephone receptionist emerges, with the message, "We've recovered more than $1 billion for thousands of hard-working people and their families."[9]

Nonprofit advocacy groups and coalitions that promote asbestos awareness and educate publics about current legislation are prevalent, but it is hard to distinguish legitimate grassroots groups from "astroturf" organizations.[10] Astroturf organizations are fake grassroots organizations, also known as "front groups," that are funded and managed by those associated with the asbestos industry. For example, the so-called asbestos education nonprofit, Coalition for Asbestos Resolution (CAR), describes itself as an advocacy organization for those who suffer from the effects of asbestos exposure. In reality, the organization is funded by the GAF Corporation to lobby the government for an asbestos industry bailout. The organization promoted a bill in 2000 creating "tax breaks for asbestos manufacturers for payments made for asbestos litigation claims" (Labaton *1999)*.

One of the most powerful coalitions lobbying in Washington DC is the Asbestos Study Group (ASG). According the Center for Media and Democracy's SourceWatch website, the organization consists of some of the largest corporations (e.g., Halliburton, Honeywell, Pfizer, Viacom, and General Motors) that face asbestos litigation due to their use of asbestos in manufactured products (Center for Media and Democracy *2008)*. This industry coalition was formed expressly to establish ceilings on asbestos-related lawsuits and protect the financial interests of coalition members. The coalition is powerful due to its money backing: "The Asbestos Study Group reported spending $5.56 million in the first half of 2003 on lobbying for legislation to end asbestos lawsuits, making the group one of the top spenders, according to disclosure forms. ... PoliticalMoneyLine.com, which tracks spending totals from lobbying reports, said the Asbestos Study Group was the 11th biggest lobby spender in the first half of 2003."[11] By the end of 2003, ASG had spent over $11 million dollars. The organization continued to spend close to $10 million dollars in lobbying in 2004 and 2005, and close to $5 million in 2006 (Center for Responsive Politics *2008)*.

[8] For example, see: http://www.asbestosnetwork.com/tools/tl_support.htm

[9] See: http://www.weitzlux.com/asbestos-lawyer_1054.html

[10] See: http://www.motherjones.com/

[11] See: CQ Moneyline website: http://www.sourcewatch.org/

Preliminary Conclusion

This case study of asbestos in society, addressed from multiple perspectives, is designed to provide an understanding of what often drives US health policy in the public interest when corporate profits are at stake. The asbestos industry effectively suppressed information about the harmful effects of asbestos for decades. Although armed with insurance company studies and internal company reports early in the twentieth century, businesses failed to protect employees and consumers from asbestos exposure until policy was passed decades later. Independent medical studies conducted as early as 1917 pointed to serious health consequences of asbestos exposure, yet this information was withheld from workers and did not become widely known to the public until the 1960s when asbestos worker lawsuits began to emerge. While scientists were busy conducting studies, few people were taking this information and advocating for worker and consumer safety. The tide turned, perhaps, once lawsuits were filed in the 1960s and the dangers associated with asbestos exposure became a matter of public record and a media focus. Lawyers may very well have been the agents for change.

The asbestos litigation served as a catalyst for public health policy, as well. At that point, mainstream media combined with consumer advocacy efforts, led to regulatory actions. Rather than science and government, grassroots efforts bolstered by the mass media accelerated the process. Once health studies became widely available to the general public with the media's coverage of asbestos victim lawsuits, asbestos' deadly effects could no longer be hidden from society. Public debates and policies regarding asbestos abatement in public schools also brought heightened attention to critical health issues. Lawyers, occupational health professionals and environmental organizations advocated for those who did not have the educational background or expertise to do so on their own.

The media aroused the public's emotions with stories of asbestos victims, incidents of deception and cover-up by the asbestos industry, and stories of Congressional members being influenced by special interest groups. These stories resonated and heightened public interest in the topic. The media also put a human face on a serious health problem that is actually quite complicated. For example, an investigative documentary on the Libby, Montana, asbestos case was released in 2005, and has received some awards at independent film festivals. Although it has not been distributed to mainstream theaters, word-of-mouth and the Internet keep it in the public eye.

Surprisingly, 40 years of national attention to asbestos dangers have not led to corresponding tough regulations to protect workers and consumers or compensate victims and future victims. Instead, the asbestos industry was able to employ the same tactics that have been successful for other special interest groups and businesses, such as the tobacco industry. As this case study reveals, several strategies were employed by the asbestos industry to protect their profits in the face of imminent regulation and lawsuits:

1. Establishment of credible doubt. The asbestos industry raised doubts about the science of assessing asbestos exposure levels and the threshold at which humans become affected. The new findings that countered decades of research came from industry-supported laboratories.
2. Strategic use of industry money. Front groups masquerading as victim advocacy groups were financed by the asbestos industry. The front groups and industry coalitions poured millions of dollars into lobbying government to shape legal reform and, again, to block efforts for more rigorous asbestos regulations.
3. Support of favorable political candidates. The 1980s, under President Ronald Reagan, represented a decade of deregulation, when businesses enjoyed greater freedoms and less controls. Asbestos industry leaders, such as W.R. Grace, enjoyed status on key political commissions.
4. Engineering of public sympathy. The asbestos industry orchestrated a shift of public sympathy away from asbestos victims toward regulatory action "victims." The industry and government created fears of an economic collapse should the class-action lawsuit be allowed to continue without placing limits on victim compensation.
5. Good lawyers. Industry lawyers have been very effective at using the law to their clients' advantage and delaying proceedings that could negatively affect their client.

In addition to industry efforts, lawmakers favorable to the asbestos industry helped gain public support for their actions by employing two key strategies:

1. Creation of a false image of government action. Minor actions that had little overall impact were touted, making it appear that government was actively working for the public's best interests in regard to asbestos policy. In reality, a backwards step was taken.
2. Discrediting asbestos victim lawsuits. The asbestos class-action lawsuit was rhetorically aligned with highly publicized litigation that the public saw as frivolous or junk lawsuits. This was an effort to gain support for legal reforms that appeared to be fair but instead rewarded corporate malfeasance.

Failings on the part of public health advocates also paved the way for industry dominance in political maneuvering:

1. Missed opportunities. In the 1990s, public health advocates failed to take advantage of a favorable regulatory political climate. While the 1990s invited greater government scrutiny of the asbestos industry, the 2000 election ushered in another decade of deregulation. Advocates failed to take advantage of the opportunity to manage asbestos policy in support of workers, consumers, and victims.
2. Lack of effective strategy to gain public support. By keeping the media and public focus on vulnerable publics such as school children, public health advocates may have been more successful at putting a sympathetic face on asbestos regulation. The public might believe that workers knowingly took risks, but children are seen as innocent victims.

Other issues continue to delay action for asbestos victims. A lack of agreement prevails within the asbestos industry, the insurance industry, and trade organizations and unions representing asbestos workers, regarding the best approach to litigate the tens of thousands of victim claims. This makes it complicated to determine which organizations have the best interests of asbestos victims in mind rather than corporate bottom lines. Corporations that have the financial backing to lobby government for their own financial benefit have successfully shaped the regulatory and legal environment for asbestos, in spite of overwhelming evidence of corporate malfeasance and criminal negligence.

Ambiguity surrounds this case; there is much to be gained or lost financially from eliminating this ambiguity, depending on what asbestos-affected party you represent. It is presumptuous to assert that if guerilla media tactics were prevalent in the early twentieth century, that asbestos regulation would have followed scientific discovery more closely. Scientists often blame media for creating alarmist public concerns over threats that actually pose little risk to human life (Ratzan *1998)*. Scientists also argue that the media's science reporting is often inaccurate or fails to provide the critical details for citizens to make an informed decision. This is partially true, due to the limited space and time afforded to such topics and the media's drive to provide news in easily digested sound bites. Many public health professionals, however, now view the media as an important tool for activating citizens to become informed participants in the public policy process.

Corporate Social Responsibility

It is difficult to determine if case studies such as this one will encourage more efficient health policy development to protect the US public. First, a culture of change must occur in both business and government. The triple bottom line (TBL), sometimes known as "People, Planet, and Profits," is a term coined by John Elkington *(1994)* to describe a standard of accountability for businesses to engage in sustainable practices. According to Elkington, business success should be measured in three areas: the economic (financial) bottom line, the social bottom line, and the environmental bottom line. The social bottom line calls on corporations not only to practice ethical conduct in their labor practices and in relationships with the communities in which they operate, but also to create objectives that maintain mutually beneficial relationships with all internal and external publics. The environmental bottom line requires businesses to "do no harm" to the environment and even to make it a better place than they found it. Finally, the economic bottom line is the traditional measure that most businesses use to gauge success, but instead it is viewed as being shared by the business, host community, and workers. Elkington suggests that following the tenets of TBL will make businesses more successful and communities healthier. Corporate critics warn, however, that the triple bottom line can be staged to give the appearance of corporate responsibility, when it is just business (and public relations) as usual. Although triple-bottom-line reporting is voluntary, it is possible to create a culture where consumers and government support corporations that follow these principles, and they should follow them both locally and globally.

As the federal lawsuit involving the Libby, Montana, mine and mill finally comes to trial this year, more attention will be drawn to asbestos. Consumers may turn to the media to explain the risk associated with living around asbestos. Perhaps the result will be the public's heightened awareness of the prevalence of corporate corruption at the expense of human lives. This, in turn, may create a more active citizenry, although lessons learned from the tobacco lawsuits suggest otherwise.

References

ABC News Productions. (2005). *Nightline, November 4, 2005: A killer in town, Part 1 and 2*. New York, NY: ABC News Productions.

Aguilar-Madrid, G., Juárez-Pérez, C. A., Markowitz, S., Hernández-Avila, M., Sanchez Roman, F. R. & Vázquez Grameix, J. H. (2003). Globalization and the transfer of hazardous industry: Asbestos in Mexico 1979–2000. *International Journal of Occupational and Environmental Health, 9*(3), 272–279.

Asbestos Resource Center. (2008). *History of asbestos*. Retrieved, December 12, 2008, from http://www.asbestosresource.com/history/

Asbestos Work Group, Department of Health and Human Services [DHHS] & National Institute for Occupational Safety and Health [NIOSH]. (1980). *Workplace exposure to asbestos: Review and recommendations*. US Department of Health and Human Services with National Institute for Occupational Safety and Health, Publication No. 81–103.

Associated Press. (2005, February 8). Charges issued over asbestos at a mine. *The New York Times*, February 8 (Online archive).

Barbalace, R. C. (2004). *A brief history of asbestos use and associated health risks*. Retrieved June 22, 2007, from http://EnvironmentalChemistry.com/yogi/environmental/asbestoshistory 2004.html

Bartrip, P. W. J. (2004). History of asbestos-related disease. *Postgraduate Medical Journal, 80*(940), 72–76.

Berg, R. (2004). When science crosses politics: the case of naturally occurring asbestos. *Journal of Environmental Health, 66*(10), 31–39.

Brodeur, P. (1985). *Outrageous misconduct: The asbestos industry on trial.* New York: Pantheon Books.

Bowker, M. (2003). *Fatal deception: The untold story of asbestos–why it is still legal and still killing us*. Emmaus, PA: Rodale Press.

Bush, G. W. (2005, January 7). Remarks in a discussion on asbestos litigation reform in Clinton Township, Michigan. *Weekly Compilation of Presidential Documents, 41*(1), January 10, 2005. Retrieved, March 10, 2009, from http://fdsys.gpo.gov/fdsys/pkg/WCPD-2005-01-10/pdf/WCPD-2005-01-10-Pg21-2.pdf

Castleman, B. I. (1996). *Asbestos: Medical and legal aspects.* Englewood Cliffs, NJ: Aspen.

Castleman, B. (2002). WTO confidential: the case of asbestos. World Trade Organization. *International Journal of Health Services, 32*(3), 489–501.

Center for Media and Democracy: Sourcewatch. (2008). *Asbestos Work Group*. Retrieved, July 10, 2008, from http://www.sourcewatch.org/index.php?title=Asbestos_Study_Group

Center for Responsive Politics. (2008). *Asbestos study group.* Retrieved July 28, 2008, from http://www.opensecrets.org/lobby/clientsum.php?year=2005&lname=Asbestos+Study+Group

Centers for Disease Control and Prevention, National Center for Health Statistics (1999–2001). *Multiple cause-of-death, 1999–2001* [Data file]. Hyattsville, MD: National Center for Health Statistics.

Congressional Research Service. (2007). *Ban Asbestos in America Act of 2007: S. 742 – 110th Congress*. Retrieved, July 15, 2008, from http://www.govtrack.us/congress/bill.xpd?bill=s110-742&tab=amendments

Consumer Product Safety Commission (1979, May 21). *CPSC accepts corrective actions from major hair dryer companies*. Retrieved, June 18, 2007, from http://www.cpsc.gov/cpscpub/prerel/prhtml79/79022.html

Consumer Product Safety Commission, Environmental Protection Agency, & the American Lung Association (2005). *Asbestos in the Home*. CPSC Document No. 453. Retrieved, March 10, 2009, from http://www.cpsc.gov/CPSCPUB/PUBS/453.html

Cooke, W.E. (1924). Fibrosis of the lungs due to the inhalation of asbestos dust. *British Medical Journal, 2*, 147–152.

Cooke, W. (1927). Pulmonary asbestosis. *British Medical Journal, 2*, 1024–1025.

Corn, J. K. (1999). *Environmental public health policy for asbestos in schools: Unintended consequences*. Boca Raton, FL: Lewis.

Creswell, J. (2006, February 9). Large and small businesses part ways on asbestos bill. *The New York Times*, sect. C, p. 3.

Djerassi, L., Kaufmann, G., & Bar-Nets. (1979). Malignant disease and environmental control in an asbestos cement plant. *Annals of the New York Academy of Sciences, 330*, 243–253.

Dodson, R. F., Williams, M. G., & Satterley, J. D. (2002). Asbestos burden in two cases of mesothelioma where work history included manufacturing of cigarette filters. *Journal of Toxicology and Environmental Health, 65*(16), 1109–1120.

Elkington, J. (1994). Towards the sustainable corporation: win–win–win business strategies for sustainable development. *California Management Review, 36*(2), 90–100.

Environmental Protection Agency. (1986). Asbestos; manufacture, importation, processing, and distribution in commerce prohibitions; Final rule (54 FR 29460). *Federal Register, 54*(132), 29460–29513.

Environmental Protection Agency. (1992). *Asbestos; manufacture, importation, processing and distribution prohibitions; effect of court decision* (57 FR 11364). *Federal Register, 57*(64), 11364–11365.

Environmental Protection Agency. (2003). Draft, exposure and human health evaluation of airborne pollution from the World Trade Center disaster. *Federal Register, 68*(44). Retrieved from ehscenter.bna.com/pic2/ehs.nsf/id/BNAP-738L5S?OpenDocument

Environmental Protection Agency. (2004). *Review of conflict of interest allegations pertaining to the peer review of EPA's draft report, 'Exposure and human health evaluation of airborne pollution from the World Trade Center disaster.'* EPA: Report No. 2005-S-00003.

Environmental Protection Agency. (2008). *Libby updates: EPA and Department of Health and Human Services launch $8 million effort to study Libby, Montana asbestos exposure*. Retrieved, March 8, 2009, from http://www.epa.gov/region8/superfund/libby/updates.html

Environmental Protection Agency, Office of Pollution Prevention and Toxics. (2007). *Fact Sheet/Q&A: Asbestos-containing vermiculite*. Retrieved June 22, 2007, from http://www.epa.gov.

Feng, Y., Liu, J., Zhang, T., & Pon, G. (2002). Asbestos in China: country report. In K. Takahashi, S. Lehtinen, & A. Karjalainen (Eds.), *Proceedings of the Asbestos Symposium for Asian Countries*. Helsinki, Finland: Finnish Institute of Occupational Health.

Franklin, B. A. (1984, March 10). Court voids emergency asbestos standard. *The New York Times*, sect. 1, p. 22.

Harris, L.V., & Kahwa, I. A. (2003). Asbestos: old foe in 21st century developing countries. *The Science of the Total Environment, 307*(1–3), 1–9.

Hensler, D. R., Felstiner, W. L. F., Selvin, M., & Ebener, P. A. (1985). *Asbestos in the courts: The challenge of mass toxic torts*. Santa Monica, CA: Rand. Retrieved, December 10, 2008, from http://www.rand.org/pubs/reports/2006/R3324.pdf

Hoffman, F. (1918). Mortality from respiratory diseases in dust trades. *United States Department of Labor Bulletin, 231*, 178.

Huncharek, M. (1993). Exporting asbestos: disease and policy in the developing world. *Journal of Public Health Policy, 14*(1), 51–65.

James, M. S. (2001, September 13). *Most WTC air tests don't show danger: Health officials stress caution, but say measured levels safe*. Retrieved from http://abcnews.go.com/Health/story?id=116630&page=1

Johnson, K. (2009, February 19). Ex-Grace officials on trial in asbestos poisoning. *The New York Times*, p. A15.

Kramer, L. (1979, December 30). The decade of the regulatory boom: Americans began to demand more safety, fairness. *The Washington Post, Business and Finance*, p. H1.

Labaton, S. (1999, October 17). How a company lets its cash talk. *The New York Times*, sect. 3, p. 1.

Landrigan, P. J. (1998). Asbestos – Still a carcinogen. *New England Journal of Medicine, 338*(22), 1618–1619.

Landrigan, P. J., Lioy, P. J., Thurston, G., Berkowitz, G., Chen, L. C., Chillrud, S. N. et al. (2004). Health and environmental consequences of the World Trade Center disaster. *Environmental Health Perspectives, 112*(6), 731–739.

Lerman, Y. (1992). Asbestos and cancer 1934–1965: and what happened thereafter? *American Journal of Industrial Medicine, 22*, 455–456.

Lilienfeld, D. E. (1991). The silence: the asbestos industry and early occupational cancer research – A case study. *American Journal of Public Health, 81*(6), 791–800.

Lilienfeld, D. E., Mandel, J. S., Coin, P., & Schuman, L. M. (1988). Projection of asbestos related diseases in the United States, 1985–2009. *British Journal of Industrial Medicine, 45*(5), 283–291.

Lowrance, W. (1976). *Of acceptable risk*. Los Altos, CA: Willaim Kaufman

Lyman, F. (2003a, September 11). *Anger builds over EPA's 9/11 report: Charges of a cover-up hit nerve with New Yorkers*. Retrieved May 12, 2007, from http://www.msnbc.msn.com/id/3076626/

Lyman, F. (2003b). *Messages in the dust: What are the lessons of the environmental health response to the terrorist attacks of September 11?* Denver, CO: National Environmental Health Association. Retrieved (last access), March 12, 2009, from http://www.neha.org/pdf/messages_in_the_dust.pdf

Maltoni, C. (2000). Re: call for an international ban on asbestos. *American Journal of Industrial Medicine, 37*(2), 230–231.

McCulloch, J., & Tweedale, G. (2004). Double standards: the multinational asbestos industry and asbestos-related disease in South Africa. *International Journal of Health Services, 34*, 663–679.

McCulloch, J., & Tweedale, G. (2008). *Defending the indefensible: The global asbestos industry and its fight for survival*. Oxford: Oxford University Press.

Meeker, G. P., Bern, A. M., Brownfield, I. K., Lowers, H. A., Sutley, S. J., Hoefen, T. M., et al. (2003). The composition and morphology of amphiboles from the Rainy Creek complex near Libby, Montana. *American Mineralogist, 88*(11–12), 1955–1969.

Millls, R. G. (1930). Pulmonary asbestosis: report of a case. *Minnesota Medicine,13*(July), 495–499.

Murray, P. (2003, February 6). *Murray questions why our government isn't warning homeowners and protecting workers from dangerous insulation*. Retrieved June 18, 2007, from http://murray.senate.gov/news.cfm?id=191239

Needleman, H. L., & Landrigan, P. J. (1994). *Raising children toxic free: How to keep your child safe from lead, asbestos, pesticides, and other environmental hazards*. New York: Farrar Straus & Giroux.

Nuhfer, E. B., Proctor, R. J., & Moser, P. H. (1993). *The citizens' guide to geologic hazards: A guide to understanding geologic hazards, including asbestos, radon, swelling soils, earthquakes, volcanoes, landslides, subsidence, floods, and coastal hazards*. Westminster, CO: American Institute of Professional Geologists.

Occupational Safety and Health Administration, United States Department of Labor. (2007). Retrieved August 10, 2007, from http:// www.osha.gov/pls/oshaweb/owadisp.show_document?p_table=STANDARDS&p_id=9997
Oliver, T. (1925). Some dusty occupations and their effects upon the lungs. *Journal of the Royal Sanitary Institute. 46*, 224–230.
Pan, X. L., Day, H. W., Wang, W., Beckett, L. A, & Schenker, M. B. (2005). Residential proximity to naturally occurring asbestos and mesothelioma risk in California. *American Journal of Respiratory and Critical Care Medicine, 172*(8), 1019–1025.
Ratzan, S. (1998). *The mad cow crisis: Health and the public good.* New York: New York University Press.
Roggli, V. L., Oury, T. D., & Sporn, T. A. (2004). *Pathology of asbestos-associated diseases.* New York: Springer.
Schepers, G. W. H. (1992). Re: changing attitudes and opinions: asbestos and cancer 1934–1965. *American Journal of Industrial Medicine, 22*(3), 461–466.
Schepers, G. W. H. (1995). Chronology of asbestos cancer discoveries: experimental studies of the saranac laboratory. *American Journal of Industrial Medicine, 27*(4), 593–606.
Schneider, A. (1999a, November 18). A town left to die. *The Seattle Post-Intelligencer*, p. A1.
Schneider, A. (1999b, November 18). Miners' search for gold led to vermiculite. *The Seattle Post-Intelligencer*, p. A11.
Schneider, A. (2000, October 4). Virginia miners at risk from asbestos: potential for serious health problem found, which would put workers in the same danger as those who died in Libby, Montana. *Seattle Post-Intelligencer*, p. A2.
Schneider, A. (2001a, February 23). Asbestos study's results alarming: exposure in Libby blamed for high rate of lung disease. *Seattle Post-Intelligencer*, p. A1.
Schneider, A. (2001b, February 28). Asbestos confusion exposing millions: search for asbestos in Crayons finds government detection methods obsolete. *Seattle Post-Intelligencer*, Retrieved, March 10, 2009, from http://seattlepi.nwsource.com/uncivilaction/talc28.shtml
Schneider, A. (2007, June 5). Big asbestos prosecution in jeopardy: US argues government ask appellate panel to overturn lower-court rulings in W.R. Grace case. *Seattle Post-Intelligencer*, p. A1.
Schneider, A. (2008, July 23). Agencies asked to ease safety rules: White House tries to push changes on asbestos, toxins. *Seattle Post-Intelligencer*. Retrieved August 3, 2008, from http://seattlepi.nwsource.com/local/371959_asbestos24.html
Schneider, A., & McCumber, D. (2004). *An air that kills: How the asbestos poisoning of Libby, Montana uncovered a national scandal.* New York: G. P. Putnam's Sons.
Scott, T. (2008, June 24). *Supreme Court won't hear W.R. Grace appeals. The Missoulian.* Retrieved July 7, 2008, from http://www.missoulian.com/articles/2008/06/24/news/local/news02.txt
Shabecoff, P. (1985, October 3). Budget Office attached over rules for asbestos exposure. *The New York Times*, sect. A, p. 32.
Smith, C. (2001, August 24). Up to 30% tested in Libby hurt by asbestos: Montana mining town residents hear findings of health survey. *Seattle Post-Intelligencer*, p. A1.
Talcott, J. A., Thurber, W. A, Kantor, A. F., Gaensler, E. A., Danahy, J. F., Antman, K. H., & Li, F. P. (1989). Asbestos-associated diseases in a cohort of cigarette-filter workers. *New England Journal of Medicine, 321*(18), 1220–1223.
Van Gosen, B. S. (2005). *Reported historic asbestos mines, historic asbestos prospects, and natural asbestos occurrences in the eastern United States* (United States Geological Survey Open-file Report 2005-1189). Retrieved December 18, 2008, from http://pubs.usgs.gov/of/2005/1189/
Vollers, M., & Barnett, A. (2000). Libby's deadly grace. *Mother Jones*, May/June. Retrieved, March 12, 2009, from http://www.motherjones.com/environment/2000/05/libbys-deadly-grace
Von Eckardt, W. (1979, November 10). Office pollution: on-the-job peril; perils on the job; your working environment may be doing a job on your health. *The Washington Post*, p. B1.

Webber, J. S., Jackson, K. W., Parekh, P. P., & Bopp, R. F. (2004). Reconstructions of a century of airborne asbestos concentrations. *Environmental Science & Technology, 38*(3), 707–714.

Webber, J. S., Getman, M., & Ward, T. J. (2006). Evidence and reconstruction of airborne asbestos from unconventional environmental samples. *Inhalation Toxology, 18*(12), 969–973.

Weissenstein, M. & Ritter, M. (2002, May 18). Hundreds of workers cleaned potentially hazardous trade center dust without standard protections. *Associated Press, New York Newsday*.

White, M. J. (2004). Asbestos and the future of mass torts. *Journal of Economic Perspectives, 18*(2), 183–204.

Wilson, R., Langer, A. M., Nolan, R. P., Gee, J. B., & Ross, M. (1994). Asbestos in New York City public school buildings – public policy: is there a scientific basis? *Regulatory Toxicology and Pharmacology, 20*(2), 161–169.

Wisconsin Department of Natural Resources (2006). *What you need to know about asbestos: a regulatory summary for contractors and building owners*. Madison, WI: Bureau of Air Management.

Wisconsin Department of Natural Resources (2007). *Asbestos: History and uses*. Retrieved from http://dnr.wi.gov/air/compenf/asbestos/asbes3.htm

Chapter 6
Risk Perception and Management of the Asbestos Industry in Korea: Rise and Fall of the Industry and Health Issues

Domyung Paek and Hajime Sato

Introduction

The term asbestosis was first used in 1924 by Cook to describe a case of chronic lung fibrosis from asbestos exposure. An association of asbestos with lung cancer and malignant mesothelioma has been reported since the latter half of the twentieth century, and since then asbestos has become one of the best known occupational carcinogens. Lung cancer and malignant mesothelioma caused by asbestos exposure will continue for many more years to constitute the majority of the occupational cancers diagnosed and compensated in the world. Almost every country regulates the use of asbestos, and many countries have banned its use completely.

Even though the carcinogenic nature of asbestos has been reported in academic journals, its use soared in many developed countries until the 1980s. Its use is still increasing in other parts of the world, especially in newly developing countries such as those in Southeast Asia. Historically, the regulation of the manufacture of asbestos products in one country has led only to the transfer of the production base to another country.

In Korea, asbestos consumption has been rising since the beginning of industrialization in the 1960s. This increase in consumption was caused by an increase in demand for asbestos products in the construction industry. The growth in manufacturing, such as automobile assembly, also contributed to the increased demand for asbestos products. Also, asbestos imports soared to more than 50,000 tons in the 1970s. This increase continued up to the 1990s, and imports peaked at 95,000 tons in 1992. Soon after the economic crisis of 1997, however, imports dwindled to less

D. Paek
School of Public Health, and Institute of Health and Environment,
Seoul National University, Seoul, Republic of Korea

H. Sato (✉)
Department of Public Health, Graduate School of Medicine, The University of Tokyo,
Tokyo, Japan
E-mail: hsato-tky@umin.net; hsato@post.harvard.edu

H. Sato (ed.), *Management of Health Risks from Environment and Food*, Alliance for Global Sustainability Bookseries 16,
DOI 10.1007/978-90-481-3028-3_6,

than one-third of the peak. The government then plans to ban both the import and consumption of asbestos completely, starting in 2009. The export of asbestos to Japan increased steadily between 1960 and 1974; this was about 10 years ahead of the expansion phase in Korea. But after 1988, the export of asbestos to Japan decreased (Morinaga et al. 2001; Furuya et al. 2003). Asbestos consumption at first rose steadily, reached a plateau, and then suddenly fell. This inconsistent progression reflects an interplay of economics, administrative policy, and politics. To determine future administrative policies and resolve any unnecessary political disputes, it is essential to understand who and what have contributed to these changes in the consumption of asbestos. In this chapter, we systematically review the unfolding of the asbestos problem in Korea, as well as the social and political reactions to it, from the perspectives of various interest groups. We then draw from this analysis constructive lessons for other developing countries, albeit somewhat belatedly.

Methods

We began by dividing the changes in asbestos consumption in Korea into distinct phases, to reflect the changing trends in consumption over time. Specifically, we compared the periods of rising and falling consumption in both Japan and Korea to see if we could find an underlying pattern of change in the industry that was shared by both countries. Each period was divided into three phases (expansion, plateau, and shrinkage) based on changes in the amounts of asbestos mined and imported. These phases were then matched to relevant events concerning health and safety. Examples of such events are the passage of the Industrial Safety and Health Act or the compensation of the first asbestos victim.

Next, we attempted to identify the key players in the rise and fall of asbestos consumption in Korea. Changes in consumption generally reflect changes in the demand for and use of asbestos products by various industries. They can be influenced by the three factors mentioned above: the economic situation, administrative policy, and the politics of health and safety. The economic situation drives the domestic and foreign demand for asbestos products and, consequently, how much asbestos is imported and exported. Administrative policy discourages or encourages certain economic tendencies by setting regulations and launching various campaigns and enforcement programs directed at the use of either asbestos products or suitable substitutes. Finally, the politics can raise health and safety concerns about asbestos products as social issues or dismiss these problems as negligible, thereby changing underlying public attitudes toward asbestos consumption. In the context of these three factors, the important stakeholders who have contributed to the changes in the consumption of asbestos, starting in the 1920s when it was first mined by the Japanese in Korea, were identified. The chapter aims to delineate the role of each stakeholder over time and draws lessons for the future.

Results

Trends and Phases in Asbestos Consumption

The consumption of asbestos or asbestos products in Korea has fluctuated over the years. The changes can be assigned to four periods: minimal consumption, expansion, the plateau, and shrinkage.

Asbestos was first mined on the Korean peninsula in the 1920s, when Korea was under Japanese occupation. At that time, the Japanese military considered it to be a strategic material, and it was mined and exported to Japan exclusively for the Japanese navy to build warships. When the occupation ended in 1945, following the defeat of Japan in World War II, the mining of asbestos in Korea stopped completely, as the demand had disappeared. It did not resume until 1960, after which the consumption has been exclusively domestic. The mining of asbestos steadily increased and peaked in 1982. In Japan, the manufacture of asbestos products relied mainly on imports of asbestos, but these imports plummeted to zero with the end of World War II. They resumed in 1949 and peaked in 1974.

In both Korea and Japan, there was a linear increase in asbestos consumption once the importation and production had resumed. After the initial expansion, asbestos consumption remained at a plateau for a while in both countries. The steady decline in asbestos consumption, especially the reduction in foreign imports, began in 1988 in Japan and in 1995 in Korea. Japan had already banned the use of all the asbestos products in 2005, and Korea will do so in 2009. Putting all these changes together, it can be seen that the asbestos industry in both countries has progressed through an initial expansion phase, an intermediate plateau phase, and a final shrinkage phase (Fig. 6.1).

The Industrial Safety and Health Act was passed as the initial expansion phase was coming to an end in both countries: 1972 for Japan and 1981 for Korea. The first mesothelioma case in Korea was compensated in 1994, at the end of the plateau phase. In Japan, the first mesothelioma case was compensated in 1978, but there were no further cases until 1984, during the plateau phase. The shrinkage phase in Japan, however, did not begin until 1988.

In both countries, the plateau phase began after the implementation of administrative regulations, and the shrinkage phase followed the appearance of repercussions from previous asbestos exposure. In Korea, asbestos consumption from 1920 to1959 was low and domestic demand almost nonexistent. From 1960 to 1982, consumption increased steadily. In 1983, consumption entered the plateau phase, which continued until 1995, with a small peak in 1992. From 1995 onward, consumption has been dwindling (Fig. 6.2).

Putting all these observations together shows that changes in the various phases of the Korean asbestos industry have been lagging about 10 to 15 years behind the corresponding change in Japan. The importing of asbestos by Japan resumed in 1949, and the mining of asbestos in Korea resumed in 1960, 11 years later. Likewise, the peak of asbestos imports occurred in 1974 in Japan and in 1992 in Korea,

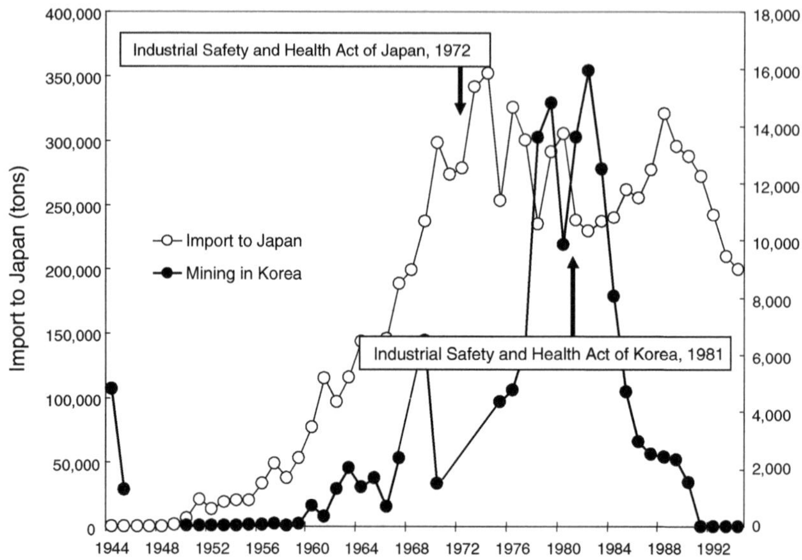

Fig. 6.1 Asbestos consumption in Japan and Korea

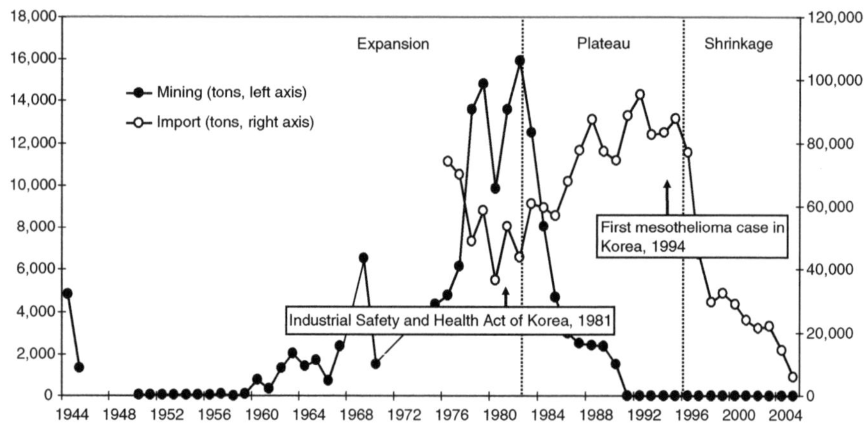

Fig. 6.2 Asbestos mining and import, Korea

a 17-year difference. The linear decrease in asbestos consumption in Japan began in 1988 and in Korea in 1995, a difference of 7 years. The first compensation for a case of asbestos-related mesothelioma occurred in 1978 in Japan compared to 1994 in Korea, a 16-year difference (Murayama et al. 2006; Takahashi et al. 2007) (Table 6.1).

Table 6.1 Chronological comparison of asbestos-related events in Japan and Korea

	Japan	Korea	Difference (years)
Start of expansion	1949	1960	11
Peak of imports	1974	1992	17
Decline of imports	1988	1995	7
Banning	2005	2009	4
First compensated mesothelioma	1978	1994	16

History of Asbestos Regulation in Korea

Period 1 (1920–1959): Minimal Consumption of Asbestos

The mining of asbestos in Korea began in the 1920s. During the Japanese occupation, the Japanese Navy was the principal consumer of asbestos, and most of the asbestos mined in Korea was exported to Japan; domestic consumption of asbestos in the Korean peninsula was almost nonexistent. When the Japanese occupation came to an end, the demand for asbestos mining in Korea had just recently disappeared, to return only in the 1970s.

In terms of regulation, no particular attention was paid to the potential (and real) health hazards of asbestos, either in its mining or its manufacture, and the government offered no protective policies or programs. In fact, during World War II, many young Korean men were conscripted and forced to work in the asbestos mines, including the mine in Kwang-Cheon, Korea, once the largest in all of Asia.

During this period, the behavior of the asbestos industry was not raised as an issue either by policymakers or the society at large. Even though cases of pneumoconiosis in coal workers had been recognized as early as the 1950s, the health problems of Korean asbestos miners under Japanese control were not looked into.

Period 2 (1960–1982): Industrial Expansion

The Korean asbestos mines were reopened in 1960 to meet domestic demand. Most of these mines, however, were very small and fewer than 10 miners worked in each one. The asbestos fibers were not very long, and because of their poor quality consumption was limited to just a few applications, such as the manufacture of asbestos cement. For this reason, Korea had to depend on foreign imports for producing its asbestos products. A major exporter was Canada, and to promote these exports the Asbestos Institute of Canada played a major role in creating the illusion among health professionals and government officials that the health hazards of asbestos could be controlled.

Since the 1960s, asbestos mining in Korea has increased incrementally. For the first 10 years, the pace was rather slow. Then in the 1970s, it started to accelerate, reaching a peak in 1982. Because data on the importing of asbestos before 1976 were not available, we defined the phases in terms of mining. Using this criterion, the expansion phase lasted from 1960 to 1982 in Korea.

Construction materials, friction materials, and textiles were the three major asbestos products. The manufacture of these products in Korea began with foreign investment and joint ventures in the 1960s and 1970s. It is noteworthy that many of these foreign companies, especially those making asbestos textiles, started their operations just after occupational safety and health regulations were tightened in their country of origin. In 1971 and 1973 respectively, Japan and Germany instituted such measures to protect workers from asbestos dust. During this same period, the Japanese and German asbestos textile companies moved their operations to Korea.

During the expansion period in the asbestos market, none of the responsible parties paid any attention to asbestos health hazards. Even though pneumoconiosis in coal workers had been studied and diagnosed since the 1950s in Korea, neither the government, the mine owners, the product manufacturers, nor the endusers of asbestos products did anything to protect workers or consumers from asbestos exposure (Paek et al. 1995).

Government policies during this period centered on economic development. With the initiation of government-led developmental projects in the 1960s, the demand for asbestos products increased. Especially after the Korean Government introduced the "New Town Movement" as a way to mobilize the rural population during the 1970s, there was a sudden surge followed by a steady increase in asbestos demand. In particular, the Movement led to the renovation of traditional townhouses by replacing the straw-woven rooftops with new asbestos slates, thereby creating a huge demand for asbestos and asbestos products. However, no particular regulations were introduced during this period. The Industrial Safety and Health Act was not enacted until 1981, and the effect of its enforcement was almost nonexistent. Even the measurement of asbestos exposure was not attempted until 1984.

Period 3 (1983–1995): The Plateau of Asbestos Consumption

Asbestos mining in Korea stopped in 1985. This was due to the fact that the asbestos mined in Korea was rather short in length, and thus of poor quality. Asbestos mining in Korea could not compete with the cheap, good-quality foreign imports, especially from Canada, which was also the biggest exporter of asbestos during this period (Table 6.2). As in the previous period, asbestos products were extensively manufactured and many of these products were exported to other countries (Fig. 6.3).

The exposure levels of asbestos fibers in Korean workplaces were measured for the first time by the Ministry of Labor in 1984, 3 years after the passage of the Industrial Safety and Health Act. Even though such measurements were legally sanctioned, enforcement was rather loose for many small, dirty shops. More serious and stringent enforcement did not occur until compensation for the first case of asbestos poisoning in 1994. It is noteworthy that legal protection of that time was restricted to the manufacture of asbestos products, which means that none of the endusers, such as construction workers or ship builders, were protected.

Table 6.2 The four phases of asbestos use and regulation

	Pre-expansion	Expansion	Plateau	Shrinkage
	1920–1959	1960–1982	1983–1995	1996–
Key events		Economic development plan (1960)	Industrial Safety and Health Act (1981)	First mesothelioma case (1994)
Economy	Mining	Mining, importing	Importing of raw minerals and exporting of manufactured products	Importing of manufactured products, abatement
Administration		No safety and health measures	Limited safety and health measures, mainly focused on measurement	Safety and health measures outside the manufacturing sector
Politics		Bureaucrats	Bureaucrats, professionals	Bureaucrats, professionals, NGOs, and victims

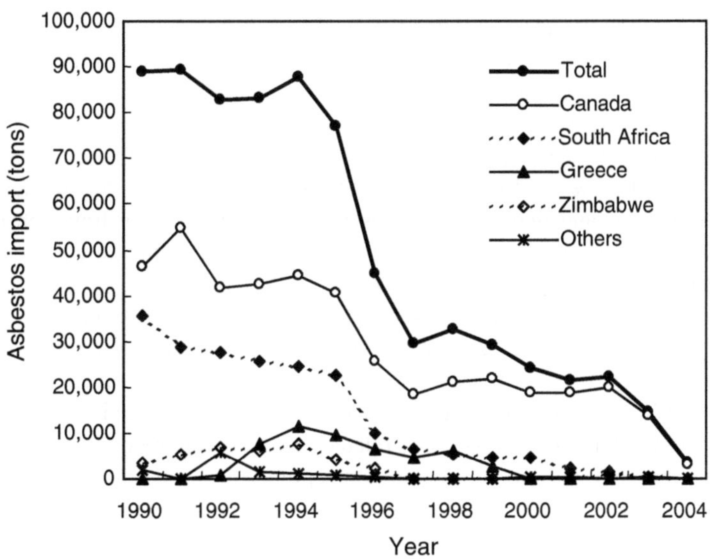

Fig. 6.3 Import of raw asbestos by exporting countries

Even though the life cycle of products manufactured with mined asbestos can only end safely if the products are buried (as waste disposals) in the ground, the protection measures covered only a small portion of this life cycle, and the most problematic portions were left uncovered. With the introduction of the Act in 1981, the infrastructure for administrative control of asbestos had been put in place and strengthened. The measures included new regulations and professional manpower.

However, the asbestos problem remained largely unnoticed during this period, even though its existence was sporadically reported. In a study of 20 mesothelioma patients from 1977 to 1987 (Kim et al. 1990), only one construction worker (5%) was found to have a definite history of occupational exposure to asbestos. The others' occupations were farmer (6), housewife (2), teacher (2), clerk (2), photographer (1), policeman (1), car driver (1), and herb doctor (1) (Kim et al. 1990).

Various administrative measures have been introduced since the early 1990s for the control and management of asbestos hazards. When the Industrial Safety and Health Act was revised in 1991, asbestos was added to those raw materials requiring prior permission before they could be used in manufacturing. From 1992 onward, those who had worked at an asbestos manufacturing site for more than 3 years became eligible for follow-up medical screening, even if they had already stopped working at the site.

The first case of asbestos-related disease in Korea was reported in 1993. It involved a former female employee of an asbestos textile company who was found to have malignant mesothelioma at the age of 46. After working for this company for 19 years, she had to stop because of thoracic pain. She was initially treated for tuberculosis, but soon thereafter malignant mesothelioma was diagnosed. The causal relationship with asbestos was immediately recognized based on her uncomplicated work history. Since that time, asbestosis and lung cancer due to asbestos exposure have also been recognized and compensated. The second study, conducted in 1993, reported that there were 12 mesothelioma patients found from 1990 to 1993; (Na 1994). Their occupational histories were also ascertained by the researchers. One of these patients (8%) had worked as an architect and served as an engineer at the construction sites, and two others had lived near either a port or train-construction sites where asbestos exposure was possible (16%). The first health-checkup record was provided to a former asbestos worker in 1995.

The mass media always have a short attention span, so the same topic is never covered repeatedly. During this period, most of the coverage about problems in the handling of asbestos was limited to news reports from foreign countries. Moreover, no responsible stakeholders, including academics, consistently paid attention to the subject, and no diagnoses of asbestos-related diseases were recognized. The use of asbestos continued with no significant interruptions.

Period 4 (1996–Present): Shrinkage of Asbestos Industry

Starting in the early 1990s, the asbestos manufacturing facilities started to move from Korea to other countries, most notably China and Indonesia. This change, in turn, reduced the manufacturing of asbestos products in Korea, a trend that became quite evident by the year 2000. This shrinkage in domestic manufacturing was compensated for by an increase in the importation of asbestos products, especially from Southeast Asia (Fig. 6.4).

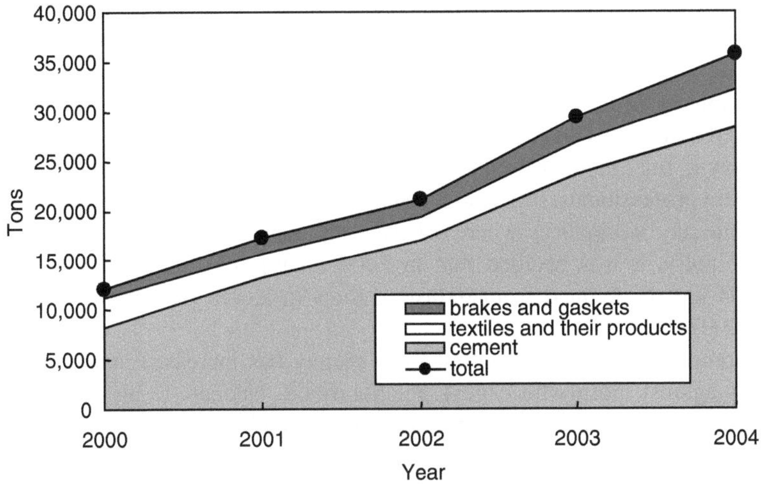

Fig. 6.4 Import of asbestos products by product category

The importation and use of crocidolite and amosite were banned in 1997, and asbestos factories that exposed workers to asbestos dust were shut down by the Ministry of Labor for the first time in 1998. The interest shown by academia in the asbestos problem was rather belated. Even after the first systematic national survey of asbestos-related diseases in 1993 (Paek 2003), it was near the end of twentieth century before academics paid attention to the need for a mesothelioma registry. Such a registry, based in this case on pathologists' reports, was finally established in 2001. It was, and has continued to be, funded by the Korean Occupational Safety and Health Agency (KOSHA) (Paek & Choi 2002). The 1988 exposure limit for asbestos (2 fibers/cc) was lowered to 0.1 fibers/cc in 2001. In 2003, actinolite, anthophylite, and tremolite were added to the list of banned asbestos types.

Public campaigns targeting the dangers of asbestos have been staged by labor unions and non-governmental organizations (NGOs) such as the Green Alliance. These efforts of the unions and NGOs, along with those of lawyers, have made a positive contribution to public awareness of the asbestos problem and have demanded attention from the public authorities as well as the researchers (Paek 2003). In early the 2000s, these efforts led to the identification of asbestos products in subway stations and trains. A health-effects survey of subway workers, requested by the labor unions, was carried out in 2007. The results revealed pleural plaques in over 30% of those surveyed.

It is noteworthy that in 2003 the scope of the protection measures was finally expanded to include asbestos workers outside the manufacturing sector, especially workers engaged in asbestos abatement. A stipulation was added that required employers to obtain permission before dismantling or removing asbestos materials from industrial facilities and office and residential buildings. The first permit was issued for two sites in 2003, and the number increased to 115 in 2005. The amount of asbestos waste in Korea fluctuated between 50 and 183 tons per year from 2000 through 2005[1] (Report of Ministry of Environment 2006). Together with the NGOs' heightened vigilance over the illegal removal of asbestos products during the renovation of old residential and office buildings, the number of facilities dedicated to asbestos abatement has increased. As the asbestos abatement industry has grown, the major Korean importers of asbestos have become asbestos abatement professionals. Protection from asbestos exposure has also been expanded to include any workplaces where asbestos products are either used or manufactured. Finally, it was decided that in 2009 all types of asbestos and asbestos products will be banned from industrial sites in Korea (Report of Ministry of Labor 2007).

Awareness of the asbestos problem in society has increased, resulting in civil lawsuits against those who expose neighborhood residents to airborne asbestos. Asbestos victims have in fact been found among members of the general population who have never worked at, but lived near, an asbestos manufacturing site.

Up to now, about 34 workers have received compensation for either mesothelioma or lung cancer. These cases, however, were mainly those of workers exposed to asbestos products, not those working in asbestos plants. The third study of mesothelioma patients listed the occupations of 18 mesothelioma victims from 2004 to 2005. Eight of the patients (44%) had positive occupational exposure histories, and 6 (33%) had probable environmental exposure at home (Jung et al. 2006). Even though the number of studies is limited, their results collectively show a clear trend of increasing asbestos exposure among mesothelioma patients compared to the general population in Korea (Kang et al. 2006). The repercussions of these data are beginning to appear. With respect to time differences in the phase changes, Korea is expected to catch up to Japan in 10 to 15 years.

Discussion

By examining the progression of the asbestos problem in Korea over time, we can say that the problem is not just medical, that is, a matter of how many cancer cases can be expected in each period and how many patients can be managed. In many ways, the asbestos problem is rooted in economic and political factors. As a way to conceptualize the stages of the asbestos problem as they have unfolded over time, we propose the Source, Exposure, Effect, and Action (SEEA) model. The model is intended to explain the status of the asbestos industry in different countries and to describe the various strategies these countries have adopted at different times to solve the many aspects of the asbestos problem (Table 6.3) (Fig. 6.5).

[1] Asbestos waste in Korea: 148.7 tons in 2000, 164.2 in 2001, 148.6 in 2002, 97.1 in 2003, 50.2 in 2004, 183.0 in 2005.

Table 6.3 The SEEA Model: the phases of the asbestos industry, safety policy, and politics

Phases	Industry	Exposure	Effect
Stages	Input-dominant	Process-dominant	Output-dominant
Objectives (why)	Politics-based Fragmented initiatives Economy, workforce	Economy-based Issue development Need for control	Health-based Repercussions Feedback
Participants (who)	Bureaucrats, employers	Professionals, unions	Victims, NGOs
Programs (what)	Technical Recommendations Guidelines	Managerial Inspection Research and info Compensation	Cultural Building of capacity Education and training
Delivery (how)	Code-based Legislation National policy	Labor-based H&S committee OSH management	System-based Social institutions Campaigns

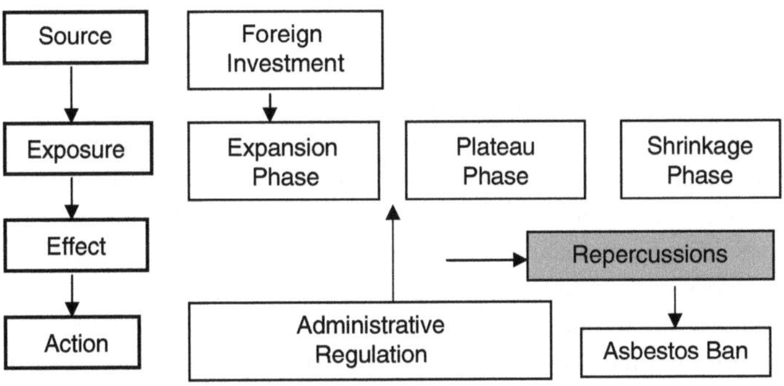

Fig. 6.5 Source, Exposure, Effect, and Action (SEEA) Model of asbestos industry

The SEEA Model as a Way to Explain the Current Status of the Asbestos Problem

The SEEA model explains the changes in the asbestos industry through an analysis of the changes in the major operating principles of the health and safety system across three phases. The model describes this system by answering a four-part question about its operation: Why (the primary purpose), by whom (the role players), with what (the programs), and how (the delivery of services) does the system operate? The phase changes in the system can also be represented as shifts in the primary emphasis of the operating principles from input dominance to process dominance, and the stages from process dominance to output dominance.

The input-dominant stage of the industry phase can be described as bureaucrats delivering primarily technical programs using code-based dictates for mainly political aims. This stage is called *input-dominant* because the operation of the system is often regulated by the amount of input, regardless of the rationale of the process or its results. Next, the process-dominant stage of the exposure phase can be described as professionals taking advantage of the dynamics of labor relations to coordinate primarily managerial programs with the major aim of economic rationalization. This stage is called *process-dominant* because the final (health) outcome is not yet appreciated (also not realized) even though the emphasis is on the rationalization of the process. Finally, the output-dominant stage of the effect phase can be described as victims systematically carrying out programs geared to changes in socio-cultural attitudes and with the aim of achieving a healthy and safe society.

In the SEEA model, the components of the various problem-solving approaches, programs, or strategies employed by a particular player are addressed across technical, managerial, and socio-cultural categories. The components of the strategy are as follows: (1) *Technical intervention*: to provide the opportunity to observe and recognize hazards; (2) *Managerial intervention:* to enable a search of alternatives for actions and protective measures; and (3) *Social-cultural intervention*: to compare and change the value systems regarding the relative importance of different health hazards.

The phase transitions between these three stages do not occur automatically unless source (S), exposure (E), and effect (E) are linked with one another sequentially through certain actions (A). The actions that lead to the input-dominant stage (source phase) are often organized by political initiatives, and politicians and bureaucrats play the major role in the shifts to this stage. Then the actions that lead to the process-dominant stage (exposure phase) are usually organized by demands from the professionals who are in charge of the preexisting technical programs intended to make the regulatory measures more efficient and effective, and streamline the institutional measures. Further actions linking the process-dominant stage to the output-dominant stage (effect phase) are organized by victims reacting to the repercussions of previous exposures. The last actions of this round (which are also the actions leading to the next round of the system-building cycle) are fueled by the feedback provided by all of the participants, but political maneuvering is still important in creating new initiatives.

In the SEEA model, the most important determinant of phase shifts between the three different stages is the presence of key role players. However, the mere presence of a key player in a stage is not sufficient for change to occur; the players must be matched to the *appropriate* stage. These stage-specific determinants of phase change in the system differ from the conventional health and safety profiles in individual countries, in which some profile components can be redundant or even counterproductive in effecting phase changes, depending on the situation.

Industry Phases and Choice of Regulatory Measures

A few Asian countries are nowadays trying to ban the use of asbestos and its products, and the other Asian countries are experiencing an increase in asbestos consumption. Overall, the manufacture and consumption of asbestos products has been shifting to Asia (International Social Security Association 2005). One-by-one comparisons of countries that are at different stages of the process may not be appropriate and feasible, because the situation in each country is different. However, some overall patterns can be detected. They could be observed also in the industrialized countries.

In general, the countries that are experiencing an increase in asbestos consumption are trying to expand their overall economy by stimulating the manufacturing and construction sectors. These countries, which are most likely in the input-dominant or process-dominant stage, have yet to experience a single case of asbestos-related cancer. Meanwhile, in the countries that are trying to ban or restrict asbestos use, asbestos-related health problems have become a major social issue. These countries are either in the output-dominant stage or starting the next round of system building.

Although all these countries have instituted regulatory measures, the effectiveness and meaning (real social functions) of the regulation vary between countries. Each country's regulations should be crafted, reflecting the experiences of that particular country in solving the occupational and environmental problem, by moving through the action sequence of source, exposure, effect, and feedback. In this sense, the measures aimed at curbing asbestos exposure and its effects vary across stages in their concreteness and comprehensiveness. The different societal stages seem to reflect the historical experiences of problem solving in that particular country.

The controlled use of asbestos, which is advocated by Canada, is not viable in developing countries because it ignores the socio-cultural conditions there. The countries that employ the controlled-use strategy limit themselves to engineering/technical solutions that are totally dysfunctional in developing countries. Paradoxically, controlled use is attractive to developing countries because it ignores the socio-cultural conditions that such countries are ill-equipped to tackle (Pandita 2006; Virta 2006).

However, if we continue to ignore these conditions, we will never be able to solve the asbestos problem successfully. To be successful, a strategy must be focused on how to eventually engage all the key role players in a network consisting of politicians, bureaucrats, workers, and victims. Each of these constituencies must play a positive and fully active role in facilitating the needed sequence of changes. Of course, there will always be some skeptics on the periphery, and it is important to turn these peripheral players into friendly activists. This approach must proceed incrementally as the situation changes. However, even in the face of such changes, priorities must be set, and the overall direction of change should be driven by sympathy and consideration of human rights. Those who seek to insert such values into the system should be recognized and their roles actively promoted. In most countries, activist researchers should become key role players.

We need an asbestos industry that fights other industries rather than victims groups. We need victims groups that fight ignorance, bias, and discrimination in the general population. But first of all we need, not engineering or managerial solutions, but sympathetic attitudes and changes in traditional or contemporary values. The roadmap for this problem-solving approach requires that the problem solvers have consciousness-raising experiences. The methods through which ethical imperatives will eventually come to override economic justifications must be devised strategically, instead of by an exclusive reliance on medical and public-health approaches.

References

Furuya, S., Natori, Y., & Ikeda, R. (2003). Asbestos in Japan. *International Journal of Occupational and Environmental Health, 9*(3), 260–265.

International Social Security Association. (2005). *Asbestos production on the increase*. ISSA (International Social Security Association). Retrieved, from http://www.issa.int/pdf/orlando05/2art-asbesto.pdf

Jung, S.-H., Kim, H.-R., Koh, S.-B., Yong, S.-J., Choi, B.-S., Ahn, Y.-S., et al. (2006). Epidemiologic characteristics revealed with a malignant mesothelioma surveillance system in Korea. *Korean Journal of Occupational and Environmental Medicine, 18*(1), 46–52.

Kang, S.-K., Kim, E.-A., & Ahn, Y.-S. (2006). *Asbestos related diseases and compensation in the Republic of Korea*. ISSA/COMTECH05-07/ATMP/SHENZHEN/2006.

Kim, C.-B., Jung, S.-H., Lee, K,-J., & Kang, J.-D. (1990). Survival analysis of hospitalized mesothelioma patients. *Korean Journal of Preventive Medicine, 23*(1), 77–86.

Morinaga, K., Kishimoto, T., Sakatani, M., Akira, M., Yokoyama, K., & Sera, Y. (2001). Asbestos-related lung cancer and mesothelioma in Japan. *Industrial Health, 39*, 65–74.

Murayama, T., Takahashi, K., Natori, Y., & Kurumatani, N. (2006). Estimation of future mortality from pleural malignant mesothelioma in Japan based on an age-cohort model. *American Journal of Industrial Medicine, 49*(1), 1–7.

Na, M. (1994). *A study on mesothelioma and asbestos exposure: MPH thesis*. Seoul, Korea: Seoul National University, School of Public Health.

Paek, D. (2003). Asbestos problems yet to explode in Korea. *International Journal of Occupational and Environmental Health, 9*(3), 266–271.

Paek, D., & Choi, J. K. (2002). Asbestos in Korea: Country Report. *Journal of University of Occupational and Environmental Health, 24*(4), 42–50.

Paek, D., Paik, N. W., Choi, J. K., Son, M. A., Im, J. G., Lee, W. J., et al. (1995). Prevalence of asbestosis in Korean asbestos industry. *The Korean Journal of Occupation Medicine, 7*(1), 46–57.

Pandita, S. (2006). Banning asbestos in Asia. *International Journal of Occupational and Environmental Health, 2*, 248–253.

Takahashi, Y., Miyaki, K., & Nakayama, T. (2007). Analysis of news of the Japanese asbestos panic: a supposedly resolved issue that turned out to be a time bomb. *Journal of Public Health*. Advance Access. Retrieved from http://jpubhealth.oxfordjournals.org/cgi/content/abstract/fdl081v1

Virta, R. L. (2006). *Worldwide asbestos supply and consumption trends from 1900 through 2003*. Retrieved (last access), March 10, 2009, from http://pubs.usgs.gov/circ/2006/1298/

Part III
Management of food risk: Cases of BSE-related human risk management

Chapter 7
Policy and Politics of BSE-related Human Disease Prevention in Japan: In Pursuit of Food Safety and Public Reassurance

Hajime Sato

Introduction

Bovine spongiform encephalitis (BSE) is a cattle disease that first emerged in the UK in 1985. A cow with BSE presents a variety of neurological symptoms, such as an abnormally stilted gait, heightened sensory perception, and anorexia, leading to the common name "mad cow disease." BSE is caused by prions that enter the food chain through the practice of feeding sheep remains to cows; these prions require a 4- to 5-year incubation period. As the infected/contaminated feed was exported and used extensively, many other countries came to discover their own BSE cases. In 1996, when eating meat from infected cows was associated with the human disease variant Creutzfeldt-Jakob disease (vCJD), the general public became extremely concerned about the safety of beef. The governments took steps to ensure the safety of beef and restore public confidence that it was safe to eat.

Cases of BSE and vCJD increased over time, as did scientific knowledge about them. However, there is still a lack of sufficient information and research evidence about the causes of the diseases, how they can be effectively and efficiently screened and prevented, precisely how much risk they pose to humans, and what the public is willing to do to avert these risks (Coulthart & Cashman 2001; Will 1999). In response, various policies have been considered, adopted, and implemented by the governments. With the benefit of hindsight, it appears that too many of the precautionary actions were nothing more than unscientific and unproductive hype, and they were seen by the public as policy failures and government negligence. These gaps between public expectations and government performance, unless bridged by effective communications efforts, can lead to spectacular political issues. This is what happened in the cases of BSE and vCJD.

The BSE issue is further complicated by the fact that the management of BSE risks requires the involvement of multiple social sectors, including the agriculture

H. Sato
Department of Public Health, Graduate School of Medicine, The University of Tokyo, Tokyo, Japan
e-mail: hsato-tky@umin.net; hsato@post.harvard.edu

H. Sato (ed.), *Management of Health Risks from Environment and Food*, Alliance for Global Sustainability Bookseries 16,
DOI 10.1007/978-90-481-3028-3_7,

and food industries, commerce, health, trade, and a variety of interest groups ranging in level of aggregation from individuals to nations. The lack of coordination in the actions of different government agencies has led to delays, and the lack of harmonization among international and domestic policies have resulted in several trade disputes. Each actor had its own orientation and stakes. With no clear and predetermined guidelines, the governments have tried hard to muddle through, while the public drifted along, listening to more and more reports of worrisome data followed by reactionary government reassurances. There have apparently been no coherent and effective communication and leadership directed at developing rational risk-management strategies, integrating the evolving scientific knowledge, and choosing among various technical options, and democratic values.

With the above as background, this article aims to examine, from an international comparative perspective, the history of the BSE problem and government responses (or lack of responses) to it, with an emphasis on communication efforts to reassure the public. Other emphases include the use of scientific arguments (always changing and accompanying some degrees of uncertainty) to justify policy decisions, the adoption of scientifically unjustified policies to reassure the public, and efforts to establish – and in some cases reestablish – public confidence in food safety. The following two questions will be addressed: (1) Ultimately, has the BSE issue helped the public and policy makers to understand the food risks and the management of these risks? (2) Have the communication efforts of the governments or the media successfully served as public fora in advancing democratic values? The chapter concludes with a discussion of how the health risks associated with BSE can be managed strategically, including through the use of communication.

Materials and Methods

To examine policy making with regard to the regulation of BSE, a comprehensive search for relevant historical documents was conducted. These documents were taken from various sources, including the archives of public records (both printed and online), libraries, government record archives, the offices of non-governmental organizations, and international organizations. Specific sources include the National Diet Library (Tokyo) and academic databases such as the Social Science Citation Index, the Science Citation Index, MEDLINE, and ICHUSHI. Other types of documents and books, such as theses, essays, and recollections, were also gathered. Government archives on BSE were thoroughly examined (Food Safety Commission; Ministry of Agriculture, Forestry and Fisheries; Ministry of Health, Labor and Welfare; Office of Prime Minister). From 2005 to 2007, supplementary interviews were conducted with bureaucrats, politicians, researchers, and some members of non-governmental organizations – both past and present.

A set of models of the political process, thus far developed from studies in different policy domains, were examined to determine if they could help explain the dynamics of policy making. Representative examples were taken from classical and contemporary textbooks on policy and politics. These examples include linear and rational policy making, bounded rationality models, models of bureaucratic incrementalism and political leadership, and the garbage-can model (see, Chapters 1 and 2). These models were applied to the event sequences reviewed in this chapter to identify and analyze the important factors that facilitate and present obstacles to policy making.

Chronology of BSE-Related Events and Policies in Japan

BSE as a Cattle Disorder

As its name implies, BSE is a disease found primarily in cattle. The first case identified as BSE occurred in the UK in 1985. The symptoms included excitability and loss of movement and control in the limbs. The brain of the diseased cow was brought to the Central Veterinary Laboratory of the Ministry of Agriculture, Fisheries and Food (MAFF), where spongiform degeneration was detected. In 1987, prion was determined to be the most likely cause of BSE. Gradually, more cases came to light in the herds of southern England. In 1988, BSE was designated to be a notifiable disease as a zoonosis in the UK (Donnelly et al. 1997). By February 1988, a commercial feed, meat and bone meal (MBM), was suspected of being infected, and therefore of the cause of the BSE epidemic. It was recommended that the feed industry voluntarily withdraw MBM from the market, that the carcasses of animals showing clinical signs of the disease be destroyed, and that the inclusion of ruminant protein in ruminant feed be banned.

Although the Southwood Committee, established on the recommendation of the Chief Medical Officer (CMO), considered it highly unlikely that BSE would become a human health hazard, it did not completely rule out the possibility. Therefore, in 1989, to prevent the consumption of material from subclinically infected animals, the UK government introduced the specified bovine offal (SBO) ban. This required exclusion from the human food chain of certain bovine tissues considered to be highly infectious, regardless of whether the cows were actually infected. On May 15, 1990, the CMO issued a statement that included the phrase, "British beef can be safely eaten by everyone, both adults and children." The MAFF first opposed taking any further protective measures that were not clearly justified scientifically or recommended by the Spongiform Encephalopathy Advisory Committee (SEAC). However, this policy was modified shortly thereafter, when it was discovered that BSE could be transferred to pigs. BSE infection was also confirmed in mice that had eaten BSE brain tissue. The British government introduced a policy called the slaughter and compensation scheme, which required the abandonment of cattle showing clinical symptoms of BSE, and compensation was provided to their producers (Packer 2006). In addition, this act officially banned the sale, supply, and use of SBO as animal feed and required government permission for its export.

Several countries soon introduced an import ban on UK cattle. In 1989, France, Germany, and the US banned cattle import from the UK, and the EU adopted a ban on the export of cattle born before July 18, 1988, or suspected calves from the UK. In 1989, France banned importation from the UK of meat and bone meal (MBM) powder to be used as animal feed. Japan had been banning the importation of beef from the UK since 1951 because of the presence of foot-and-mouth disease, but in 1989 it extended the ban to include the importation of live cattle and tightened its regulation on the importing of MBM. In 1990, France banned the use of animal protein (except from milk, eggs, and fish) as feed. The EU and France subsequently designated BSE as a notifiable infectious disease in animals, and France established a national network for clinical surveillance for BSE. This latter action was in response to recommendations from the Office International des Epizooties (OIE, the World Organization for Animal Health). The US introduced an active surveillance program targeting abnormal cattle before they are slaughtered.[1] Despite all the bans listed above, the first case of BSE was reported in Switzerland in 1990 and in France in 1991.

Meanwhile, the number of BSE-infected cattle in the UK was increasing, reaching a peak of 37,000 in 1991 (Office International des Epizooties, OIE 2003). Pressed by the EU and its member countries, in 1991 the UK government banned bovine offal derived from cattle older than 6 months, as well as the use of SBO (MBM) as fertilizer. It also started regulating the transfer of SBO-protein materials from farms of origin. Although the prevention of BSE had been expected, thanks to the feed ban of 1988/1989, on March 22, 1991, some animals born after the feed ban, which came to be known as BABs, were found to be infected. It was suspected that some quantities of contaminated feed or feed ingredients might remain on farms because of loose and ineffective regulation. The discovery of BSE in the UK led the Japanese Ministry of Agriculture, Forestry and Fisheries (MAFF) to tighten its import regulations on beef products and MBM from France.

In August 1992, the UK Creutzfeldt Jakob Disease Surveillance Unit, established in 1990, reported that a probable case of CJD had been identified in a farmer who had reported a case of BSE on his farm (Sawcer et al. 1993; Will et al. 1996). A second probable case of the disease in a farmer was reported in 1993, which prompted the Department of Health (DoH) to begin issuing repeated statements of reassurance. Then, in 1994, a 16-year-old girl was suspected of having contracted CJD (Brown et al. 2001). Prompted by reports of such cases and the results of animal experiments, in 1994 the EU and the UK introduced a ban on feeding mammalian proteins to ruminants.

The number of BSE cases in the UK was becoming substantial. By the summer of 1994, it had become clear that the presence of BSE in a high proportion of BABs could not be attributed to the carry-over effects of feed given to cows after July 1988 – or even after October 1988, when the sale of ruminant feed containing ruminant protein actually stopped. Many believed instead that the contamination resulted from the loose enforcement of the regulations. It was discovered in this connection that some operators were ignoring the rules regarding the treatment of SBO. Consequently,

[1] Passive surveillance denotes reporting of clinically suspect cases, while active surveillance is the testing and reporting based on the birth and health condition of the cattle pre-determined.

a new national body, the Meat Hygiene Service, was established in the UK on April 1, 1995, to take over responsibility for enforcing the regulations in slaughterhouses. In the same year, France greatly improved its cattle registration and tracking system, which was originally established in 1978 (Declaire 95/276).

From Cattle Disorder to Human Disease: Fear Turning into a Threat to Human Health

On March 20, 1996, the UK announced that the SEAC had found a new form of CJD in young people. They concluded that the patients had most likely been exposed to animals that transmitted the disease to them 1 year before the symptoms appeared (Collinge et al. 1996). This announcement triggered a major food scare and resulted in a substantial decline in beef consumption. The Prime Minister as well as officials from the MAFF and the DoH recognized the necessity of increasing consumer confidence. Based on the SEAC recommendation, on April 3, 1996, the government adopted the so-called “30-month scheme,” which excluded animals over 30 months of age from the food chain. More precisely, the sale of beef from cattle slaughtered after March 1996 was banned, unless it could be proven that they were less than 30 months old. Furthermore, the sale or use of feed containing mammalian MBM was banned (Food Standard Agency, The BSE Inquiry 2000).

Immediately thereafter, on March 25, 1996, the EU banned the importation of live cattle, bovine semen and eggs, all meat, and MBM from the UK. France ordered the removal of SBO from cattle fed with infected MBM and required the incineration of all animal carcasses. In the US, beef producers and academics recommended the voluntary cessation of ruminant-protein feeding. In Japan, the Ministry of Health and Welfare (MHW) issued a notice and recommended a voluntary cessation of the importing of beef products (including MBM) from the UK, and the MAFF banned the same imports officially. Soon after the release of the “Report of a WHO Consultation on Public Health Issues related to Human and Animal Transmissible Spongiform Encephalopathies (WHO 1996),” the MAFF convened a panel of experts, including officials from the Milk and Meat Section and the Environmental Health Bureau of the MHW. The MAFF then issued a directive mandating that ruminants’ organs, such as MBM, not be used in ruminants’ feed. The MHW held a meeting of its own panel of experts, to which MAFF officials were invited, and issued a notice to local governments that bovine organs already imported should not be consumed, and it subsequently launched a BSE surveillance program.

The next year, 1997, the EU banned the use of Specified Risk Materials (SRM). They also issued a more general ban of any materials considered to be a BSE risk, but this directive was not implemented. The EU commission also required each member country to introduce a cattle tracking system or improve its existing system (EC 820/97). Despite a paper in *Nature* predicting that BSE would virtually disappear by 2001 – even without selective culling – the UK announced the establishment of a cattle-tracing organization for such culling (Anderson et al. 1996; Ferguson et al. 1998). The government subsequently required the removal of bones from beef before delivery to consumers, as well as the incineration of SRM and of cattle older than 30 months. Furthermore, as political criticism of the government's policies mounted, a BSE Inquiry under the direction of Lord Phillips was announced in December 1997. The report was published on October 26, 2000. Across the Atlantic, the US government banned the use of mammalian protein as ruminants' feed, as well as the importation of live ruminants and ruminant products from Europe.

In Japan, BSE was placed on the agenda of the Diet, both the House of Representatives and the House of Councilors. A few politicians raised questions about the effectiveness of certain measures already taken against BSE, calling for regulations that were stricter than the previous administrative directives. They also wanted to prohibit the use of MBM for cattle feed. MAFF officials from the Animal Industry Bureau responded that such measures might not be necessary, as MBM was not widely used in Japan, and any value it once had almost entirely vanished after the official notice of the previous year (The Exploratory Committee on the BSE Issue 2001a). The Domestic Animal Infectious Disease Control Law was partially revised on April 11, 1997, but the revisions only served to designate BSE as a notifiable animal disease.

In 1998, the EU started an evaluation of the status of the BSE problem, and it issued guidelines for the epidemiological surveillance of BSE. At the same time, it lifted the trade ban on beef produced in Northern Ireland, although still requiring that the cattle be traceable. In response to the EU Act of 1999, the UK adopted the Food Standard Act in 1999, and established the Food Safety Agency in 2002, which had the primary role of coordinating the functions of the MAFF and the DoH. As a result, the EU lifted its import ban on UK beef from cattle born after August 1996. In 1998, France legislated the monitoring of food safety and tightened the related inspections. Subsequently, France established its own Agency for Food Hygiene and Safety (Millstone & Zwanenberg 2002).

Despite all these efforts, the incidence of BSE in EU countries other than the UK was not effectively contained. In November 2000, the (re-)emergence of BSE cases sent the European beef market into turmoil. Both the retail and food-service sectors felt the impact of consumers dramatically reducing their beef consumption. In the same year, the EU published its "White Paper on Food Safety," recommending the removal and abandonment of SRM and calling for comprehensive food-safety regulations, including the establishment of agencies (food authorities) to enforce these regulations (European Commission, DG Health and Consumer Protection, 2000). The following year, it was required that all cattle older than 30 months be tested for BSE if they were to be slaughtered for food. It also mandated random-sample testing of cows that had died on farms or during transport (later changed to testing of all cattle older than 30 months). The decision of whether live cattle could be imported from a country had to be based on that country's BSE-related status. Finally, the OIE adopted a BSE status-evaluation plan.

In accordance with EU recommendations, the UK introduced in 2001 a surveillance program for cattle on farms. The specific targets were dead or injured cattle older than 24 months, a random sample of cattle older than 30 months (later changed to 24 months), and all descendants of BSE cattle older than 30 months

(Packer 2006). In 2000, France implemented a BSE surveillance program targeting cattle older than 24 months that had died from unnatural causes (abnormal deaths). In 2001, BSE testing was extended to all cattle older than 30 months that were to be slaughtered for food. However, shortly thereafter a BSE case was identified in a cow born after July 1996. Consequently, BSE surveillance was introduced for at-risk cattle older than 24 months, and the age at which healthy cattle must be tested for BSE was lowered from 30 months to 24 months.

In 1998, the US banned the importation of live ruminants and their products from countries it considered to have inadequate testing procedures. The Food Safety Council was established under the Office of the President, comprising the Secretaries of the Department of Agriculture (USDA), the Department of Health and Human Services (DHHS), the Environmental Protection Agency (EPA), and others. Simultaneously, an evaluation of the effectiveness of the USDA's risk-avoidance policies was initiated. In 1999, BSE surveillance was expanded to include dead and abnormal cattle. In 2001, a BSE steering committee was established (USDA 2004a). Presided over by the Food and Drug Administration (FDA), the committee included representatives from the FDA, The USDA, and the Department of the Treasury. In its report entitled "Food Safety and Security," the General Accounting Office (GAO) recommended the establishment of a new agency focusing on food safety.

Watching the course of events overseas unfold, Japan's Ministry of Health, Labor, and Welfare (MHLW, formerly the MHW) called in December 2000 for a voluntary cessation of the importing of bovine brains and spinal cords. Based on recommendations from its panel of experts, the MAFF issued directives banning the use of ruminants' MBM as animal feed or the use of materials from ruminants to produce animal drugs. In February 2001, the MHLW banned the importing of beef products from the EU. In March of the same year, the MAFF banned the importing of animal drugs made from ruminants' materials and initiated a surveillance program for abnormal prions in neurologically abnormal cattle older than 24 months.

Both Japan's domestic beef producers and the Japanese government publicly boasted about the safety of domestically produced cattle and beef. Labels with the logo "J-beef" soon appeared on beef sold at retail stores as part of a campaign started in 1994, shortly after the liberalization of the beef market, to promote sales of domestic beef products. The Japan Meat Information Service Center also advertised domestic beef, implicitly suggesting that, unlike imported beef, it was safe.

Emergence of Domestic BSE Cases in Japan: Disruption of the Illusion of Safety About Domestic Beef

On August 5, 2001, a veterinarian declared a cow in the Chiba prefecture to be diseased. The next day, the prefecture Meat Inspection Section diagnosed the cow as septic and ordered that it not be used for food. The laboratory sent its specimen to the National Institute of Animal Health for BSE testing. On August 24, through a pathological examination, the cow was found to have had a spongiform brain; on September 10, an immunopathological test confirmed that it had abnormal prions. The test results were immediately reported to both the Minister and Vice-Minister of the MAFF and the Inspection and Safety Division of the Department of Food Safety of the MHLW. The latter informed the Minister and Vice-Minister of MHLW.

The MAFF immediately publicized the case, explaining that the cow had been incinerated and not used for food. It also set up a BSE taskforce, chaired by the Vice-Minister. The MHLW ordered that beef shipped from the farm where the BSE-infected cow had been raised must no longer be sold. The next day, the MHLW established a taskforce for BSE-related food safety, calling on the BSE research group. Meanwhile, the MAFF held a briefing session for consumers. It subsequently initiated a complete cattle survey for BSE (called as an emergency census) (Ministry of Agriculture, Fisheries and Food, & Ministry of Health 1989, p.3). Having discovered that the BSE-infected cow had been rendered to produce MBM for aquaculture, the authorities ordered on September 14 that the MBM not be moved from the stock location.

On September 19, the MHLW decided to upgrade the BSE screening tests, requiring that they be applied to all cattle older than 30 months (24 months if they had neurological abnormalities or were generally ill). On September 28, the MAFF implemented an import ban on MBM from all countries if it was to be used for feed or fertilizer. The MHLW then asked food manufacturers and processors to start checking on October 5 for the use and withdrawal (calling-in) of SRM. On October 9, the BSE project team, consisting of seven vice-ministers from different ministries, was formed. On the same day, the MHLW solicited the cooperation of the prefectural governments in launching a program to screen all cattle for BSE. This testing, which was to take place at the slaughterhouses, began on October 18. On October 10, the same day the MHLW announced its screening program, a meeting of the project team, headed by the Vice-Minister of the MAFF, was held. On the next day, the MAFF directed farms to postpone all cattle shipments until the screening program started.

The first meeting of the MAFF panel of experts on BSE was held on October 5. The panel members expressed their concern that the general public was worried about the safety of domestic food in general – not just Japanese beef – because prior to the BSE case, the authorities had been stressing that domestic food was safe (Food Safety Commission 2004a). It was argued that before (and at the latest when) safety measures are announced, the scientific basis for the claims regarding the safety of the food and the rationales for those measures should be clearly presented to the public. A member of the National Union of Consumers' Groups claimed that the union was asking the MHLW to test all cattle. A member of the Food Safety Department of the MHLW explained that, since the present BSE tests are not supposed to be capable of detecting prions in cattle younger than 24 months, the EU had decided to test only cattle older than that. He then questioned the scientific justification for the testing of all cattle. In response, a member from the farming sector, while acknowledging that scientific judgments are important, insisted that because the country was in a crisis, it was imperative that all cattle be tested so the public could be put at ease and harmful rumors averted. A member from the fertilizer industry stated that the ban on the use of MBM in fertilizer production is not necessary and should be lifted soon (The Exploratory Committee on the BSE Issue 2001b).

In the end, the panel found testing all cattle to be justified, not only as a way to reassure the public, but also because it was impossible to determine the precise ages of all the cattle. Suppliers were concerned that public confidence would not be restored if there was untested beef in the marketplace. Shortly thereafter, on October 15, the MAFF issued a ban on the production, sale, and use of animal feed containing MBM, although the use of such feed was allowed if it was shipped before October 31. At the same time, compulsory labeling of the MBM ingredients in fertilizers was initiated. The MHLW adopted a policy requiring the removal and incineration of SRM. The BSE screening program started as planned, on October 18, 2001, using the Western blotting method, which the MAFF considered to be standard.

On November 21, the second BSE case in Hokkaido Prefecture was confirmed. Approximately 10 days later, a third case was reported in a cow from Saitama Prefecture. The MAFF decided that all domestic beef produced before the BSE screening program began (October 18) must be incinerated. Regulations targeting the importation of beef products were also adopted. On December 22, the MAFF announced a ban on the importation of beef products from EU countries, and the MHLW called for a voluntary cessation of the importing of products containing bovine organs coming from the EU. The Japan Meat Information Service Center upgraded its "J-beef" campaign, a main function of which was the advertising of the claim that all domestic beef products had been thoroughly examined for safety (Fig. 7.1).

The total number of BSE cases in the EU – excluding the UK – increased from 482 in the year 2000 to about 1,000 in 2001. The numbers of confirmed and suspected vCJD cases also increased in several European countries (United Kingdom

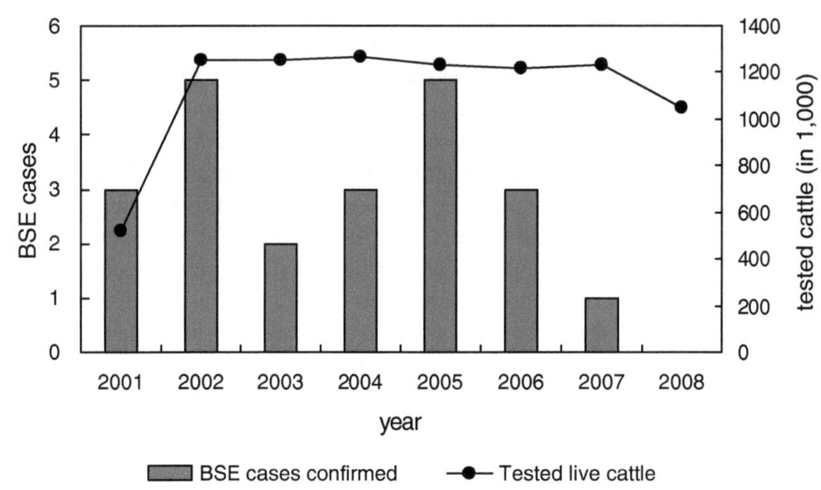

Fig. 7.1 BSE cases in Japan

Department of Health 2006). The enforcement of Regulation 999/2001 regarding BSE began on July 1, 2001. The EU Scientific Steering Committee began an international BSE risk assessment, which was called the Geographic BSE Risk Assessment (GBR) (Matthews 2003; Ricketts 2004). The Olsson Report on BSE in February 2002 confirmed that the use of MBM in animal feed had been suspended, that BSE testing of all healthy cattle older than 30 months and at-risk or dead cattle older than 24 months had begun, that the list of SRMs had been expanded to include, most notably, the spine and the whole intestine, and, finally, that meat recovered mechanically from the bones of ruminants had been banned. A panel of EU agriculture ministers established the Food Safety Authority. Finally, to improve policy implementation, on January 31, 2002, the Japanese MHLW issued a directive requiring the removal of the spinal cord before the spine was split open.

To match trends in the EU, the Japanese authorities called for the establishment of a food-safety organization. On November 6, 2001, the Exploratory Committee on the BSE Issue was established as a private consulting body for MHLW and MAFF ministers. On April 2, 2002, it published its report. Then, on June 11, 2002, the Ministers' Panel on Food Safety published its report, "On the Future Administration for Food Safety," which recommended the enactment of the Basic Food Safety Law (Shokuhin Anzen Kihon Ho) and the establishment of the Food Safety Commission (MAFF 2002a). This report also noted that, through the administrative reform of the MAFF, the section of the MAFF devoted to risk management should be a separate entity from the section(s) with the mission of promoting the agricultural industry.

On June 14, 2002, the Special Provisional Act for BSE Mitigation was enacted. The act was intended to protect public health as well as to promote the development of the cattle, beef, and related food industries by preventing the occurrence and spread of BSE. It stated that the mitigation of BSE was the responsibility of both the national government and local governments, and it called for them to commit themselves to a number of action plans, including a ban on the use of MBM for cattle feed, compulsory examination and notification of dead cattle beyond a certain age, BSE inspections at slaughterhouses, measures against the economic problems with respect to compensation and subsidies caused by BSE, efforts to communicate with and educate the public, and the promotion of research; all of these measures were defined as necessary. The act also made the government responsible for establishing a tracking system for all cattle. Given the failure so far to prevent BSE, the appendix of the act called for a thorough overview of all administrative systems for food safety – from production to consumption.

The MAFF provided a number of organizations, such as the Japan Center for Meat Consumption, the National Union of Food Meat Cooperatives, the associations of the livestock industry, and the Japanese Association of Veterinarians, with a variety of subsidies to help increase beef consumption. This support resulted in TV programs, symposiums, cooking seminars and the like that went beyond compensation for the economic losses incurred by the confiscation and incineration of suspected cattle, beef, and beef byproducts. The money allocated for these

BSE-related programs amounted to 20 billion yen. On the downside, there were numerous reports of attempts to swindle money from these government compensation programs.

The consumption and market price of beef in Japan, as well as the sale of Japanese barbeque restaurants, hit bottom in February 2002, before gradually returning to the previous year's level in September (Daiichi Life Research Institute 2004). On September 10, the Minister of the MAFF expressed his satisfaction with the measures so far adopted, requesting that the public calmly wait for the government's efforts against BSE to fully succeed and that people avoid harmful rumors and panic. He was thus promising the successful BSE containment by the government's efforts, namely, the enactment of the provisional act, the establishment of cattle tracking system and the introduction of a thorough BSE testing mechanism (targeting all cattle). At the same time, he announced to drastically shift agricultural policy from a producer orientation to a consumer orientation, by making a new law and a special commission on food safety (MAFF 2002b).

On May 23, 2003, the Diet enacted the Basic Food Safety Law, which was designed to protect and promote food safety. As required by this legislation, on July 1, 2003, the Food Safety Commission (FSC) was established under the Cabinet Office to promote food safety by gathering information, assessing and communicating food risks, and developing crisis management procedures, in cooperation with the MHLW and the MAFF. The commission was charged with conducting scientific assessments of food risks, assigning risk management to the MHLW and the MAFF, and facilitating risk communications. BSE was an important agenda item from the outset. On December 1, 2003, a tracking system for cattle – from production to the slaughterhouse – was introduced, and on the same day in 2004 a new system for beef tracking was introduced. The purpose of the cattle tracking system was to prevent the BSE epidemic, while that of the beef tracking system was to maintain consumer confidence in food safety.

The incidence of BSE overseas continued to trigger reaction in Japan in the form of official statements, regulations, and import bans. The first BSE cases in Canada and the US were reported on May 21, 2003, and December 24, 2003, respectively. On the same days, the importation of beef and beef products from these countries was tentatively banned. The MHLW, The MAFF, and The FSC immediately dispatched a team of experts to learn more about the situation.

The top exporter of beef to Japan had been the US. A series of Japanese-US consultations on BSE began on December 29, 2003, when officials from the MAFF, the MHLW, the Cabinet Office, and the Ministry of Foreign Affairs (MFA) sat down with their counterparts from the USDA. The Japan–US BSE Working Group, composed of experts and working-level officials from both countries, exchanged opinions and conducted inspections of relevant facilities. In its final report, published on July 22, 2004, it stated that, "the United States will establish a marketing program that enables a resumption of some trade for an interim time period. The operational details of the Beef Export Verification (BEV) Program managed by the USDA Agricultural Marketing Service will be further worked out by Japanese and US experts." A number of actions, most notably SRM removal, cattle tracking (traceability), and age verification of cattle, were also agreed upon (Japan–United States BSE Working Group 2004).

The US BEV Program was to be reviewed and certified according to OIE and WHO standards. The section on the Prevention of Trade Disruption stated that both Japan and the US have (already) had sufficiently robust food safety systems in place so that the identification of a few additional BSE cases would not result in market closures and the disruption of beef trading patterns in the absence of scientific justification. The working group's memos and report stated that Japan and the US had come to a mutual understanding of the scientific and technical aspects of the BSE problem, as well as the similarities and differences in the countermeasures implemented by both countries.

The Soothing of Panic and a Path to a More Rational Approach: The Resurgence of Scientific Arguments

In February 2004, the FSC started to assess Japan's BSE-related policies and programs. The Prion Expert Panel established by the FSC discussed a report prepared by an international team of experts that had been dispatched to the US. The panel concluded that BSE surveillance should be strengthened, all at-risk cattle older than 30 months should be tested, and SRM should be completely eliminated (Food Safety Commission 2004b). Although the panel recognized that testing all cattle older than a particular age is a meaningful BSE measure, they considered its cost-effectiveness to be minimal. Finally, it concluded that the BSE risk was not completely clear and that further risk communications efforts were important to make the most of the limited resources.

Six months later, in August, the panel published an interim report on Japan's BSE risk management (The Prion Expert Panel of the FSC 2004). At that time, 11 BSE cases had been reported in Japan, and 1.9 million cases had been reported worldwide in 23 countries (1.84 million in the UK; 1,400 in Ireland; 905 in France; 894 in Portugal, etc.). In addition, 157 vCJD cases had been reported outside Japan. The panel concluded that the obex sample test detects BSE only in the later stages of the latency period. They also stated that it is not clear whether organs other than the SRM have no infectivity and that the exposure threshold limit of prion intake to cause/not to cause BSE remained unknown. Although the panel determined that the BSE tests at slaughterhouses and SRM removal had reduced the risk of BSE to humans, it argued that the examination of at-risk cattle, tracking systems, and feed regulations were necessary if BSE is to be abolished. The panel reconfirmed that the tests cannot completely eliminate BSE-infected cattle, since they have detective limits and there was no way of knowing the risk of infection in the latent stage of those infected cattle. In conclusion, the group stated that as long as SRM removal remained compulsory, the absence of active surveillance of cattle under the threshold age for detection of BSE did not increase the risk of vCJD. Though it was said that the available BSE tests could not detect BSE among cattle not older than 20 months, the exact (and optimal) threshold age remained unknown and needed to be determined.

The year 2004 witnessed another epidemic, that of the highly pathogenic avian flu. One farm failed to report the incidence of massive chicken deaths, fearing possible economic losses. This incident seemed to add more fuel to the fire regarding the communication of potential risks. Programs to inform the public about the BSE problem and related policies continued to be implemented. The FSC, along with the MAFF and the MHLW, occasionally held meetings on the topic "Risk Communication on Food." These meetings were open to – and sometimes targeted at – the public. As public hearings and education sessions, they were intended to promote the public's understanding of the BSE problem, and at the same time, to assess public perceptions and opinions concerning the issue and the past and present actions of the government to deal with it. The meetings included lectures by invited experts, both domestic and foreign, question-and-answer periods, and concluding remarks by an FSC representative. Opinions on the FSC interim reort were also solicited, both explicitly and implicitly.

Before the FSC report was finalized on October 15, 2004, ministers from the MHLW and the MAFF again consulted with the FSC on the compliance of industry practices with the current regulatory standards as well as the rationale for testing cattle regardless of age. The point in question was whether the scope of the BSE testing could be safely reduced from all cattle to cattle over 20 months old. Some members wondered why they had been repeatedly consulted, starting almost immediately following the presentation of their interim report. They assumed that this consultation was requested because of the trade conflict with the US. The MHLW responded that the intent was simply to ensure scientific rationality. Later, the reasons behind the consultations were elaborated as follows: The BSE mitigation measures introduced in 2001 were emergency measures taken in response to the fact that exact ages of all cattle could not be known and that the public was seriously concerned about beef safety; therefore, the assessment of those emergency measures was well warranted.

Risk-communication programs, especially in the form of public hearings, were continually being implemented at 50 locations nationwide. The voices at these meetings revealed that opinions were divided (Food Safety Commission 2004c); some supported the continuation of the testing of all cattle, whereas others supported limiting the testing to cattle older than 20 months. More people came to question the effectiveness of and justification for the testing of all cattle. Some even argued that such tests were confusing and that the government should frankly admit that, as demonstrated in other countries, the risk of vCJD can be, and already had been, lowered to a reasonable degree by the feed regulations and the removal of SRM. Some supported testing of all cattle based on the risk of insufficient compliance with or effectiveness of the feed regulations. Others expressed concern about the safety measures taken in the US, as compared to the measures taken in Japan. Summary reports of these opinions and concerns were presented to the FSC's Prion Expert Panel and disclosed shortly thereafter on the FSC's official websites.

The FSC panel drafted its report in December 2004 (Food Safety Commission 2004a). It presented a quantitative documentation and summary of the effectiveness of past policy measures and estimated the future incidence of BSE. It did not mention vCJD, possibly because of the small rate of its incidence. Although the FSA report called for further scientific research on BSE and tests for its detection, it stressed that the enforcement of the designated measures was critical to the effectiveness of the policy. It simultaneously noted that abnormal prion protein was not detectable when its level reached the detection threshold – regardless of the age of the cattle. In response, the MHLW and the MAFF promised to increase their efforts to enforce BSE testing, facilitate the complete removal of SRM, increase the effectiveness of the feed regulations, and promote research on BSE. International difference in BSE-related policy measures was not directly addressed then (Matibag et al. 2005) (Table 7.1).

One item on the trade agenda, the USDA Export Verification (EV) program, started to specify product requirements for beef exported to Japan. Beef or beef offal eligible for export to Japan must, according to the Food Safety and Inspection Service (FSIS) website, be processed using the same standard operating procedures that the facility employs for the Hazard Analysis and Critical Control Point (HACCP) as regards the cow's head, spinal cord, and distal column of the spine (Ministry of Foreign Affairs 2004). In addition, the evaluation of all carcasses must conform to proper age verification and the carcasses cannot be marketed until they are tested and approved by proficiency-tested USDA evaluators.

The FSC report, "BSE-related Health Effects of Food," was finalized on May 6, 2005 (Food Safety Commission 2005a). It reviewed the forecasts of future BSE incidence as well as the importance of policy enforcement. It further pointed out that no estimates can be final; even though the data available for estimating BSE risks in Japan had been accumulated throughout the testing of all live cattle, it did not include the dead cattle, some of which could have been BSE-positive. It was only after April 2004 that the testing of all cattle, including dead cattle, became compulsory. A decisive answer to the question consulted about was that switching from the testing of all cattle to the testing of cattle older than 20 months did not significantly increase the risk of BSE in humans, if it had any effect at all. However, at the end of the report, the arguments for the continuation of testing all cattle were noted, based on the reasons that a change in the testing program should occur only after additional risk-mitigation measures were taken, and that the testing of all cattle is important for the future assessment of more sensitive detection methods.

On May 24, 2005, the MHLW and the MAFF consulted with the FSC on the equivalence of the BSE-related risk from beef and cattle organs on the domestic market and the risk from the corresponding products imported from the US and Canada and regulated in these countries. The FSC first presented itself as a risk-assessment organ, clearly separated from the kind of risk-management organizations

Table 7.1 Difference in BSE testing programs in 2004

Test targets	Japan	EU	US
Healthy cattle	All	All cattle older than 30 months	Samples
Abnormal cattle	All	All cattle older than 24 months	Samples older than 30 months
Dead cattle	All cattle older than 24 months	All cattle older than 24 months	Samples older than 30 months

exemplified by the MHLW and the MAFF. It then asked why these ministries consulted on the topic before starting substantial discussions. The official reply from the MHLW and the MAFF stated that both ministries were reconsidering a change of policy as a result of the FSC report. Furthermore, they noted that the import ban on US and Canadian beef products was a provisional action, and it was necessary to examine its appropriateness in light of the latest scientific knowledge.

On July 1, 2005, without waiting for the FSC verdict, the MHLW and the MAFF changed the enforcement regulations for the BSE Mitigation Provision Act. One month later, on August 1, BSE testing of all cattle was replaced by testing limited to cattle older than 20 months. It was also decided that financial supports by the MHLW for the testing of cattle below this age limit would be discontinued within 3 years. However, some prefectural governments, especially those producing cattle, declared their intention to continue testing all cattle. On October 23, 2005, Japanese and US officials revealed that they had come to an agreement that bilateral trade in beef and beef products should be resumed, under specified conditions and modalities, following approval by the respective countries of their own processes and decisions, so long as these were based on sound scientific judgments. The two countries issued a joint press release stating the above agreement.

The consultation by the Japanese and US experts with each other, as well as with experts from the OIE and WHO, continued to help both sides gain a further understanding of the pathogenesis and patterns of BSE disease. The USDA Agricultural Marketing Service (AMS) further promised to conduct a special study in which steers and heifers of known ages (within a month) would be slaughtered and evaluated for physiological maturity. The purpose of the study was to determine an expected endpoint of maturation to ensure the exclusion of steers or heifers older than 20 months from the certification program for export to Japan (USDA, APHIS, *Newsroom*).

Six months later, in November 2005, the draft report entitled "Risk assessment concerning the comparability between risks of consuming beef and internal organs regulated by the beef export verification program of the United States/Canada and risks of consuming beef and internal organs of Japanese cattle" was prepared by the FSC's panel of experts on prion and shortly thereafter released to the public through the internet. Its final conclusion was that the difference in risk between domestic beef and beef from the US and Canada is minimal, so long as the regulations, which were drawn up specifically for export to Japan, were enforced as promised. After a short period of time for public comments, the final verdict allowing the importation of US and Canadian beef products was submitted on December 8 (Food Safety Commission 2005b). The FSC at this time also stressed the need for effective communication with the public.

Four days later, on December 12, the Japanese government agreed to conditions for trade with its US and Canadian counterparts and decided to lift the import ban. Only beef from cattle younger than 20 months could be exported, provided that the rigorous standards for production and processing were followed. Immediately after the decision to resume trade, MHLW officials were dispatched to the US and Canada to judge compliance with the designated standards. The MHLW simultaneously decided that until the end of March 2006 it would conduct full rather than random inspections of the beef products from these countries. Public meetings were held in nine locations nationwide to explain the government plans and actions concerning the importation of beef. Upon their return to Japan on December 26, the inspection teams reported that the factories were complying well with the designated standards.

Persistence of the BSE Issue as a Source of Trade Conflict: Pressure Toward International Harmonization

On January 20, 2006, a calf vertebra – prohibited by trade standards – was found in an airborne package from the US during a quarantine inspection. The beef products shipped with it were incinerated, the products already imported were checked by local governments and trading companies, and the approval of imports was immediately stopped until the US government could provide a report on the incident. Two days later, on January 22, a second inspection mission was dispatched again to the ten US factories. In the meantime, officials from Japan's MFA, MAFF, and MHLW initiated a series of meetings with officials from the USDA.

The USDA report came on February 17, and a Japanese translation was submitted to the Japanese public on March 3. On March 1, in response to the Japanese investigative report on exports, the USDA's Food Safety and Inspection Service (FSIS) issued directives and notices listing measures to ensure that exporters comply with the requirements of the Export Verification (EV) Program (Embassy of the United States, Japan 2006). On March 6, Japanese officials sent a detailed set of queries to the USDA concerning specific points raised by the MHLW and the MAFF. These latter agencies demanded a further exploration and explanation of the causes for the failure to comply with the trade regulations.

In May 2006, the USDA AMS audited the 35 beef-processing facilities to ensure that Japan's country-specific requirements were met and to ensure factory compliance with the EV Program. One month later, officials from the MHLW and the MAFF visited those same facilities to (re-)confirm the compliance. They also visited five farms, two feed factories, and one factory for rendering works. Two additional facilities were inspected by a Japanese team in August. Likewise, from November 26 to December 13, 2006, the Japanese inspection team revisited the US factories participating in the Beef Export Program for exports to Japan. With the exception of a few minor violations, their compliance with the export standards put into effect following the resumption of trade on July 27 was confirmed, although a more rigorous implementation of the export rules was requested. Consequently, the trade ban on exports from the US was again lifted on December 26.

Reactions to the official resumption of beef trade were not always the same among restaurant chains and food retailers. Some welcomed it, but others denounced both the US and Japanese governments, claiming that they were placing politics and economy over health and safety. A restaurant chain even declared that it would not use US beef until the US government introduce complete cattle testing and ban MBM use (*Hochi Shimbun*, July 27, 2006).

In 2007, the communications programs for food risks began to be implemented continuously in Japan. The first annual public meeting was scheduled for January 15, 2007, to explain domestic BSE-related policies and trade policies involving beef products, especially those coming from the US. The consumption of US beef remained low – about one tenth of what was before the import ban 6 months earlier. However, Japanese retailers hesitated to stock up, because they were worried that concerned consumers would be reluctant to buy. The US cattle and beef industries repeatedly complained about Japan's beef-safety regulations, particularly the 20-month age limit, which was stricter than the OIE limit of 30 months. They also complained that the strict monitoring of cattle's ages was too costly. On January 19, US Trade Representative Schwab and Agriculture Secretary Johanns formally requested that Japanese Agriculture Minister Matsuoka relax the import regulations. Matsuoka replied that such a relaxation would depend on the public understanding and acceptance of BSE-related risks in Japan. The MAFF and the MHLW again conducted inspections of beef-processing plants in the US.

On February 16, a package of beef imported from Lexington, Nebraska, contained plates not listed in the EV. The factory admitted that the non-verified plates were from cattle older than 20 months, violating the limit of 20 months agreed to by the US and Japan. Exports of products from that factory were immediately suspended. In tandem with this incident – and during a series of diplomatic talks between the governments – beef exports from the US continued to be a hot item on the political agenda and repeatedly came to the fore in discussions. In March, the OIE tentatively classified the GBR of the US beef as "managed risk," the second safest level; the classification was made formal in May (Embassy of the United States, Japan 2007). It allowed the US to export its beef, regardless of the age of the cattle. In response, the USDA began requesting that the Japanese government abolish the age limit on beef entirely, which the Japanese were not inclined to do, insisting that a more thorough inspection of the beef processing factories was necessary.

On June 13, the MAFF and the MHLW submitted their inspection report to the Liberal Democratic Party (LDP), the majority party in Japan. The report stated that all the conditions had been met at all the beef processing plants in the US. Thus, the loosening of the beef import regulations was back on the agenda and remains under discussion to this day. Even since then, violation of the EV Program by some US beef exporters were occasionally reported, attracting media attention. Supermarket chains and restaurants then recalled the beef shipped from the same exporters (*Yomiuri Shimbun*, April 15, 2008).

On August 31, 2007, the MHLW issued its notice that requested prefectural governments not to provide financial support for the testing of cattle older than 20 months. The reason for this was explained: consumers could get confused about safety and the beef market would be disoriented if some beef products are labeled with "all cattle tested" and others not. However, nine prefectural governments then declated the continued support for the testing of all cattle (*Asahi shimbun*, September 11, 2007), saying that consumers demand such testing. The Japan Consumers' Federation manifested its objection to the MHLW's policy, requesting the continuation of all-cattle testing. An opinion poll, conducted in February–March

2008, disclosed 58% of consumers still considered the testing of all cattle desirable (*Mainichi Shimbun*, July 18, 2008).

On July 18, 2008, the Fair Trade Commission announced that it would be illegal (violating the truth-in-advertising laws) if producers are to market their beef products coming from cattle below 21 months old with a label "tested to be safe." On July 31, the MHLW discontinued its financial support for the testing of cattle below 21 months. However, all the prefectural and municipal governments continue the testing of all cattle with their own money. They justified this for the reason again that consumers demand the testing of all cattle, and for the possible loss of market competitiveness of local beef brands unless the testing is changed uniformly (*Yomiuri Shimbun*, July 11, 2008; *Mainichi Shimbun*, July 28, 2008).

Discussion

Although different countries responded differently to the BSE/vCJD incidents, they all sought to increase food safety as well as reassure the public. An examination of the past policies of these countries reveals that none of these policies were adopted without criticism. At first glance, the important question appears to be whether the policy decisions were sound in light of what was known at the time, and whether those concerned were sufficiently frank about the risks to the public. For example, in the UK it was suggested that the crisis stemmed from serious policy failures, and that government secrecy, cover-ups, and mendacity were the main causes of the whole wretched saga (Packer 2006, p. 1). This indicates that the problem arose from technical ineffectiveness, policies that could not be administered efficiently, and a failure to adequately communicate the risks of BSE to the public.

In France and Japan, people loudly voiced their concern about the safety of their domestic beef products even before BSE cases were reported in their own lands. At the same time, precautionary actions were taken, such as cattle surveillance and bans on beef import from certain countries with BSE. Despite such measures, when BSE incidents did occur in these countries, people feared possible infection and consequently decreased their beef consumption. In Japan, public opinion seemed divided; on the one hand, some people considered that the government's actions should be protective so that science warranted, whereas others, on the other hand, thought that the government should adopt every possible measure to decrease the possibility of infection. As a result, scientifically unreliable BSE testing continued to be conducted and scientifically unjustified regulations continued to be enforced. Once such measures were introduced, the government found it hard to repeal them.

In more general terms, the real issue is the discordance between the government's risk management and its fulfillment of the public's expectations about this management. Undoubtedly, risk communications with the public were intended to change their perceptions, attitudes, and behavior regarding food risks, and at the same time, to adjust government policies in accordance with them. The public's expectations about the policies were a key ingredient in how it judged the success or failure of those policies. The performance of the government, the public's expectations of this performance, and how the government presented itself in the public arena, were all possible targets in the assessment of the risk management. The history of BSE-related events provides a valuable opportunity to examine the factors that influence policy making and the fates of the resulting policies.

Science and Policy Making

The generation and utilization of scientific knowledge are not free from their social and political contexts. Rather, science and politics are intertwined. Because of the possible manipulation of research (its agendas, funding, and even its findings), the promotion of science, and the secrecy of its components, science is much less a matter of autonomous rational discourse and much more a matter of politics (Miller 1999). As evident in the BSE case, a centralized system in which government agencies control science for the government's benefit is inherently vulnerable to alliances of scientific experts and interest groups. Such alliances could undermine the credible assessment of public health and safety risks (Beck et al. 2005).

An important issue concerning science and science-based policy making is the presence of ambiguity in scientific data and how that ambiguity is interpreted – in terms of both the uncertainty and pluralism. The BSE crisis was partly the result of a failure to acknowledge the mutable and controversial nature of the disease. Several scientists have suggested that during the BSE crisis a scientific pluralism existed, at least in the medium term (Butler 2001; Krebs et al. 2002); yet scientific knowledge claims are often accompanied by uncertainty (Hinchliffe 2002), and it is this uncertainty, when appropriately presented, that allows for more discretion in policy choices and provides a space in which political leadership can be exercised, in addition to protecting the legitimacy of science itself.

The horizons of policy making – the wide range of policy alternatives and the ways of deciding which ones to choose – are inevitably limited. As classic theses, the bounded rationality model of policy making and the concept of satisficing behavior stress the importance of limited human capacities and organizational environments in structuring bureaucratic attitudes and behavior (Simon 1957). Therefore, in addition to the limitations posed by scientific uncertainty, policy making must always be pragmatic and adaptive. Although scientists can wait as long as necessary for definitive answers to a problem, policy makers must make timely decisions, even when the information is incomplete (Stallones 1982). Decisions cannot be delayed; they must be made in light of the information available at the time and of the theories that best account for the known facts. The resulting decisions may not always be optimal from an historical vantage standpoint. Such a situation is inevitable, and history abounds with examples of catastrophes caused, in effect, by ignorance. Scientists have two main preoccupations: the desire to wait until all the angles are thoroughly checked and the desire to take their proper share of the credit. In both the public and private domains, those who are eager to criticize any delay in the announcement of scientific findings will not stand up if things go wrong and explain that the odd error is a price worth paying for complete transparency.

In addition, value judgments play a substantial role in the assessment of scientific information and the planning of policies, specifically with regard to the interpretation of uncertainty and that of possible outcomes. The limitations of scientific judgments and technological choices are not always made clear (Jensen & Sandoe 2002). Governments usually balance different objectives in managing risks. The same evidence can be used to inform or influence different policies. Decision making is different than ascertaining the hard facts of science (Weinberg 1972). Even when some committee of scientific experts is called in, they have to make their own "political" judgments about the policy goals and the acceptability of possible risks, or whether to endorse the judgments of government officials and/or ministers on such matters.

Specific societal conditions further encourage risky or opportunistic behavior in policymakers. Such behavior is conducive to delays and inaction until such time as the evidence of a health risk becomes overwhelming (Beck et al. 2005). Organizations and individuals, both private and public, construct risks in close relation to their places and visions, including their horizons of globalization (Tacke 2001). This might be always the case, as every organization is the mobilization of a bias (Pigg 2002; Schattschneider 1975). Policy makers' value orientations can also be biased; organizational processes are inevitably constrained by the routine filtering and reinterpretation of information and problems in ways that fit with the organization's values, beliefs, and world views. Consequently, certain issues and problems may be routinely distorted because of how organizations define their missions and responsibilities (Hawkins & Thomas 1989). Organizations that have an economic stake in the beef industry may act so as to avert the regulation of commercial and trade activities.

Apart from the particular organizational structures and their preferences, the concept of the risk-opportunistic policy maker can be found in the literature on short-term and opportunistic policy making. It indicates that so long as the issues are sufficiently complex, politicians are able to act in ways that opportunistically favor special interests without their behavior being detected (Lowi 1979). An inherent compatibility exists between the modern bureaucratic state and a science that is conditioned by the opportunistic behavior of both sides. This compatibility leads to the formation of different types of alliances (Lowi 1992). When opportunistic policy makers are able to pick from a number of possible scientific explanations of a phenomenon that seem to be equally credible, they will choose the explanation that best serves their own interests or those of their network or peer group (Beck et al. 2005).

All these factors can lead to institutional attenuation, i.e., the institutional processes that diminish inspectors' awareness of a risk and/or the policy importance they attach to the associated regulations (Rothstein 2003). Alternatively, when these socio-political factors maintain or increase the awareness and impact of risks, this phenomenon is called risk amplification (Kasperson et al. 1988; Krimsky & Golding 1992).

It has been sometimes suggested that in the UK, policy changes made by a conservative government committed to deregulation led to the spreading of BSE.

The UK MAFF from the outset equated the public interest with that of the meat industry and always put the financial interests of the dominant firms ahead of public health. The health risks to plants, animals, and humans, as well as the financial risks to industry and the market, were to be considered in policy making (Caswell 2006). The government then committed itself to reassuring the public, insisting that beef was safe to eat. In risk assessment and policy making, the economic well-being of the food industry generally takes precedence over human safety and health. Whenever the stability of the beef market or government expenditures was threatened, the preferred policy was to gamble that the risks to public health would turn out to be acceptably low.

Different agencies had different value orientations and policy primacy. The initial problem definition in expert and policy discourse as an animal health problem played an important role in shaping official responses. The lack of coordination between and within institutions and the misapplication of experts' advice raised serious questions about the government's ability to manage serious BSE crises (Gerodimos 2004). The problem was left mainly to the MAFF to solve. The delay in informing the UK Department of Health (DoH) about BSE resulted from the lack of concrete information about BSE, as well as the perception that BSE was primarily an animal health problem. The DoH was largely excluded from decision making, and the MAFF took the lead in addressing BSE (Miller 1999). Thus, some of the difficulties in tackling BSE were attributed to confusion – and sometimes conflict – in the allocation of administrative responsibilities for food production and safety among government departments, as well as between their local and national tiers (Winter 1996).

Consequently, in the early years, government actions in the UK were based on the assumption that BSE was unlikely to have serious implications for human health, although due to the uncertainty the government also considered some precautionary measures. The optimistic scenario that BSE would not cross over from cattle to man, which had been expressed by the Southwood Committee in 1989, was adopted by policy makers and introduced on many subsequent occasions (Zwanenberg & Millstone 2003, p. 32). Thus, the government did not dare to introduce rigorous measures. The threat to the livestock industry, the desire to reassure the general public about the safety of British beef, and an unwillingness to increase public expenditures were also key factors influencing policy formation.

In Japan, the situation seemed to be more or less the same as in the UK, in that the MAFF took the lead in handling the BSE issue before 2001. However, prior to 2000, the experts – especially those in other countries – had suggested the possibility that Japanese cattle had been infected by the contaminated cattle feed imported from the UK. Yet, the response of the Japanese government to a ban on the use of MBM as feed (MAFF 1996), the designation of BSE as a notifiable disease (MAFF 1997), the diagnostic criteria for vCJD (MHW 1997), and the disease registry for vCJD and other prion disorders (MHW 1999) were all fairly late in coming. The fact that these measures were introduced in the UK and other European countries before they were introduced in Japan indicates that Japan benefited from the experience and knowledge accumulated elsewhere. It might also suggest that the Japanese government lacked sufficient scientific information on BSE, perceiving only the "potential" threat of the disease, and therefore had no strong incentive to act more proactively. Most of the efforts to address the BSE problem, including the funding of research, did not start until the connection of BSE to vCJD was confirmed.

Food Safety and Public Reassurance

The judgments about risks, including their acceptability and horribleness, are not usually based on probabilities, but rather substantially influenced by their properties which are called "outrage factors" (Sandman 1993). This is especially the case when it comes to food safety (Fischler 2000). As a result, the behavior of the general public has not always been rational from the technical standpoint. For example, although consumers were concerned with food safety in the abstract, they rarely justified their food choices in terms of how safe they considered foods to be and how well they thought the government was monitoring those risks. Instead, they adopted rules of thumb based on the dichotomy of "good" versus "bad" as well as the trust in food categories and that in food retailers (Green et al. 2003; Ilbery & Kneafsey 2000).

In many countries, the incidence of BSE – whether domestic or foreign – eroded consumers' confidence in food safety as well as in the regulatory agencies. As a result, beef consumption dropped significantly (Harvey et al. 2001; Latouche et al. 2000; Pennings et al. 2002).

The UK policy history provides insights on these topics. For several years, the controls were made progressively more stringent. The animal SBO ban was introduced in 1990, after initially being declared unnecessary. New controls for calves were introduced in 1994. The production of MBM from cattle vertebrae had been permitted for many years before it was banned. It gradually became clear that the measures taken between 1988 and 1990 did not completely eliminate the source of infection from animal feed in the UK. This revealed the inadequacies of the current regulations, in terms of both their content and their enforcement.

Drawing on the reports of advisory committees, UK government officials during the mid-1990s continued to emphasize the safety of British beef. Although scientists outside government circles expressed concern over the spread of the infection, the officials maintained their position that beef was absolutely safe, and they were unwilling to discuss the possibility of a threat to humans. With no contingency plans presented by the government, the public got the message that BSE was not and had never been a danger (Packer 2006, p. 124). The original conclusion of the Southwood panel that the transfer of BSE to humans was highly unlikely obviously needed to be qualified significantly when it was shown that BSE could be transmitted to other species – most notably cats. Throughout the ordeal, the official pronouncements on BSE and the safety of beef were focused too heavily on the possible reactions of the public.

In response to the increase of public uneasiness in the early 1990s, and particularly in light of the increasing evidence that the controls in place to eradicate BSE among cattle had not been properly implemented or enforced, there was a series of incremental adaptations and extensions to the existing legislation in the UK. Finally, as the scientific evidence for the human risks and potential disaster resulting from BSE became increasingly compelling, the government was forced to reverse its stance – they now maintained that BSE could be transmitted to humans, so beef was not safe to eat after all. By 1996, given the increasing number of infected cattle and the 10 reported cases of vCJD, the government's previous position had become impossible to sustain.

As a first step, the UK government used a new report by the advisory committee to justify the announcement that cattle more than 30 months old had to be deboned, and that MBM could not be used in animal feed. As this announcement failed to quell public concerns, the government's next step was to proclaim a complete ban on cattle more than 30 months old (Hornsby 1998). This campaign of reassurance eventually proved to be a mistake, as the public felt betrayed (Phillips et al. 2000, Vol. I, xviii–xxi). As it became more and more evident that BSE had important implications for human health, the policy prescriptions were adapted accordingly.

Retrospectively, over time, as more became known about BSE, better decisions could have been made. At each stage, the UK government tried to legitimize its decisions by appealing to scientific reports and recommendations. However, this approach had an inherent flaw: When science could not produce the necessary level of certainty for decisive actions of any kind, a crisis ensued. At every stage, a government's actions must be thoroughly justified so that they are not to harm the legitimacy of policy makers, but this is a difficult goal to achieve. Any action can incur criticism, especially when it leads to failure. Having articulated its conclusion that the science was robust and the risks were either negligible or non-existent, the UK MAFF found it extremely difficult in the early 1990s to accept and respond to the new evidence that implied that its assumptions about the risks, and the policies that followed from those assumptions, might need to be revised. Accordingly, it became progressively less reassuring (this will be discussed further in the following chapter).

The Japanese government received a lucky break when some official measures barely preceded the first "report" of the domestic BSE crisis. The enactment of the MBM ban (MAFF, December 2000) and the ban on the importation of beef products from the EU (MHLW, February 2001) preceded the first reported case of BSE in August 2001 by many months. At the time, it seemed somewhat reassuring to stress the safety of domestic beef products as compared to the imported alternatives. This reassurance also satisfied the economic interests of the domestic beef producers by maintaining, and even encouraging, the sale of their products in the market. Furthermore, the cattle surveillance program (MAFF, March 2001) was introduced 5 months before the first reported case. This sequence of events certainly helped the government maintain its legitimacy.

On the other hand, the government's actions after the first BSE incident appeared to be very quick and thorough in Japan. The MAFF and the MHLW started working closely together, and in June 2000 the political leadership enacted a special law for the mitigation of the BSE threat; a year later, the Basic Food Safety Law was passed, and in July 2003 the FSC was established. Instead of trivializing the risks of BSE to humans, the officials plainly stated – and even stressed – the possibility of BSE/vCJD risks before taking concrete steps to soothe the concerns of the public – even if these measures were not scientifically justified and necessary. In other words, they chose to err on the side of caution. The government then endeavored to persuade the public that it had taken every possible measure, and by doing so had successfully protected its own position. In 2004, the FSC initiated a critical review of previous BSE-related policies.

Those activities hastened in the latter phase were induced by the occurrence of domestic BSE cases, at which point the public discovered that both imported and domestically produced beef products might be unsafe. More than ever, the politicians, bureaucrats, beef industry experts, and retailers became concerned about a possible decrease in beef consumption, the maintenance of the cattle-to-beef chain, and the economic distress imposed on the domestic cattle and beef industries. In public, the government agencies, such as the Cabinet Office, the FSC, the MHLW, and the MAFF, meticulously pushed food safety to the foreground and proclaimed it to be the most important, if not the sole, objective of government policy, downplaying the concurrent agenda of protecting the cattle industry and beef producers. Beef consumption dropped briefly but rebounded in a little over a year, and some consumers continued to believe that domestic beef was safer than imported beef.

The Instrumental Use of Science and Safety Testing

In 1996, faced with public panic following the announcement of a link between BSE and vCJD that collapsed confidence in the British beef industry and the export ban introduced by the EU, policy makers in the UK were forced to go beyond what the scientific community deemed "proportionate action." For example, they introduced a measure to prohibit of the use of certain SBOs from cattle more than 6 months old, even though the concerns previously raised by the Southwood Committee applied only to the use of SBOs in baby food. They also banned the use of all bovine vertebrae in mechanically recovered meat, and they extended the bovine offal ban to include the thymus and the intestines of calves under 6 months of age. Target slaughtering was introduced, which ended up preventing relatively few cases of BSE. Policy makers justified these steps as necessary to restore public confidence in the beef industry, which had been severely affected by the media response to the announcement (Greer 1999). Public confidence in the government's handling of BSE substantially eroded when it became known that BSE could be transmitted across species and to humans. In justifying its decisions to the public, the government represented them as science-based and consumer-oriented, even when they were in fact based on political and economic considerations (Zwanenberg & Millstone 2003). Therefore, scientific endorsement was sought for such decisions.

On the European continent, several policies were adopted to reassure the public, but they lacked rigorous scientific backing. In 1996, when the risk of BSE transmission to humans was confirmed, the European Commission adopted emergency measures to protect against BSE and CJD by imposing a ban on exports of cattle and beef products from the UK to the other EU members or third countries (McNelis 2001). Concerned about a dramatic loss of confidence in beef products, they sought to sustain the beef chain and protect the industry. The veterinary authorities in these countries were well aware that this would achieve very little in terms of protecting the public, but that was not why the governments of the EU member states banned imports from the UK – the intent was to reassure the public, even if there was no rigorous assessment of the ban's scientific grounds and effectiveness (Packer 2006, p. 164). The cost of reassuring the EU public in their own markets was borne by the exporters.

Reversing its non-compliance policy, the British government later adopted safety measures corresponding to those of other countries. In 1996, it ruled that only cattle under 30 months old could be eligible for human consumption. Among scientists, this age limit is still disputable. In 2003, the UK FSA/SEAC stated that the BSE risk to public health is so low that the over-30-months (OTM) rule should be replaced by rapid testing, the policy adopted by the rest of the EU. Based on the assumption that there would be 5,000 vCJD deaths in total, they concluded that replacing the OTM rule with testing would lead to a 0.04% increase in vCJD deaths over the subsequent 60 years. The FSA board reported to the ministers that this policy change would not significantly increase the risk of vCJD. The policy was in fact revised: quick tests would be performed on healthy animals 30 to 42 months old and animals that are casualties during the age of 24 to 30 months, but not on any animals younger than 24 months and not on healthy animals between the ages of 24 and 30 months.

In Japan, after the domestic BSE cases appeared, the government adopted a set of health-protection measures without delay. Some of these lacked rigorous scientific justifications – the most obvious example being the introduction of BSE tests on cattle regardless of age. The scientific evidence indicates that the BSE tests currently available can detect BSE (abnormal prions) only in cattle older than 24 months. Although a few cases have been reported in which such tests revealed abnormal prions in younger cattle, the reliability, sensitivity, and specificity of such tests is considered to be quite low in such cases; thus, they are considered to be ineffective, inefficient, and impractical when applied to cattle younger than 24 months. Although there is no justification for expecting serendipitous detections, the testing of all cattle, regardless of age, was presented to the public as justified for purposes of reassurance.

Discussions of testing, including its overall usefulness and scientifically established limitations, took place at meetings of the expert panels. These meetings provided the public with symbolic reassurance, and the records of the discussions were posted on internet websites. However, when the policy of testing all cattle was introduced, neither safety nor reassurance was presented as its rationale. The policy was just stated, and its interpretation was left to the public. The government's ambivalent attitude toward the policy helped maintain its reassuring effect, but it found itself in a difficult position when it became clear that the policy had a weak scientific basis. The adoption and subsequent termination of such measures could no longer be justified by rigorous scientific arguments.

Communications and Public Participation

The communication of risk is an interactive process involving exchanges of information and opinions among individuals, groups, and institutions. There are multiple messages, some describing the nature of risk and others expressing concerns, opinions, or reactions to risk messages or to the legal and institutional arrangements for risk management (National Research Council 1989). Failure of communication can result in a state of civic dislocation, which is a mismatch between what government institutions are expected to do for the public and what they in fact accomplish (Jasanoff 1997). In this disoriented state, trust in the government vanishes and people look to other institutions. The legitimacy of public institutions and the body politic depends substantially on meeting the public's need for credible reassurance.

The perception gap between the British authorities and the British public over the risk of BSE has left a legacy of frustration, distrust, and conflict (Schwing & Albers 1980). Since 1988, the British government, as well as some of its advisors, repeatedly stated that beef was safe, implying that transmission of BSE from cows to people was impossible. The reason for this statement was the fear of overreaction on the part of the public. Certainly, there was a tension between openness and the desire to avoid exacerbating the situation by provoking undue public alarm. Government officials considered that irresponsible or ill-informed publicity would likely be unhelpful, as it might lead to hysterical demands for immediate, draconian measures and/or lead other countries to reject UK exports (Greer 1999, p. 603). In contrast, the ethos of the DoH was that the public could react intelligently to sound information and that it was the duty of ministers to provide the electorate with objectively presented facts on which they could make an informed choice.

The policy of reassuring the public by stressing safety (and trivializing, or at least not highlighting, the possibility of risks) was counterproductive, because the risk communications did not then appear trustworthy (Jensen 2004). In the UK, the "bland assertions" about safety convinced non-MAFF experts and the public that the ministers were covering something up, especially as such actions occurred repeatedly, slowly, and usually after some damaging revelation. Once it became obvious that a risk to human health existed, the previous failure to communicate reduced public trust even more. In contrast, the private domain responded more effectively to the public fear of BSE/vCJD. For example, McDonalds and Burger King ignored the MAFF reports about the safety of

British beef and turned to long-used routines of risk management that had been successful in the business world. Their strategy was to reassure the public that they took its fear seriously.

The new conventional wisdom is that all risk must be communicated – that a policy of openness is the correct approach. If doubts are openly expressed and publicly explored, the public is capable of responding rationally, and it is more likely to accept reassurances and advice if and when they come (Phillips et al. 2000, Vol. I, para. 1301). However, the road to open, democratic, and consensual policy-making is not straightforward. First, risk-regulation policies based on public opinion have a very insecure foundation, even when the government is trying to be responsive to public opinion. Public preferences and attitudes may be polarized or fail to distinguish between means and ends (Sato et al. 2005; 2006). Public opinion can be volatile, and differences in knowledge and expertise between regulators, other players, and the lay public may mean that snap survey responses are unrealistic, simplistic, or tainted by misunderstandings (Hood et al. 2001).

Second, although some social scientists recognize the need to conceptualize risks in ways other than their probability of occurrence and their consequences, no clear and standardized procedure has emerged for gauging and incorporating the contextual (e-)valuation of risk (Kunkel et al. 1998). In participatory procedures, the various stakeholders become involved in the risk analysis early on, so they can evaluate the risks for themselves even before they receive a formal assessment from elsewhere. This process helps elicit the values and the perspectives of the community, so that the multiple dimensions of the risk can be taken into account early in the assessment (Amendola 2002). Such measures, however, are still experimental. Third, the mutual trust between policymakers, experts, and the public is not always well established; a study in Ireland indicates that most experts have little confidence in the public's understanding and assessment of food risks, their ability to deal with scientific information, and their food-safety practices (Boer et al. 2005).

In Japan, government officials, academics, and industry representatives all muddled through to reach the point of acknowledging the importance of communication in the management of risks and crises. The Basic Food Safety Law and the Food Hygiene Act state that communication efforts are indispensable to the achievement of the government's objectives. The interim and final reports of the discussions held by the panels of experts, as well as the government's position papers, were promptly disclosed. Furthermore, input from the public was sought through postings on official internet websites. From July 2003 to June 2006, the FSC, the MHLW, and the MAFF held 229 public hearing nationwide that focused on BSE. The FSC simultaneously formed a committee of experts on risk communications, which has met almost every month since September 2003.

In July 2004, a year after it was established, the FSC publicized its report entitled "Present status of and agenda for food safety and risk communications." The report first summarizes a set of communication problems present at the time of the FSC's inception (Food Safety Commission 2004d). They include insufficient communication among administrators (especially between the MAFF and the MHLW) and between administrators and scientists/experts; insufficient disclosure of information and incorrect announcements by administrators, as well as insufficient explanations on emerging (risk) events and ongoing policy discussions; a lack of precision and clarity in messages to the mass media, combined with insufficient knowledge about food safety on the mass media's part; and insufficient efforts to facilitate consumers' understanding of food safety (issues), as well as inattention to the effects of the messages the government does send out on the public's perception. The report then examines the actions taken by the FSC, other administrative organs, local governments, and private institutions to remedy the above problems. The FSC's second report, released in November 2006, further discusses the situations that needed to be improved and those under study, as well as the ongoing efforts of the FSC. These include the release, transmission, and mutual exchange of information, the efficiency of communication media and associated technologies, the monitoring of food safety, and the education of communication experts.

One question has emerged from all this: Did the advent of the precautionary principle affect the BSE problem or vice-versa? O'Riordan and Cameron (1994) describe the precautionary approach to environmental and science policy as follows: If unambiguous scientific proof of cause and effect is not available, it is necessary to act with care; if the benefits of early action are judged to be greater than the likely costs of delay, it is appropriate to be proactive and inform the public why such action is being taken; if there is a possibility of irreversible damage to natural life-support functions, precautionary action should be taken irrespective of the abandoned benefits. Moreover, policy makers should always listen for calls to change course, include those making such calls in deliberative forums, maintain transparency throughout the process, never shy away from publicity, and never try to suppress information, however unpalatable. Finally, if there is public uneasiness, such uneasiness should be responded to decisively by means of extensive discussion and deliberation.

In practice, however, the implementation of precaution in policy is a very complex exercise (Cross 1996). The application of the precautionary principle to the BSE problem has been considered, at least in theory (Jacob & Hellstrom 2000). The fact is, however, that even when precaution is applied in dealing with such risks, one has to justify it in terms of the existing information about the threat, as well as balance the benefits of such procedures against their costs. These requirements make it difficult to determine whether mere precaution is an effective policy tool. Communication, defined as the exchange of a wide range of information among policymakers, experts, and the public, could facilitate the application of the precautionary principle in policymaking, by making it clearer how the principle should be used practically.

It remains to be seen in the BSE case whether precaution has been the main driver of safety policy in any country. This question is closely related to how political and social considerations have been coupled with scientific information that always seem to be changing, accompanying some degree of uncertainty, and susceptible to multiple interpretations.

Trade Conflicts as an Opportunity for Policy Adjustments

International collaboration is undoubtedly required to decrease and eradicate BSE and vCJD, both nationally and internationally. Throughout history, many epidemics have traveled the trade routes (Kastner et al. 2005). Cattle, beef, and beef products are heavily traded internationally and, likewise, people move across borders. In such a highly integrated market, the problems in one country can have significant effects on the production systems and markets of other countries. A ban on economic activities, even for food safety purposes, can incur huge economic losses and limit the effectiveness and efficiency of independent, diversified national policies. In the case of BSE, however, there have been substantial policy differences across countries. The divergence in the types of BSE risk management (e.g., feed bans, surveillance, and quarantines) perhaps reflects differences in the domestic BSE risks and the public's attitudes towards them, as well as the economic costs of BSE. These costs reflect the occurrence of BSE, the public response to that occurrence, the impact that the preventative measures (and countermeasures) have on agriculture and other industries, and the political configurations that affect government action (Kellar & Lees 2003).

Problems arising from the failure to harmonize international policies become most evident when one looks at trade, especially in terms of its social, economic, and political impact. As happened in Europe in 2000, the markets for live cattle and beef products in Canada, Mexico, and the US, which had been closed due to The NAFTA, were severely disrupted in 2003 by the confirmation of single cases of BSE in Canada and the US. The US imports many live animals from Canada and Mexico (United States Department of Agriculture, Economic Research Service, USDA 2004b), and to a lesser extent trades in beef and veal products (the US and Canada rank second and third, respectively, in global beef exports). The BSE cases in Canada in May 2003, and in the US in December 2003, caused a total sealing of international borders against the importation of beef and cattle. The Canadian domestic market was too small to absorb the Canadian beef produced for export (Boame et al. 2004). The closure of its border initially had a negative spillover effect on its cattle trade with Mexico, because the US cut off imports as well until a full assessment of the risk was made. The prices of cattle and beef products drastically fell in Canada, while prices in the US surged for more than a year.

Arguably, this disruption could have largely been avoided if these countries had done a better job of coordinating their risk-management programs in accordance with international standards (Sparling & Caswell 2006). However, this was not so easy. Although the BSE cases in 2003 revealed major weaknesses in the animal tracking systems, the industry continually opposed stricter controls, as it was concerned about higher costs (Acord & Feldman 2004). From the trade perspective, public programs for tracking and relief (i.e., government investment in systems to prevent BSE) could be regarded as a form of subsidy, providing an unfair trade advantage to the subsidizing country (Sparling & Caswell 2006). Thus, agreement on the necessary conditions and the measures to attain them is vital.

Around 2000, about 65% of US beef exports went to non-NAFTA countries – primarily Japan and Korea. When the BSE cases in the US and Canada were reported, Japan and Korea responded with a ban on the import of all US beef products. The economic impact of these bans on the US cattle and beef industries was quite significant. The USDA repeatedly pressured the Japanese government to open its market to beef products and it launched a safety program that specifically targeted Japan. At the same time, it pressured Japan to change its import regulations, e.g., by placing limits on the age at which cattle could be traded. The trade talks certainly helped to stimulate discussion on the appropriateness of safety regulations in the Japanese market, especially the requirement of cattle testing regardless of ages or – more precisely – the scientific ineffectiveness and inefficiency of such testing.

The US government argued that the safety regulations for the Japanese market should comply with international standards. The GBR standards, for example, link BSE risk levels to OIE judgments of the degree to which a country can impose trade restrictions that are consistent with protecting the health of the animals and the public (Heim & Mumford 2005; Matthews 2003; World Organisation for Animal Health 2003). A trade conflict and its resolution generate valuable opportunities for the reexamination of domestic and international policies intended to promote safety and health. These opportunities include the reassessment of policy and policy goals and their rationales (e.g., scientific, economic, and social).

An extension to the trade agreements and attempts to harmonize the necessary safety measures are currently underway. In 2005, Canada, the US, and Mexico jointly publicized the Report of the North American Chief Veterinary Officers (CVOs) on the Harmonization of a BSE Strategy (North American Chief Veterinary Officers 2005). The WTO decided that when member countries change their importation policies in response to an emerging disease, they must file notifications of such changes through WTO's Sanitary and Phytosanitary Committee (Kimball et al. 2004; Hart & Babcock 2002; World Trade Organization 1994, 2000, 2003). In addition, the international harmonization of the risk assessment methods for food safety came to be realized, as general guidelines, starting – for example – as part of the necessary activities of Codex and the WTO (Hathaway 1997).

Lessons Learned

In the face of public criticism of their BSE-related policies, the past governments of several countries have reviewed the activities of their agencies and/or government-appointed task forces. Generally, the failure to adequately manage BSE risks has been attributed to an incomplete understanding of the nature of the risks and a lack of maturity in risk management as a discipline (Beck 2004).

The Phillips Report concluded that the BSE epidemic happened because contaminated animal protein was recycled as feed for ruminant animals (Phillips et al. 2000). The UK government then erred by attempting to reassure the public that there was no risk in eating beef when it had insufficient evidence for that claim. In addition, there were unacceptable delays in reaching decisions and putting policy into practice. Specifically, arrangements needed to be in place to facilitate a synchronized approach to the common problems of animal health and human health. Finally, the communication of the relevant government agencies with the public was poor (Goldwater 2001). In Japan, although no clear policy fault was acknowledged in the appraisal report, it was pointed out that divided – and sometimes overlapping – jurisdiction over the BSE issue within the administration created problems for managing the problem. This is a classic example of administrative dispersion and diffusion, in which the responsibility for addressing different aspects of a problem is scattered across a range of government departments, agencies, and advisory bodies.

With the advent of the BSE problem and several other food safety issues (e.g., foot-and-mouth-disease), there was a serious loss of public confidence in the safety of foods and the policy-making institutions that were supposed to deal with the problem. Consumers reportedly were looking for greater transparency or traceability and were even willing to pay for it (Latouche et al. 1998). In many countries, new agencies were created, dividing the responsibilities for regulating and sponsoring the agricultural and food industries, conducting risk appraisals, making open, democratic decisions, and drawing on experts representing a wide range of interests (Millstone & Zwanenberg 2002; Smith et al. 2004). In 1999, France created its *Agence Française de Sécurité Sanitaire des Aliments* (AFSSA, French Agency for Food Safety). In 2000, the UK set up its Food Standards Agency, which was charged with protecting the public from health risks that might arise from the consumption of certain foods and other related problems (Food Standards Act, 1999, Section 1(2)). In 2002, the EU established its European Food Safety Agency.

Triggered by the BSE issue, especially public criticism of the government's efforts to address it, the Japanese government made several institutional changes and introduced a new set of measures to keep food safe and reassure the public. These measures included the enactment of the Basic Food Safety Law, the establishment of the FSC (an independent body under the Cabinet Office specializing in food safety), the reorganization of the MAFF (clearly separating the food safety department from the agricultural development department within the ministry), and a series of "risk communication" efforts. The merits of some of these measures, especially the establishment of the FSC, have been debated since the late 1990s, but its establishment was certainly propelled by the 2002 report "On the Future Administration for Food Safety." Based on this report, it was decided that neither the MAFF nor the MHLW should ask the FSA to deal with issues beyond its competence – specifically, they should not be asked to make risk-management decisions.

Throughout the period of institutional and procedural reforms, a commitment to openness and public participation was stressed (Krebs 2001; Murphy-Lawless 2004). It has increasingly come to be expected that the public be immediately informed of just about everything. However, broadening of participation per se does not necessarily lead to more democratic policies or better outcomes, although it may have some limited value in improving public confidence in the regulatory regime. The potential benefits of the stakeholder process were mitigated by several factors, such as interpretative flexibility in representing consumer interests and the concept of precaution, restrictions on openness and the exclusion of key stakeholders, and the supranational context of regulation (Rothstein 2004). The impact of participation on policy robustness depends on the actual participants, the information available to them, and their ability to handle this information.

Risk opportunistic decision making could be constrained by requiring openness, transparency, and accountability in handling scientific information (Lobstein 2001). Thus, openness and accountability in government can be expected to result in policies that reflect the values of the general public and democratic principles. A potential but critical question would be who (whose values) represents the public or the citizen at large, as the public is in reality not a uniform subject in terms of interest, knowledge, preference, opinion, or mobilization (Neuman 1986). A similar issue is the integration of experts and the public, particularly as there is no value-free interpretation of, and therefore no value-free use of science in policy making (Majone 1989).

The last important question is whether the policy makers, the public, and the media really understand the concepts of risk and safety, the available policy alternatives, and the possibilities and limitations of managing them in regards to their "rationality." Special attention should be paid to the functions of the mass media, which do not simply reflect controversy or help shape its portrayal in the public sphere (Goodell 1987). Media reporting, public responses, and experts' opinions are the context in which policy making occurs, and they must be taken into account by all the participants in the policy-making process. The media can function as a forum for facilitating public discussion and the advancement of public understanding of food safety and risks.

References

Acord, D., & Feldman, B. (2004). Meat industry says USDA grossly underestimates cost of new BSE rules. *Food Chemical News, April*, 13–14.

Amendola, A. (2002). Recent paradigms for risk informed decision making. *Safety Science 40*(1–4), 17–30.

Anderson, R. M., Donelly, C. A., Ferguson, N. M., et al. (1996). Transmission dynamics and epidemiology of BSE in British cattle. *Nature, 382*, 779–788.

Beck, M. (2004). Obstacles to the evolution of risk management as an academic discipline: some tentative thoughts. *Risk Management, 6*(3), 13–21.

Beck, M., Asenova, D., & Dickson, G. (2005). Public administration, science, and risk assessment: a case study of the UK bovine spongiform encephalopathy crisis. *Public Administration Review, 65*(4), 396–408.

Boame, A., Parsons, W., & Trant, M. (2004). *Mad cow disease and beef trade: an update* (11-621-MIE2004010). Ontario, Canada: Statistics Canada.

Boer, M. D., McCarthy, M., Brennan, M., Kelly, A. L., & Ritson, C. (2005). Public understanding of food risk issues and food risk messages on the Island of Ireland: the view of food safety experts. *Journal of Food Safety, 25*: 241–265.

Brown, P., Will, R. G., Bradley, R., Asher, D. M., & Detwiler, L. (2001). Bovine spongiform encephalopathy and variant Creutzfeldt-Jakob disease: background, evolution and current concerns. *Emerging Infectious Diseases, 7*, 6–16.

BSE Inquiry. (2000). *The inquiry into BSE and variant CJD in the United Kingdom: Volume 1–16*. Retrieved (last access), March 8, 2009, from http://www.bseinquiry.gov.uk/pdf/index.htm

Butler, D. (2001). Brain mix-up leaves BSE research in turmoil. *Nature, 413*, 760.

Caswell, J. A. (2006). A food scare a day: why aren't we better at managing dietary risk? *Human and Ecological Risk Assessment, 12*(1), 9–17.

Collinge, J., Beck, J., Campbell, T., Estibeiro, K., & Will, R. G. (1996). Prion protein gene analysis in new variant cases of Creutzfeldt-Jakob disease. *The Lancet, 348*(9019), 56.

Coulthart, M. B., & Cashman, N. R. (2001). Variant Creutzfeldt-Jakob disease: a summary of current scientific knowledge in relation to public health. *Canadian Medical Association Journal, 165*(1), 51–58.

Cross, F. B. (1996). Paradoxical perils of the precautionary principle. *Washington and Lee Law Review, 53*(3), 851–925.

Daiichi Life Research Institute. (2004). *Effects of US mad cow disease on Japanese economy*. Retrieved, September 20, 2007, from http://group.dai-ichi-life.co.jp/dlri/news/pdf/nr2003_26.pdf

Donnelly, C. A., Ferguson, N. M., Ghani, A. C.,Woolhouse, M. E. J.,Watt, C. J. & Anderson, R. M. (1997). The epidemiology of BSE in GB cattle herds: I. Epidemiological processes, demography of cattle and approaches to control by culling. *Philosophical Transactions of the Royal Society, B 352*, 781–801.

Embassy of the United States, Japan. (2006). *USDA news release (no. 0255-06)*. Retrieved, August 22, 2007, from http://japan.usembassy.gov/j/p/tpj-j20060724-50.html

Embassy of the United States, Japan. (2007). *GOJ-USG joint press statement: Japan EV program verification period now ended*. Retrieved, September 10, 2007, from http://japan.usembassy.gov/e/p/tp-20070613-79.html

European Commission, DG Health and Consumer Protection. (2000). *Food safety: from the farm to the fork*. Retrieved (last access), March 16, 2009, from http://ec.europa.eu/food/fs/bse/legislation_en.html

Ferguson, N. M., Ghani, A. C., Donnelly, C. A., et al. (1998). BSE in northern Ireland: epidemiological patterns past, present and future. *Proceedings of the Royal Society London, 265*, 545–554.

Fischler, C. (2000). Securite sanitaire des aliments et consommateurs. *Dossier, 7*(5), 443–448.

Food Safety Commission. (2004a). *Measures against bovine spongiform encephalopathy (BSE) in Japan: an interim report*. Retrieved (last access), March 8, 2009, from http://www.fsc.go.jp/sonota/measure_bse_injapan.pdf

Food Safety Commission. (2004b). *Deliberation of the Food Safety Commission on the revision of BSE measures in Japan*. Retrieved (last access), March 8, 2009, from *http://www.fsc.go.jp/sonota/revision_bse_measures170524.pdf*

Food Safety Commission. (2004c). *Food-related risk communication: results of the questionnaires*. Retrieved, September 20, 2007, from http://www.fsc.go.jp/koukan/risk160420/anketosyukei_160420.pdf

Food Safety Commission. (2004d). *Present status of and agenda for food safety and risk communications*. Retrieved, July 27, 2007, from http://www.fsc.go.jp/iinkai/riskcom_genjou.pdf

Food Safety Commission. (2005a). *BSE-related health effects of food*. Retrieved, October 10, 2007, from www.fsc.go.jp/iken-bosyu/point_measures_bse170513.pdf

Food Safety Commission. (2005b). *Risk assessment concerning the comparability between risks of consuming beef and internal organs regulated by the beef export verification program of the United States/Canada and risks of consuming beef and internal organs of Japanese cattle*. Retrieved, October 2, 2007, from www.fsc.go.jp/sonota/bse-risk-assessment-concerning.pdf

Food Safety Commission. *BSE oyobi vCJD ni tsuite* [On BSE and vCJD]. Retrieved (last access), March 8, 2009, from http://www.fsc.go.jp/sonota/bse1601.html

Food Standard Agency. *Over thirty months review archive*. Retrieved (last access), December 20, 1997, from http://www.food.gov.uk/archive/bsearchive/otmreview/

Gerodimos, R. (2004). The UK BSE crisis as a failure of government. *Public Administration, 82*(4), 911–929.

Goldwater, P. N. (2001). Bovine spongiform encephalophathy and variant Creutzfeldt-Jakob disease: implications for Australia. *Medical Journal of Australia, 175*(3), 154–159.
Goodell, R. (1987). The role of the mass media in scientific controversy. In H. T. Engelhardt & A. L. Caplan (Eds.): *Scientific controversies: case studies in the resolution and closure of disputes in science and technology.* Cambridge, UK: Cambridge University Press.
Green, J. M., Draper, A. K., & Dowler, E.A. (2003). Short cuts to safety: risk and rules of thumb in accounts of food choice. *Health, Risk and Society, 5*(1), 33–352.
Greer, A. (1999). Policy coordination and the British administrative system: evidence from the BSE inquiry. *Parliamentary Affairs, 52*(4), 598–615.
Hart, C. E., & Babcock, B. A. (2002). *U.S. farm policy and the World Trade Organization: how do they match up?* Iowa State University, Center for Agricultural and Rural Development, Working Paper 02-WP 294. Ames, IO: Iowa State University.
Harvey, J., Erdos, G., & Challinor, S. (2001). The relationship between attitudes, demographic factors and perceived consumption of meats and other proteins in relation to the BSE crisis: a regional study in the United Kingdom. *Health, Risk and Society, 3*(2), 181–197.
Hathaway, S. C. (1997). Development of food safety risk assessment guidelines for food and animal origin in international trade. *Journal of Food Protection, 60*(11), 1432–1438.
Hawkins, K., & Thomas, J. (1989). *Making regulatory policy.* Pittsburgh, PA: University of Pittsburgh Press.
Heim, D., & Mumford, E. (2005). The future of BSE from the global perspective. *Meat Science, 70*(3), 555–562.
Hinchliffe, S. (2002). Indeterminacy in decisions: science, policy and politics in the BSE (Bovine Spongiform Encephalopathy) crisis. *Transactions of the Institute of British Geographers, 26*(2), 182–204.
Hood, C., Rothstein, H., & Baldwin, R. (2001). *The government of risk: understanding risk regulation regimes.* Oxford, UK: Oxford University Press.
Hornsby, M. (1998, November 7). Public may have been misled on beef threat. *The Times*, p. 2.
Ilbery. B., & Kneafsey, M. (2000). Producer constructions of quality in regional speciality food production: a case study from south west England. *Journal of Rural Studies, 16*, 217–230.
Jacob, M., & Hellstrom, T. (2000). Policy understanding of science, public trust and the BSE-CJD crisis. *Journal of Hazardous Materials, 78*(1–3), 303–317.
Japan–United States BSE Working Group. (2004). *Final report, Japan–US BSE Working Group: July 22, 2004.* Retrieved, September 22, 2007, from http://www.mhlw.go.jp/kinkyu/bse/yunyu/dl/040723-1b.pdf
Jasanoff, S. (1997). Civilization and madness: the great BSE scare of 1996. *Public Understanding of Science, 6*(3), 221–232.
Jensen, K. K. (2004). BSE in the UK: why the risk communication strategy failed. *Journal of Agricultural and Environmental Ethics, 17*(4–5), 405–423.
Jensen, K. K., & Sandoe, P. (2002). Food safety and ethics: the interplay between science and values. *Journal of Agricultural and Environmental Ethics, 15*(3), 245–253.
Kasperson, R. E., Renn, O., Slovic, P., Brown, H. S., Emel, J., & Goble, R. (1988). The social amplification of risk: a conceptual framework. *Risk Analysis, 8*(2), 177–187.
Kastner, J., Powell, D., Crowley, T., & Huff, K. (2005). Scientific conviction amidst scientific controversy in the transatlantic livestock and meat trade. *Endeavour, 29*(2), 78–83.
Kellar, J. A., & Lees, V. W. (2003). Risk management of the transmissible spongiform encephalopathies in North America. *Revue Scientifique et Technique de l Office International des Epizooties, 22*(1), 201–225.
Kimball, A. M., Plotkin, B. J., Harrison, T. A., & Pautlet, N. F. (2004). Trade-related infections: global traffic and microbial travel. *EcoHealth, 1*(1), 39–49.
Krebs, J. R. (2001). Science, uncertainty and policy: food for thought. *Toxicology Letters, 120*(1–3), 89–95.
Krebs, J. R., May, R. M., & Stumpf, M. P. H. (2002). Theoretical models of sheep BSE reveal possibilities: but we must remember that these theories are based on speculation, not on fact. *Nature 415*, 115.

Krimsky, S., & Golding, D. (1992). *Social Theories of Risk.* Westport, CT: Praeger.
Kunkel, H. O., Thompson, P. B., Miller, B.A., & Skaggs, C. L. (1998). Use of a competing conceptions of risk in animal agriculture. *Journal of Animal Science, 76*(3), 706–713.
Latouche, K., Rainelli, P., & Vermersch, D. (1998). Food safety issues and the BSE scare: some lessons from the French case. *Food Policy, 23*(5), 347–356.
Latouche, K., Rainelli, P., & Vermersch, D. (2000). Quel prix pour la securite alimentaire? Une evaluation contingente suite a la crise europeenne de la vache folle. *Canadian Journal of Agricultural Economics, 48*(3), 325–340.
Lobstein, T. (2001). Crisis in agriculture: are we learning from disasters? *Consumer Policy Review, 11*(3), 78–85.
Lowi, T. L. (1979). *The end of liberalism: The second republic of the United States.* New York: WW Norton.
Lowi, T. L. (1992). The state of political science: how we become what we study. *American Political Science Review, 86*(1), 1–7.
Majone, G. (1989). *Evidence, argument, and persuasion in the policy process.* New Haven, CT: Yale University Press.
Matibag, G. C., Igarashi, M., & Tamashiro, H. (2005). BSE safety standards: an evaluation of public health policies of Japan, Europe, and USA. *Environmental Health and Preventive Medicine, 10*, 303–314.
Matthews, D. (2003). BSE: a global update. *Journal of Applied Microbiology, 94*(Suppl. 1), 120–125.
McNelis, N. (2001). The role of the judge in the EU and WTO: lessons from the BSE and hormones cases. *Journal of International Economic Law, 4*(1), 189–208.
Miller, D. (1999). Risk, science and policy: definitional struggles, information management, the media and BSE. *Social Science & Medicine, 49*(9), 1239–1255.
Millstone, E., & Zwanenberg, P. V. (2002). The evolution of food safety policy-making institutions in the UK, EU and Codex Alimentarius. *Social Policy and Administration, 36*(6), 593–609.
Ministry of Agriculture, Forestry and Fisheries. (2002a). *BSE mondai ni kansuru chosa kento iinnkai* [The exploratory committee on the BSE issue]. Retrieved (last access), March 12, 2009, from http://www.maff.go.jp/j/syouan/douei/bse/b_iinkai/index.html
Ministry of Agriculture, Forestry and Fisheries. (2002b). *Summary of the year after the first BSE incidence and the future agenda.* Retrieved, September 20, 2007, from http://www.maff.go.jp/mlet/213/2.pdf
Ministry of Agriculture, Forestry and Fisheries (MAFF). *BSE kankei* [About BSE]. Retrieved (last access), March 8, 2009, from http://www.maff.go.jp/j/syouan/douei/bse/index.html
Ministry of Foreign Affairs. (2004). *Joint press statement for the resumption of trade in beef and beef products, by the Government of Japan and the Government of United States.* Retrieved, September 10, 2007, http://www.mofa.go.jp/region/n-america/us/economy/joint0410.pdf
Ministry of Health, Labor and Welfare (MHLW). *BSE ni tsuite* [On BSE]. Retrieved (last access), March 8, 2009, from http://www.mhlw.go.jp/kinkyu/bse.html
Murphy-Lawless, J. (2004). The impact of BSE and FMD on ethics and democratic process. *Journal of Agricultural and Environmental Ethics, 17*(4–5), 385–403.
National Research Council. (1989). *Improving risk communication.* Washington DC: National Academy Press.
Neuman, W. R. (1986). *The paradox of mass politics: knowledge and opinion in the American electorate.* Cambridge, MA: Harvard University Press.
North American Chief Veterinary Officers. (2005). *Report of the North American Chief Veterinary Officers on Harmonization of a BSE Strategy.* Retrieved February 28, 2009, from http://canada.usembassy.gov/content/can_usa/pdfs/madcow_bseharmonization.pdf
O'Riordan, T., & Cameron, J. (1994). *Interpreting the precautionary principle.* London: Earthscan Publications.
Office International des Epizooties (OIE). (2003). *Number of reported cases of BSE worldwide.* Retrieved, November 21, 2007, from http://www.oie.int/eng/info/en_esbmonde.htm

Office of Prime Minister. *BSE taisaku* [BSE mitigation measures]. Retrieved (last access), March 8, 2009, from http://www.kantei.go.jp/jp/osirase/kyougyubyou.html

Packer, R. (2006). *The politics of BSE.* New York and London: Palgrave MacMillan.

Pennings, J. M. E., Wansink, B., & Beulenberg, M. T. G. (2002). A note on modeling consumer reactions to a crisis: the case of the mad cow disease. *International Journal of Research in Marketing, 19*(1), 91–100.

Phillips, N., Bridgeman, J., & Ferguson-Smith, M. (2000). *The BSE inquiry: return to an order of the Honourable the House of Commons dated October 2000 for the report, evidence and supporting papers of the inquiry into the emergence and identification of Bovine Spongiform Encephalopathy (BSE) and variant Creutzfeldt-Jakob Disease (vCJD) and the action taken in response to it up to 20 March 1996.* Norfolk, UK: The Stationary Office.

Pigg, K. E. (2002). Three faces of empowerment: Expanding the theory of empowerment in community development. *Journal of the Community Development Society, 33*(1), 107–123.

Prion Expert Panel, Food Safety Commission. (2004). *Record of the 15th meeting.* Retrieved, September 20, 2007, from http://www.fsc.go.jp/senmon/prion/p-dai15/161026_dai15kai_prion_gijiroku.pdf

Ricketts, M. N. (2004). Public health and the BSE epidemic. *Current Topics in Microbiology and Immunology, 284,* 99–119.

Rothstein, H. F. (2003). Neglected risk regulation: the institutional attenuation phenomenon. *Health, Risk and Society, 5*(1), 85–103.

Rothstein, H. F. (2004). Precautionary bans or sacrificial lambs? Participative risk regulation and the reform of the UK food safety regime. *Public Administration, 82*(4), 857–881.

Sandman, P. M. (1993). *Responding to community outrage: strategies for effective risk communication.* Fairfax, VA: American Industrial Hygiene Association.

Sato, H., Akabayashi, A., & Kai, I. (2005). Public appraisal of government efforts and participation intent in medico-ethical policymaking in Japan a large scale national survey concerning brain death and organ transplant. *BMC Medical Ethics, 6*(1), 1–12.

Sato, H., Akabayashi, A., & Kai, I. (2006). Public, experts, and acceptance of advanced medical technologies: the case of organ transplant and gene therapy in Japan. *Health Care Analysis, 14*(4), 203–214.

Sawcer, S. J., Yuill, G. M., Esmonde, T. F. G., Estibeiro, P., Ironside, J. W., & Bell, J. E. (1993). Creutzfeldt-Jakob disease in an individual occupationally exposed to BSE. *The Lancet, 341,* 642.

Schattschneider, E. E. (1975). The semisovereign people: a realist's view of democracy in America. New York, NY: Harcourt Brace College.

Schwing, R. C., & Albers, W. A. (1980). *Societal risk assessment: how safe is safe enough?* New York: Plenum.

Simon, H. (1957). *Administrative behavior.* New York: Free.

Smith, E., Marsden, T., Flynn, A., & Percival, A. (2004). Regulating food risks: rebuilding confidence in Europe's food? *Environment and Planning C: Government and Policy, 22*(4), 543–567.

Sparling, D. H., & Caswell, J. A. (2006). Risking market integration without regulatory integration: the case of NAFTA and BSE. *Review of Agricultural Economics, 28*(2), 212–228.

Stallones, R. A. (1982). Epidemiology and public policy: Pro- and anti-biotic. *American Journal of Epidemiology, 115*(4), 485–491.

Tacke, V. (2001). BSE as an organizational construction: a case study on the globalization of risk. *British Journal of Sociology, 52*(2): 293–312.

The Exploratory Committee on the BSE Issue. (2001a). *Record of the second meeting: December 7, 2001.* Retrieved, September 20, 2007, from http://www.mhlw.go.jp/shingi/0112/txt/s1207-1.txt

The Exploratory Committee on the BSE Issue. (2001b). *Record of the first meeting: November 19, 2001.* Retrieved, September 20, 2007, from http://www.mhlw.go.jp/shingi/0111/txt/s1119-1.txt

United Kingdom Department of Health. (2006). *Monthly Creutzfeldt Jakob disease statistics – March 2006.* Retrieved March 1, 2009, from http://www.dh.gov.uk/en/Publicationsandstatistics/Pressreleases/DH_4131172

United States Department of Agriculture, Animal and Plant Health Information Service (USDA, APHIS). (2004a). *BSE: USDA actions to prevent bovine spongiform encephalopathy*. Retrieved, May 25, 2007, from http://www.aphis.usda.gov/lpa/issues/bse/bsechron.html

United States Department of Agriculture, Animal and Plant Health Information Service (USDA, APHIS). *Newsroom: bovine spongiform encephalopathy*. Retrieved (last access), March 10, 2009, from http://www.aphis.usda.gov/newsroom/hot_issues/bse/inde.shtml

United States Department of Agriculture, Economic Research Service (USDA). (2004b). *Background statistics on US beef and cattle industry*. Washington, DC: United States Department of Agriculture.

Weinberg, A. M. (1972). Science and trans-science. *Minerva: A Review of Science, Learning and Policy, 10*(2): 209–218.

Will, R. (1999). New variant Creutzfeldt-Jakob disease. *Biomedecine & Pharmacotherapy, 53*(1), 9–13.

Will, R. G., Ironside, J. W., Zeidler, M., Cousens, S. N., Estibeiro, K., Alperovitch, A., Poser, S., Pocchiari, M., Hofman, A., & Smith, P. G. (1996). A new variant of Creutzfeldt-Jakob disease in the UK. *The Lancet, 347*, 921–925.

Winter, M. (1996). Intersecting departmental responsibilities, administrative confusion and role of science in government: the case of BSE. *Parliamentary Affairs, 49*(4), 550–565.

World Health Organization. (1996). *Report of a WHO consultation on public health issues related to human and animal transmissible spongiform encephalopathies*. Retrieved, October 4, 2008, from http://whqlibdoc.who.int/hq/1996/WHO_EMC_DIS_96.147.pdf

World Organisation for Animal Health. (2003). *OIE 23/05/03 Résolutions sur le statut de l'ESB*. Retrieved March 1, 2009, from http://www.oie.int/fr/info/fr_statesb.htm

World Trade Organization. (1994). *The WTO agreement on the application of sanitary and phytosanitary measures* ("Notification procedures" in Annex B: transparency of sanitary and phytosanitary regulations). Retrieved February 18, 2008, from http://trade.wtosh.com/english/tratop_e/sps_e/spsagr_e.htm#Annexb

World Trade Organization. (2000). *Notification of emergency measures*. Geneva: Author.

World Trade Organization. (2003). *Notification of emergency measures*. Geneva: Author.

Zwanenberg, P. V., & Millstone, E. (2003). BSE: a paradigm of policy failure. *Political Quarterly, 74*(1), 27–37.

Chapter 8
BSE in the United Kingdom

Andrew Webster, Conor M.W. Douglas, and Hajime Sato

Introduction

The BSE Crisis in the UK

The bovine spongiform encephalopathy (commonly known as BSE) crisis that exploded in the UK in the mid-to-late 1990s has been regarded as a classic case of failure in the management and communication of health risks to the public. Academic disciplines such as political science, social policy, management studies, sociology of science, science and technology studies, and others, have provided a range of analyses to explain what happened leading up to the crisis, helping us to understand why the management of the crisis was such a failure. One of the motivating factors behind this academic work, as well as the official inquiries, hearings, and reports of the British government, was to learn from this failure, with the hope that such a crisis will not be repeated.

This chapter provides an overview of some of the key findings from these literature sources and places them in a British context. We begin by briefly describing BSE, providing its history in the UK, and relating how it is associated with risks to human health. The rest of the report is divided into the following three parts that reflect the distinct stages in which the management and communication of health risks related to BSE took place in the British context. They include: Period 1 (1986 to 1996): BSE is identified, *but only* as a cattle and agricultural problem; Period 2 (1996 to 2000): BSE is linked to humans and leads to a health crisis; and Period 3 (2000 to the present): Institutions are transformed and the health risks are ameliorated.

In each part, the government institutions responsible for BSE risk management are identified and discussed, as are the committees of experts, formal inquiries, and

A. Webster and C.M.W. Douglas
Science and Technology Studies Unit (SATSU), Department of Sociology, University of York, UK

H. Sato (✉)
Department of Public Health, Graduate School of Medicine, The University of Tokyo, Japan
e-mail: hsato-tky@umin.net; hsato@post.harvard.edu

H. Sato (ed.), *Management of Health Risks from Environment and Food*, Alliance for Global Sustainability Book Series 16,
DOI 10.1007/978-90-481-3028-3_8, © Springer Science+Business Media B.V. 2010

advisory panels that played such a central role in communicating the risks involved with BSE. Then, the policies implemented by the relevant organizations and institutions are described in detail. At the end of each part, the successes and failures of the risk management and communication strategies are reviewed. With this review as a background, we are then in a position to discuss the convergent and divergent approaches to risk management and communication, in particular their distinctively British features.

The Background of the Disease in the UK

The first UK case of BSE was recorded in November 1986 by the Central Veterinary Laboratory (CVL), which was then under the umbrella of the Ministry of Agriculture, Fisheries and Food (MAFF) (Phillips et al. 2000, Vol. 16, chap. 1; Maxwell 1999 p. 97). Prior to 1986, a similar disease – scrapie – had existed for many years in sheep. BSE and scrapie are two of the many fatal brain diseases known as transmissible spongiform encephalopathies (TSEs). The commonly held theory is that most TSEs, including BSE, share a particular infectious protein called prions, the origin and function of which is still not completely understood (United States Department of Agriculture 2001, p. 2). It is believed that BSE in Britain was either transmitted through cattle herds by feed containing by-products infected with scrapie, or the prions in the cattle originated from "a novel source early in the 1970s, possibly a cow or other animal that developed disease as a consequence of a gene mutation." It is also possible that the use of organophosphorus pesticides increased the susceptibility of cattle to the disease (Phillips et al. 2000, Vol. 1).

TSEs are not limited to cattle and sheep. Classic Creutzfeldt-Jakob disease (CJD) is the most common TSE in humans. It has been known about since the early twentieth century, and it is *not related* to BSE. There are several types of CJDs that affect humans, although they are very rare. They include Iatrogenic CJD and Kuru, which are acquired through the consumption of human brain, offal, and nerve tissue, as well as three rare genetic disorders: familial CJD, Gerstmann-Sträussler-Scheinker syndrome (GSS), and fatal familial insomnia (FFI) (Maxwell 1999, p. 99).

Only one form of CJD is believed to be related to BSE, namely "new variant Creutzfeldt-Jakob disease" (vCJD or nvCJD). Although the relationship between these two prion diseases is not fully understood, it is believed that in the UK BSE "jumped species" to humans through the consumption of contaminated beef and bovine products during the 1990s. This consumption resulted in a public health crisis in 1996. By June 2007, nvCJD had killed 165 people in Britain and 6 elsewhere (National Creutzfeldt-Jakob Disease Surveillance Unit 2007). Although this crisis was tragic, the number of deaths in the UK could have been much larger had 179,000 BSE-infected cattle not been destroyed; subsequently, another 4.4 million were slaughtered as a precautionary measure (Brown 2001).

A major cause of the BSE-nvCJD crisis was the early belief that BSE could not be transmitted to humans. As noted in the section on Period 1, during the 10 years following the discovery of BSE, this belief caused the problem to be framed as a farming and agriculture risk.

Sources and Methods

This chapter is based on a systematic survey of numerous academic and policy sources. Two of the most detailed analyses have recently been provided by Millstone and van Zwanenberg (2001) and van Zwanenberg and Millstone (2003), who systematically reviewed the relevant empirical data. In addition, UK government policy documents were examined, including the minutes from meetings of the committees charged with overseeing and dealing with the BSE problem. Other sources included academic papers discussing how the media handled the BSE story and, in particular, how the certainty or uncertainty surrounding BSE was communicated to the public.

The major foci of the chapter are the relationship between the state and the private sector, how the risks were handled, how the expertise that framed the decisions was represented, and the eventual emergence of the new forms of governance to ensure the more effective management of such crises in the food/public health sector. The conclusions on the latter point are based on analyses of the so-called "regulatory state" (Moran 2001) and the move towards a more explicit "precautionary principle" allied to more stringent regulatory policies (Vogel 2001). These frameworks help to identify the lessons that can be learned from the cases examined.

History of the BSE Issue and Risk Management in the UK

Period 1 (1986 to 1996): BSE Identified as a Cattle and Agriculture Problem

The first case of BSE in Britain was identified in November 1986. From that point on, an important sequence of events took place that must be taken note of to understand the processes by which the BSE health risk was managed and communicated to the public. Many of these events involve the actions of various government institutions, most notably the Ministry of Agriculture, Fisheries and Food. This agency was primarily responsible for risk management and communication in the early stages of the BSE saga in the UK.

Government Institutions

The Ministry of Agriculture, Fisheries and Food

The first government office to identify and deal with BSE-infected cattle was the Central Veterinary Laboratory (CVL), which at that time fell under the umbrella of what was then the Ministry of Agriculture, Fisheries and Food (MAFF). Once the CVL identified BSE and reported to its governing body, and because it was believed at this time that BSE could not be transmitted to humans, the MAFF became the main government institution responsible for dealing with the BSE problem from 1986 to 1996. It is well documented that the centrality of the MAFF in the BSE saga was one of the principal reasons for the scope and scale of the crisis (Maxwell 1999; Millstone & van Zwanenberg 2001; van Zwanenberg & Millstone 2003). The MAFF's involvement was considered to be detrimental to the management of the BSE problem because of its organizational role and mandate. At this time, the MAFF had conflicting responsibilities and roles, which included the promotion of the economic interests of British farmers (both crops and livestock) domestically and abroad, while at the same time protecting the public's health from food-borne hazards. This situation proved to be unworkable, because the MAFF consistently placed a higher priority on the interests of British meat and dairy farmers than public health, a point that we return to in the section on "successes and failures."

Given this situation, it is perhaps not surprising that the MAFF did very little from the time it learned about BSE in late 1986 until the establishment of the Southwood Working Party in April 1988.[1] As discussed below, what little they did do was not a policy action per se, but rather the formation of a committee of eminent scientists to give advice on potential policy routes. Many commentators have suggested that this inaction on the part of the MAFF was motivated by its desire to protect the British meat and dairy industries from the reductions in their profit margins that would result from the fear experienced by foreign and domestic beef consumers if they learned about BSE. The same commentators suggested that, likewise, the MAFF was motivated by a desire to keep potentially large expenditures on BSE relatively low. It is again important to note that at this stage there was no evidence to suggest that BSE could be transmitted to humans.

The Southwood Working Party (1988–1989)

The Southwood Working Party (SWP) was the first scientific committee assembled to address the risk presented by BSE. Its members, a handful of eminent scientists, were appointed by officials at the MAFF and the Department of Health (DoH) in April 1988. It is important to note that none of these scientists had any experience or expertise concerning prion diseases or TSEs. The group was led by Richard Southwood, Professor of Zoology at Oxford, who had previously chaired both the Royal Commission on Environment Pollution and the National Radiological Protection Board (Southwood Working Party 1989).

The SWP was ostensibly charged with providing scientific advice on the BSE issue, which at that stage mainly concerned the animal food chain. Its purview consisted

[1]During this period from November 1986 to April 1988, a number of meetings were held inside the MAFF, between the MAFF and the farming industry, and between the MAFF and the Department of Health. However, these meetings produced no policy actions.

of infected animals, bovine food and non-food products, as well as other potentially related risks. However, similar to the MAFF, the SWP found itself in a dual role that complicated its proceedings. On the one hand, the SWP was asked to provide scientific advice on BSE, yet on the other hand – and at the same time – it was also being asked to provide the MAFF with policy recommendations on how to deal with the risk. This tension, and the processes by which it was handled by the MAFF, has been identified by Millstone and van Zwanenberg (2001, 2003) as a central problem in the management of BSE risks. The main reason for this tension is that the policies the SWP could recommend were severely limited by the MAFF's priorities of saving money and protecting the British cattle-related industries and, most of all, by its desire not to create a panic among local and foreign beef consumers.

After deliberations, the SWP Report ultimately concluded in February of 1989 that it is "most unlikely that BSE will have any implication for human health" (Report of the Working Party on Bovine Spongiform Encephalopathy 1989: 21). According to van Zwanenberg & Millstone (2003), "… that was, however, a provisional and partial judgment. It was based on the assumption that BSE had derived from scrapie in sheep, and on a further assumption, namely that BSE would behave exactly the same way as the scrapie agent did, once it had jumped species into cattle" (p. 32).

Despite its conflicting roles as advisor and policymaker, the SWP was nonetheless able to offer policy recommendations as to how BSE should be dealt with, based on the findings of their research (to be discussed below). Again, it is important to note that the SWP – the MAFF's expert scientific advisors – had concluded that BSE was not a hazard to humans. As a result, the SWP's conclusions – as partial and provisional as they may have been – served as the foundation on which the policymakers built their risk-management and communication strategies. This process continued until March 1996, when BSE was linked to nvCJD.

Clearly, it is evident that a partial and provisional set of scientific conclusions drawn at this stage led to a policy decision that was always likely to be lagging behind the evidence-base, especially when the latter became more and more clear. Such a lag is bound to create a policy regime that is slow-moving and likely to shift its position only when – as with the link to nvCJD – a potentially serious risk is recognized.

The Consultative Committee on Research into Spongiform Encephalopathies, and the Spongiform Encephalopathy Advisory Committee

One of the recommendations of the SWP Report was the establishment of a permanent and expert committee, which included biomedical expertise, to continue the study of BSE and follow up some of the work begun by the SWP. This meant the immediate formation in February 1989, on the same day the SWP Report was published, of the Consultative Committee on Research into Spongiform Encephalo pathies (CCRSE). This Committee was chaired by Dr. David Tyrrell, and thus it became known as the Tyrrell Committee. The CCRSE was replaced in April 1990, not long after the Tyrrell Report was issued, by an advisory committee known as the Spongiform Encephalopathy Advisory Committee (SEAC). The SEAC had the same members and chair as the CCRSE. It is still functioning today, but with a different committee composition (Spongiform Encephalopathy Advisory Committee 2007).

This constitution and reconstitution of scientific and expert committees reflects only some of the institution creation that took place in the early stages of the BSE saga. In Britain, numerous other working groups, committees, and advisory groups were formed. These include the CVL's BSE Research & Development Group, which first met in November 1988, and the Medical Research Council (MRC) Coordinating Committee on Spongiform Encephalopathies (the Murray Committee) in October 1990. In 1991, the MRC Coordinating Committee created the Clinical Subcommittee (the Allen Subcommittee). The MAFF's Expert Group on Animal Feedingstuffs (the Lamming Committee) was established in February 1991.

This explosion of advisory committees in Britain is worth noting in and of itself, and there was even more to come. These various committees exemplify the institutional and biological complexity of the BSE crisis, as well as the government's inexperience in dealing with this kind of health crisis. It seems as though each ministry or office involved with BSE felt it necessary to form its own special advisory committee, and that each particular issue that BSE touched upon also required that such a committee investigate whether the problem involved animal feed, clinical practices, or spongiform encephalopathy more generally. The majority of these committees and advisory groups are not dealt with in this chapter, because they played a lesser role regarding the major risk-management and communication policies. Worthy of note is the – perhaps particularly British – tendency for expert advisory committees to be formed when risk-laden situations are presented to the government and regulators (Jasanoff 2005).

Policy Actions

Slaughtering and Compensation Policies

We have seen that one of the first major risk-management strategies employed by the MAFF was the establishment of a committee of eminent scientists to deliberate and advise the government about BSE. However, this was not a policy action per se. From the time that BSE was discovered to the time that the Southwood Committee was formed, "clinically diseased animals were routinely slaughtered and processed for human consumption alongside healthy cattle" (van Zwanenberg & Millstone 2003, p. 29). One of the first concrete risk-management policies proposed to address the BSE crisis was called "slaughter and compensation." This policy had been suggested to John MacGregor, the Minister of the MAFF, in February 1988.

However, without a risk assessment to identify the transmissibility of BSE to humans, a "slaughter and compensation" policy could not take hold, since it was the MAFF's procedure for the farming and agricultural industries to cover the costs associated with the destruction of diseased animals (Millstone & van Zwanenberg 2001, p. 103). Here we see how the priority of keeping public expenditures to a minimum directly impacted the risk management of BSE. A "slaughter and compensation" policy was introduced half a year later, in August 1988, after the SWP was in place to carry out the desired risk assessment (Phillips et al. 2000, Vol. 16, chap. 1).

In June 1988 – based on the recommendations of the SWP – it was decided that cattle with clinical symptoms of BSE would be slaughtered and banned from the human food chain, with compensation to be paid at 50% for confirmed cases of BSE and 100% for slaughtered cattle that turned out not to have BSE (negative cases) (Phillips et al. 2000, Vol. 16, chap. 1). There was, however, no diagnostic test to identify cattle that were asymptomatic but actually had BSE, and this unfortunately remains the case (Millstone & van Zwanenberg 2001, p. 106).

The Ban on Slaughterhouse Waste

In early 1988, before the Southwood Working Party had been formed, the MAFF did ban the use of potentially contaminated slaughterhouse cattle and sheep waste for feeding cows and sheep. However, in an attempt to protect the rendering industry, pigs and chickens, the primary consumers of slaughterhouse waste in the UK, could still be given feed that might contain BSE-contaminated products. At this early stage in the BSE saga, the other priority of protecting the British farming and rendering industries came into focus. Not only did this partial ban on slaughterhouse waste fail to protect the rendering industry, it also confused the farmers, who were unable to understand why one kind of feed was suitable for pigs and chickens, but not for cows and sheep. This inconsistency ultimately jeopardized and undermined the slaughterhouse waste ban of 1988, as spot checks found that farmers were not complying with it and continued to use potentially contaminated feed for all their livestock – including cows and sheep.

In late 1988, soon after it was formed, the SWP tabled a preliminary policy recommendation to the MAFF that entailed a total ban on slaughterhouse waste and meat and bone meal (MBM) from cows and sheep in the food chain of all herbivores. However, in line with their priority of protecting the rendering industry, the MAFF discouraged the SWP from including this recommendation in their final report; the SWP complied (Southwood Working Party 1989).

The Ban on All Brain and Central Nervous System Products

With the "slaughtering and compensation" policy in place, as well as the partial ban on slaughterhouse waste in the food chain of cows and sheep, in the spring of 1989 the major British pet food manufacturer and the food trade associations began to consider a voluntary ban on all cattle offal (brain and central nervous system products).

This move towards voluntary self-regulation by the industry proved to be crucial, as the MAFF did not want to have its regulatory role usurped by an industry whose interests they had been mandated to promote. Although the movement of the pet-food industry into the policy and regulation arenas forced the MAFF to act, it took a further 9 months for it to order and implement the ban on all cattle brain and central nervous system products.

Safety Issues

The risk-communication strategies concerning non-food products and activities involving infected cattle were devised so as not to alarm the British public. Consequently, the SWP's concerns about the safety of vaccines made from bovine materials were kept quiet by the Department of Health (DoH), because the DoH was afraid that disclosing these concerns would disquiet the public. The same applied to the SWP's concerns about the safety of slaughtermen and abattoir workers.

The SWP also brought their concerns about workers who might come in contact with infected cattle to the Health and Safety Executive (HSE) – the agency responsible for attending to hazardous working conditions. In response, the SWP was told by the Chief Veterinary Officer that "... new precautions would escalate the issue unnecessarily when we were saying there was no hazard to man from BSE" (Phillips et al. 2000). The belief was that sufficient precautions had already been taken by both the DoH and the HSE, and that the SWP's concerns had already been expressed to the relevant bodies. Ironically, part of the reason why the DoH and HSE felt they were able to dismiss the SWP's concern was that they felt their existing measures were sufficient, and they felt they did not want to unduly raise the public concern because "other agencies were persuaded by the Working Party's [SWP's] *published* advice that the risks were already remote" (Phillips et al. 2000).

Optimistic messages about the non-transmissibility of BSE to humans continued to be sent via the Tyrrell Report (from the proceedings of the CCRSE) and then through the publication of the SEAC's findings. Throughout the late 1980s and early 1990s, ministers and officials were presented on numerous occasions with formal reports reassuring them that British beef could be eaten safely. Kewell et al. (2007) argued that the "overall tone [of the Tyrrell Report] was one of providing mild criticism of the Southwood Committee's approaches, *whilst affirming the legitimacy of the science behind Southwood's recommendations*" (pp. 22–23, emphasis added). As they highlight from that report:

> We need to be reassured that further spread in cattle or to new species will not take place. Most importantly we seek reassurance that the Southwood group was correct in their belief that this disease would not have implications for human health, say through food, through occupational exposure or through medicinal products that use bovine ingredients. (Ministry of Agriculture, Fisheries & Food, and Department of Health 1989, p. 3).

In July 1990, Dr. Tyrrell – now chair of the SEAC – contacted the Chief Medical Officer, essentially confirming the findings of the SWP about the safety of beef consumption by humans (Phillips et al. 2000, Vol. 16, chap. 1). The SEAC's more formal report in 1994 was in line with this foundational message that came to be used in one form or another in all subsequent risk communications, which concluded that:

> … our scientific assessment is that the risk to man and other species from BSE is remote because the control measures now in place are adequate to eliminate or reduce any risk to a negligible level. We do, however, point out that any species exposed already and before any bans were effective could be incubating disease, and therefore continuous monitoring is very important until any possible incubation period has been exceeded (Spongiform Encephalopathy Advisory Committee 1995, p. 72).

Successes and Failures of Strategic Management and Communication of the BSE Risk in the UK from 1986 to February 1996

There are numerous lessons to be learned from the approaches that the MAFF took to managing and communicating the risks associated with BSE during the first decade of the disease's history in the UK, but very few of these measures could be considered successful. It is perhaps not surprising that the failures of the risk management and communication strategies – which were numerous – are closely associated with one another. They are nevertheless dealt with in turn below, and their connections are made clear.

Lethargic Policy Actions and Implementation

The previous discussion has documented the various approaches and policy actions that were taken in regard to the management and communication of risk in the wake of the discovery of BSE. As we have seen, starting at the time of the BSE discovery, the CVL, DoH, HSE and MAFF, as well as the meat and rendering industries, held various meetings and sent out various communications. However, it is also true that it took the MAFF well over a year to put their feed ban in place and adopt their "slaughtering and compensation" policies. It took the MAFF even longer to issue an all-out ban on brain and offal tissue from cattle.

This lethargy in policy action was **not** due to the inaccessibility or unavailability of the risk management procedures that the MAFF would ultimately follow. On the contrary, MAFF officials had been provided with recommendations that they simply chose not to follow, or to follow at their own pace. Therefore, not only were there delays in the creation of such policies, but the pace at which policy recommendations were translated into practice also was dangerously slow. Dr. Steven Dealler, a British microbiologist who was working on the BSE issue and who had testified at the formal BSE Inquiry, argued that the time lag from the first SWP recommendation to ban all offal to the point in which it became practice allowed for an estimated 25,000 infected cattle, that were at least halfway through the average incubation period for BSE, entered the human food chain (Dealler 1996). The lack of conclusive evidence on the transmissibility of BSE to humans notwithstanding, there is no doubt that this delay was a key element in the mismanagement of the BSE health risk in the UK.

Upstream Framing Assumptions: Giving the Interests of the Meat and Dairy Industries Priority over Public Health and the Minimization of Expenditures

As was noted at the beginning of the "successes and failures" section, the delay in making and implementing policies regarding risk management of BSE in the late 1980s is not an isolated problem. In fact, the primary reason for this lethargy is that MAFF officials did not want to take actions that could potentially harm the British beef and dairy industries or that would be expensive and ultimately unnecessary. The MAFF's records document that its priority was to ensure a stable market for British beef, not to assuage the concerns over public health.

One can see further evidence of the MAFF's prioritization of interests not only in their policy (in)actions to BSE risk management, but also in terms of their risk-communication strategies. From relatively early on (1987), the MAFF chose to enter into discussions with farmers and industry representatives about BSE and the available information about the new disease and its risks. There were no such discussions with representatives of food-safety public interest groups, and not until an article on BSE was published in the *Telegraph* could the BSE issue be even remotely considered to have entered the public arena. The 1989 SWP Report ultimately brought the BSE issue to the attention of the MPs and the public; however, the content of the report also reflects a very particular set of risk-management and communication failures.

Thus, the institutional arrangements for (food) risk management certainly predisposed policymakers to adopt swift and effective counterproductive measures under high scientific uncertainty (Jacob & Hellstroem 2000).

Managed Science

Prior to 1996, one of the central criticisms leveled at the MAFF regarding its risk-management and communication strategies was its close involvement with the Southwood Working Party, the scientific committee that was supposed to provide independent advice to the MAFF. A Department of Trade and Industry (DTI) report that outlined the role of the expert committees addressing food safety states that "historically, it has not always been clear where the responsibilities of the advisory committee end and those of Government begin" (Department of Trade and Industry [DTI] 2000, p. 4). On a number of occasions, the SWP issued preliminary findings and expressed concerns that were either disregarded or played down by the MAFF staff. Such examples in which the advice and concerns of the scientific committee were fettered include early concerns by Richard Southwood himself about the transmissibility of BSE to humans, the importance of removing BSE-contaminated cattle waste from **all** the animal food chains (DTI 2000, p. 5), and "an early draft from SEAC [that] drew attention to the fact that infectivity was still entering the food chain ... [in which] MAFF officials simply deleted the inflammatory paragraphs" (van Zwanenberg & Millstone 2003, p. 33).

The MAFF's role in managing scientific advice did not end with its direction of the SWP and SEAC, as it also made risk-management decisions internally to their own department that reflected rather unscientific choices. For instance, when a member of the Food Sciences Division of the MAFF suggested in May 1989 that an assessment of infectivity levels of beef products be compared to the previous risk assessments conducted on sheep food products in the case of scrapie, this was blocked. Colin Whitaker, the veterinarian who had identified the first cases of BSE in 1985, reported in 1989 that "[The Ministry] didn't seem to want publicity with the disease ... The word scrapie was deemed to be emotive and I was asked not to use it" (BBC Radio Four, *Face the Facts*, May 18, 1989).

Through the establishment of the SWP, the MAFF had sought to create independent scientific advisory committees and, later, a more expert committee with the SEAC. However, due to the MAFF's all-too-close involvement with these committees, the results were conservative: A BSE risk assessment that took the form of the SWP Report, which unfortunately would be the foundation for risk-management and communication strategies until BSE was linked to nvCJD in March 1996.

The Path-Dependent Legacy of Southwood

Perhaps the greatest shortcoming in the risk-management and communication strategies in the early 1990s was the fact that these strategies were totally wed to the general tone of the SWP Report: British beef was safe to eat and there was no risk of transmissibility to humans. We have thus far described in detail some of the problems surrounding the SWP and its findings and operations that were fettered by the MAFF. It is crucial to note that given these problems, and irrespective of them, the SWP Report became the legacy from which opinions and future policies would follow. Both the Tyrrell Report and various documents from the SEAC reaffirmed the SWP's findings, and this was also the ostensible purpose of the Tyrrell Report. In reference to the alignment of the early scientific committee reports on BSE, Kewell et al. (2007) even went so far as to label the SWP Report, the Tyrrell Report, and the first two publications from the SEAC as "the Southwood Series." The SWP Report made it very difficult for ministers to enact policies or to communicate with the public in ways that could have ultimately aided the management of the BSE risk. This is because such policies or communications would have contradicted the SWP

Report. As Kewell et al. (2007) noted, "In doing so, the Southwood Series helped to create a *path dependant legacy* (Treece et al. 1997) in which BSE had come to be treated, first and foremost, as an animal health risk."

The concept of path dependency implies a form of emergent policy locking-in or entrenchment of views that become difficult to challenge or dislodge, even with countervailing evidence. The ministers believed that it was crucial to maintain the approach that had been endorsed by the SWP, not simply because the SWP's findings were meant to reflect those of an independent scientific committee (which, as we have seen, is in doubt), but because "introducing new regulations threatened to make explicit the fact that those risks were not, and had never been, negligible or certain" (van Zwanenberg & Millstone 2003, p. 34). There was concern that such "flip-flopping" would produce the same kind of fearful public reaction that the MAFF and all the other relevant agencies had been trying to avoid from the start:

> Throughout the BSE story, the approach to the communication of risk was shaped by a consuming fear of provoking an irrational public scare. This applied not merely to the government, but to advisory committees, to those responsible for the safety of medicines, to Chief Medical Officers and to the Meat and Livestock Commission (Phillips et al. 2000, Vol. 1, para. 1294).

This is a crucial finding. All the expert, policymaker, and political advice was leaning in the same direction (Forbes 2004, p. 349).

An Interim Conclusion

In summary, there were four key factors that shaped the communication and management of the threat posed by BSE: (1) procrastination by the principal committees involved, (2) too close a relationship of the regulatory or advisory committees with both industry and the government, such that the market of beef took priority, (3) framing of the BSE issue in terms that understated the risk and the evidence for it, thereby discouraging serious public debate, and (4) a strong culture within the British science-policy establishment that endeavored to sustain a position over time that, from today's perspective, was driven not by an evidence-based philosophy, but rather by a desire to eliminate any possibility of public anxiety.

Period 2 (March 20, 1996, to 2000): The Link of BSE to Humans and the Resultant Health Crisis in Britain

March 20, 1996, marks the crucial turning point in the risk-management and communication strategies concerning BSE. It was on that day that the Secretary of State for Health, Stephen Dorrell, announced to the British House of Commons that the

SEAC was now suggesting that BSE was most probably linked to a new variant of Creutzfeldt-Jakob disease (nvCJD), and that 10 cases of this nvCJD had been diagnosed within the UK by the CJD Surveillance Unit (Jasanoff 1997, p. 222). This announcement sent British government agencies, food retailers, and public interest organizations into policy crisis mode, as the prevailing communication line until that day had been that BSE is not transmissible to humans. Although there was no **conclusive** evidence of a link between nvCJD and BSE, the Philips Inquiry later concluded that "the most likely explanation is that these cases are linked to exposure to BSE before the introduction of the SBO ban in 1989" (Phillips et al. 2000, Vol. 16, chap. 1).

This phase of the management and communication of the BSE risk can therefore be most accurately characterized as a crisis, and new information was needed to restore public trust in the safety of beef products. The central risk-management and communication question was, thus: How then can this be done? This part therefore explores some of the policy actions taken by both Britain and the European Union (EU). The EU was not the only new player to become involved in the risk management and communication strategies for dealing with BSE. The retail food industry, various public interest organizations, and the media also adopted a new and more open role in risk management. Following a detailed discussion of the European and British risk-management approaches, we proceed to evaluate the relative successes and failures of these strategies.

Key Institutions and Players in the UK

The Government

Once the House of Commons was informed that BSE was most likely to be responsible for nvCJD, several interest groups joined the debate about the risk-management, and especially the risk-communication, strategies. Nonetheless, the MAFF remained the institution primarily responsible for BSE policy. Given its previous prominence in dealing with BSE, it was also the main target of criticism now that BSE had become a human health risk. The policies it developed to manage the BSE/nvCJD risk are dealt with in detail in the sections below. The Spongiform Encephalopathy Advisory Committee (SEAC) also retained its central role in assessing the domestic risk of BSE, but what is of particular interest during this period is that other institutional players came to the fore, particularly with regard to risk communication.

The Food Industry

When on March 20, 1996, the MAFF did a U-turn on the issue of BSE transmissibility to humans, its credibility – and even that of the SEAC – became tattered. Not only were British consumers now confused about the relative safety of beef and other bovine products, but various sections of the retail food industry began to panic as the potential for catastrophic profit losses loomed. As a result of the government's discredited and unreliable risk-management approaches, as well as the desire to reassure frightened consumers, a number of major – as well as many minor – players in the food industry entered into the fray, in terms of both risk management and risk communication. Shelia Jasanoff (1997) gives numerous examples of interest groups throughout the food industry getting involved. They ranged from local pubs that mounted signs about the potential risks to restaurant giants such as McDonalds and the chief supermarket chains in the UK (i.e., Tesco and Sainsbury's). These companies took risk communication into their own hands; their strategies are discussed in detail below.

Public Interest Organizations

Various public interest organizations also stepped into the void of trust that had been created by the MAFF and their advisory committees. Organizations such as the Consumers' Association not only provided information to the public about which foods were safe to eat, but they also ultimately played a crucial role in brainstorming about the organizational changes that would be needed in the food-regulatory bodies. As noted earlier, one of the main criticisms leveled at the MAFF regarding BSE was its conflicting roles of promoting the business interests of British farmers and agriculturalists and being responsible for public health as it pertains to food. The Consumers' Association, as well as key academics in the Rowett Research Institute in Aberdeen, later became key figures in proposing the establishment of an independent Food Standards Agency (FSA) (Jasanoff 1997, p. 225). The creation of the FSA came to be seen as one of the few bright spots in the doom and gloom that surrounded the BSE/nvCJD problem, and an institutional initiative that will be discussed later in detail.

The Media

The print media also played an increasingly important role, not only in risk communication, but also in forcing regulators' hands in certain areas of risk management. According to Miller (1999), coverage of the BSE problem spiked twice in the 1990s (see Fig. 1 on page 1247 of this document). The first spike occurred in 1990, "when it became clear that TSEs could jump the species barrier, when 'Max' the cat was revealed to be infected in May." The second spike followed the 1996 announcement of BSE transmissibility to humans. These spikes would prove to have ramifications on consumers' behavior, and the BSE Inquiry ultimately concluded that the fear of drastic public responses (by way of the media) would be an overwhelming factor for the MAFF's secrecy of BSE leading up to the crisis.

This does not mean that the media is to blame for the MAFF's communication strategy; rather, it highlights the fact that the media not only played a role in communicating the risk, but also that its treatment of the risk was a central concern of the government in implementing its risk-management and communication strategies, a point we return to later.

The European Commission, the European Union, and the Standing Veterinary Committee

The most influential player in the risk-management domain at this stage of the BSE crisis was the European Commission (EC) and the EU, which sanctioned a ban on British beef and bovine products on March 27, 1996, just a week after the announcement of the link between BSE and nvCJD. As we shall see, this EC/EU ban had huge ramifications on the British beef industry, as well as an effect on the risk-management policies and approaches of the British government. It is important to note that this EC/EU ban was initiated by the EU's Standing Veterinary Committee (SVC), then ratified by the EC Commissioners (Packer 2006, p. 165).

Policy Measures and Risk Management Strategies

It is perhaps rather artificial to divide the policy measures and risk-management strategies into domestic (UK) and European (EU) components, as decisions taken by one body inevitably had knock-on effects on the other. This period in the BSE/nvCJD saga is characterized by an interplay between these two levels of government. Nevertheless, for the sake of clarity, each of these is dealt with in turn, with connections between the policies made evident where possible.

Domestic Policy Actions

In the wake of the announcement of March 20, 1996, concerning the likely transmissibility of BSE to humans in the form of nvCJD, a number of risk-management measures were almost immediately put into place. These policies and strategies were intended to serve the dual purpose of protecting British citizens from the possible health risks associated with BSE and reassuring them that actions were being taken. On top of this, these initial policies were meant to reassure the public that the government was indeed doing something and to salvage the beef industry (House of Commons Records 2007). Therefore, it was stated at this time that some British beef was safe to eat. For instance, on March 20 it was announced that the SEAC did not recommend an umbrella ban on all beef consumption by humans; instead, it was suggested that only older cattle (over 30 months) should be deboned, and all their trimmings, waste meat, and MBM should be kept out of both the human and animal food chains. Many of these recommendations were taken up later in March, but the vast majority of the EU member states saw these management strategies as inadequate. Thus, and the EU moved swiftly – on March 27 – to ban **all** exports of beef and bovine products from the UK.

The details of the EU ban are discussed in detail below. Suffice it to say here that the UK was reluctant to condemn the whole British beef industry to the same fate faced by infected cattle, by adopting the EU's policy of forbidding domestic consumption of **all beef**. Thus, the UK's policies were directed at controlling the spread of the disease, while at the same time suggesting that certain parts from certain cattle were still safe for human consumption. To a degree, this approach was precautionary, but it acknowledged the importance of greatly increasing the degree of regulation.

The first attempt to halt the spread of BSE was announced on March 28, 1996, as consisting of "new BSE controls, the calf slaughter scheme and financial aid for the rendering industry" (Phillips et al. 2000, Vol. 16, chap. 1). This enhanced "slaughter and aid scheme" was eventually complemented at the beginning of July by a cattle tracking system or the "The Cattle Passport Order 1996." From that date forward, all movement of all cattle born in the UK had to be documented (Phillips et al. 2000, Vol. 16, chap. 1). This was not only intended to help prevent the spread of BSE; it was also a precautionary measure to track future outbreaks of BSE in the UK to their origin, as well as to contain and minimize them. The Cattle Tracing System, which also required that all cattle be ear-tagged within 36 hours of birth, was officially launched on September 28, 1998 (Department for Environment, Food, and Rural Affairs [DEFRA] 2004).

A new set of risk-management policies was introduced with the goal of removing all infected cattle, while not banning consumption of all beef in the UK. On March 29 several different kinds of cattle bans were put into place. This implementation occurred only 9 days after the March 20 announcement, which is rather swift in light of the lethargy of the first period. Many of these actions by the government and the MAFF were taken in response to SEAC recommendations. For instance, the "Specified Bovine Material Order 1996" stated that the entire head of all cattle over 6 months of age had to be destroyed. Another example is an amendment to the "Bovine Spongiform Encephalopathy Order 1996" that banned MBM from all the animal food chains (Phillips et al. 2000, Vol. 16, chap. 1). Ironically, while this was also a SEAC recommendation such a ban (on MBM from all farm animal food chains) had first been suggested informally, but never seriously considered, back in 1988 by the Southwood Working Party.

Perhaps the most stringent risk management-measure taken in the UK during this period was the banning for human consumption of beef from older cattle. Although not acquiescing to the total ban of beef implemented by the EU 2 days previously, along with the other policies issued by the British government on March 29, the Beef (Emergency Control) Order 1996 prohibited "the sale for human consumption of any meat from bovine animals showing more than two permanent incisors" (Phillips et al. 2000, Vol. 16, chap. 1). The caveat of "two permanent incisors"[2] was meant to discriminate between older cattle deemed to be unsafe from younger cattle, which the MAFF – in light of the measures to control the spread of BSE – considered to be safe for human consumption. Based on the later recommendations of the SEAC, this discriminating safety marker was replaced on April 3, 1996, by a ban of all bovine animals from the human and animal food chains, as well as the slaughter of all cattle over 30 months of age.

By 1997, the "total number of recorded BSE cases reaches 179,087 in the European Union, of which 99.5% are in the United Kingdom … and as a result, between 1996 and 1999, 3.3 million cattle were destroyed" (Beck et al. 2005, p. 402). May 1997 also marks the end of John Major's Conservative government, which was replaced by Tony Blair and the Labor Party. This would prove to be a relatively significant development for the British approach to risk management and communication regarding BSE. As Jasanoff (1997) argued, "[The] MAFF had followed the Thatcherite policy of deregulation [that] had allowed for relaxation in rendering practices including treatment at lower temperatures, shorter treatment periods, and elimination of some solvents" (p. 226). She even suggested that such deregulation could have been at the root of BSE. Further, the change in government almost certainly led to a new, and ministry-independent, Food Standards Agency (FSA). The creation of this agency had been an item on the Labor Party's campaign agenda. Although Labor proved to be sluggish in establishing the FSA, it would eventually signal the end of an era when responsibility for public health in regard to food was held by the same ministerial body that was responsible for promoting the agriculture business.

As the UK continued to enforce its control measures throughout 1997 and 1998, there was a movement within the EC towards lifting the total ban (Beck et al. 2005, p. 402). However, the actual enactment of changes in the European policy would not be possible until more significant management strategies were put in place domestically in the UK. On August 1, 1999, the Date-based Export Scheme (DBES) was launched. This scheme, coupled with more stringent documentation for cattle, allowed the UK to export beef to the EU only if it was registered and produced under the DBES (DEFRA 2004). The key to the plan was that only beef products from cattle born after August 1, 1996 (BABs), could be exported. It was believed that the feed and slaughter policies adopted and implemented by the UK would effectively protect against the spread of BSE to cattle born after August 1, 1996. Therefore, cattle born after this date would be free of the disease. Farmers, nevertheless, had to register for the plan through the MAFF.

The good news that the door was now slightly open for British beef exports to Europe was tempered by the SEAC's announcement on December 22, 1999, that the government should not change the "over 30 month rule" that had been in place since 1996 (DEFRA 2004). All beef was still not safe to eat in the UK.

[2]The first kind of tooth in heterodont (multi-toothed) mammals.

EU Risk Management Strategies and Policies

The EU and the EC were central players in the adoption of risk-management strategies for BSE in this period, for two reasons: first, they determined whether the UK could export its beef anywhere abroad; second, the policies they adopted eventually exerted political pressure on the UK's domestic policies (Packer 2006, p. 165). The fact that the EU (excepting the UK, which voted no) had banned British beef from all their member states and citizens did not breed confidence among British citizens, who were being told that some British beef was still safe to eat.

It should be noted that, according to the Philips Inquiry, from late May 1990 through early 1991, France, Austria, Italy and what was then West Germany had all issued import bans on British beef. These national bans were issued well before the official EC/EU decision, which was announced 5 or 6 years later, on March 27, 1996. These ad hoc national bans on British beef presented a fundamental problem for the EC. The bans eventually proved to be unacceptable, because "all trade involving the member states should be governed by EU rules. The existence of some dozen national bans on UK beef was for them an unacceptable affront to EU doctrine as well as being illegal" (Packer 2006, p. 165). It is therefore somewhat surprising that the EC did not deal sooner with the national bans adopted by France, Italy, and others. In light of the transmissibility announcement of March 20, 1996, it became the top priority to present a unified front on the status of British beef.

Within 5 days of the transmissibility announcement, a meeting of the Standing Veterinary Committee (SVC) was held in which the safety of British beef and bovine products was ruled on. Because beef products were still available domestically, the British representative argued that they should also be available throughout Europe: "UK consumers were entitled to the same protections as consumers anywhere else – neither more nor less." However, the legality of the national bans was not the concern of the SVC, and it quickly recommended a total ban of British beef and bovine products. This recommendation did not become policy, because the SVC vote – a qualified majority, with only the UK in opposition – had to be circulated to the Commissioners before publication (Packer 2006, pp. 165–166).

The objection of the British Commissioners temporarily blocked the circulation of the SVC vote. This delay provided a short window for all kinds of UK officials to lobby the EC and the SVC to reconsider their stance (Packer 2006, p. 166). The British Prime Minister (then John Major) contacted the President of the Commission, and in an attempt to avoid the ban, the Chairman of the SEAC and the Chief Medical Officer flew to Brussels to testify in front of the SVC at its next meeting on March 26. All these efforts failed to persuade the SVC to make any revisions to their vote to recommend a ban.

The EU/EC ban contained a number of items that Richard Packer (Permanent Secretary of the MAFF from 1993 to 2000) found particularly troubling, if not plainly unjust. For instance, he claimed that the ban on all beef exports to **all destinations** was "unnecessary and that it gratuitously increased his [John Major's] political difficulties for no good reason" (Packer 2006, p. 165). The EC's rationale for its total ban on exports was that beef products sold to third – non-European – parties could potentially be redirected to EU member states. Britain was isolated: "what was good enough for British citizens was not good enough, apparently, for citizens of any other country in the world" (Packer 2006, p. 165).

The other problem Packer had with the export ban on British beef was that there was no expiration date; therefore, any amendment to the ban would require a qualified majority vote of the EU member states (Packer 2006, p. 166). In his view, this measure "unnecessarily constrained flexibility." He was quite right in thinking that it would make revising or lifting of the ban difficult – but that was the point. Whether or not this EC export ban of British beef reassured and protected EU consumers is discussed later.

As explained above, the British government and its ministries were busy during this period proposing and enacting risk-management strategies. On May 31, 1996, they presented their "BSE Eradication Programme" to the EC. It consisted of all the basic policies adopted by the UK to prevent the spread of the disease, as well as to eradicate the known cases. Therefore, it mandated the removal of all cattle over 30 months old from all the food chains. The incineration and rendering policies also were tightened and refined.

The UK was not the only country being subjugated to a tightening of controls over beef and bovine products during this period. The outbreak of BSE, and its link to nvCJD, led the EC to issue the Bovines and Bovine Products (Trade) Regulations. Adopted in late April 1998, these regulations strengthened the controls on "exports of meat, meat products, meat preparations, 'other products of animal origin' and pet food derived from bovines slaughtered outside the UK" (DEFRA 2004).

Britain had hoped that with risk-management procedures such as the Eradication Program in place, and the Bovine Trade Regulations now being applied throughout Europe, the ban on British beef would be reviewed. However, in early May 1998, the European Court of Justice (ECJ) "gave its final decision, which upheld the validity of the export ban on UK beef" – and UK beef alone (DEFRA 2004). This official ruling forced the British ministers and regulators to produce new and more extensive controls, so the European market could once again be opened to beef from the UK. This ultimately led to the Date-Based Export Scheme (DBES), which allowed "exports of deboned fresh beef and beef products from cattle born after 1 August 1996" (DEFRA 2007a).

Although the creation of the DBES was a major step in the recovery of the British beef industry, countries such as Germany and France continued to block exports of British beef. This policy flew in the face of the EC jurisdiction and laws, and it was not until March 29, 2000, after the EC threatened that the European Court of Justice (ECJ) would formally reprimand it, that Germany formally lifted its ban (DEFRA 2004). France, on the other hand, continued to flout the DBES, and threats from the EC to take the case to the ECJ if it did not reopen its borders to British beef did not cause France to budge. This trade dispute lasted until October 25, 2002, at which time France formally lifted its ban on DBES beef, but only because the "European Court of Justice impose[d] a financial penalty of 158,250 Euros per day on France for non-compliance with the ECJ ruling that its ban on the import of UK DBES beef was illegal" (DEFRA 2004). The ECJ dropped its objection in November of the same year (DEFRA 2004).

Risk Communications

In addition to the institutional diversification described above, this period of BSE risk communication can also be characterized by an increased visibility (widening and opening-up) of actors employing a variety of strategies. When BSE was confined to animal health issues, risk communication was monopolized by the relevant governmental ministries and their advisory committees. With the about-face announcement that BSE was in fact transmissible to humans in the form of nvCJD, these ministries (namely, the MAFF) and their advisory committees came under scrutiny for their risk-communication strategies. These critiques worked to discredit the MAFF, and as a result the British public turned to other institutions and actors (or other institutions and actors stepped in, or were called upon) for information about BSE/nvCJD. The parties and their output included Lord Philips and the BSE Inquiry he would lead; the James Report, which proposed the formation of an independent Food Standards Agency; the food industry, which took upon itself the task of assuaging the fears of the British public about beef consumption; and the media, which played a prominent role in informing the public about these institutions, as well as assuming a more direct role in risk communication.

The Phillips Inquiry and Report (1997/8 to October 2000): The BSE Inquiry

The Phillips Inquiry and Report (PIR) are crucial to understanding the communication of the BSE risk in the UK. It played such a central role for two reasons: first, it did not shy away from criticizing the approach taken by the MAFF and its scientific advisory boards. In fact, the PIR points to poor risk communication as the central failure of the MAFF during the BSE crisis (more on that below). Second, the PIR represents the government's attempt to understand, and report, what had gone wrong in its efforts to deal with BSE. It was therefore vital that the Inquiry be thorough, open, and honest. If it was not, any remaining trust the public might otherwise have in the government's risk management and communication policies would be jeopardized.

The Inquiry was led by Lord Justice Phillips of Worth Matravers, who had a lengthy legal résumé and was appointed to the House of Lords in 1998; Malcolm Ferguson-Smith, who was (and still is) a professor in the Department of Clinical Veterinary Medicine at Cambridge University; and June Bridgeman, a senior civil servant who had been Deputy Chair of the Equal Opportunities Commission before joining the Inquiry (BSE Inquiry 2000).

The BSE Inquiry was announced in Parliament on December 22, 1997, and Phillips was appointed as Chair on January 12, 1998. The Inquiry's official mandate was "to establish and review the history of the emergence and identification of BSE and new variant CJD in the United Kingdom, and of the action taken in response to it up to 20 March 1996; to reach conclusions on the adequacy of that response, taking into account the state of knowledge at the time; and to report on these matters to the Minister of Agriculture, Fisheries and Food, the Secretary of State for Health and the Secretaries of State for Scotland, Wales and Northern Ireland" (BSE Inquiry 2000).

The final report was returned to the House of Commons at the beginning of October 2000. It was extensive indeed. The 16 volumes, which consumed 412 pages, are the following:

Volume 1: Findings & Conclusions
Volume 2: Science
Volume 3: The Early Years, 1986–88
Volume 4: The Southwood Working Party, 1988–89
Volume 5: Animal Health, 1989–96
Volume 6: Human Health, 1989–96
Volume 7: Medicines and Cosmetics
Volume 8: Variant CJD (vCJD)
Volume 9: Wales, Scotland and Northern Ireland
Volume 10: Economic Impact and International Trade
Volume 11: Scientists after Southwood
Volume 12: Livestock Farming
Volume 13: Industry Processes and Controls
Volume 14: Responsibilities for Human and Animal Health
Volume 15: Government and Public Administration
Volume 16: Reference Material

All of these volumes, along with the evidence that the Inquiry drew upon to formulate the report, are publicly accessible via their online archive. Making this final report, and the evidence used to inform its conclusions, available to the public represents a marked departure from the practices of the previous committees and working parties, and indeed the PIR criticized these previous groups for their lack of transparency. Further, the PIR contains 139 days of public hearings, including the oral evidence that was presented there. This evidence is also available online in the form of written transcripts. The hearings were indeed open, as the public could attend them (BSE Inquiry 2000).

The central finding of the PIR is that the most serious failures in response to the BSE/nvCJD outbreak "concerned communication, rather than more fundamental problems with the substantive business of assessing or managing risk" (van Zwanenberg & Millstone 2003, p. 28). Some of these communication problems were already highlighted in the preliminary findings from Period 1. In their summary of the PIR findings, Beck et al. (2005) characterized the risk-communication problems as "including time delays, secrecy regarding scientific research, and the paternalistic attitudes of some ministers" (p. 402). More importantly, as later reiterated by Beck and colleagues, Phillips concluded that "government officials should have created a consistent risk communication and management strategy rather than relying on containment strategies" (Beck et al. 2005, p. 403).

The PIR successfully identified the failures of the British risk-communication strategy – a crucial outcome. However, as a piece of risk communication itself, the PIR did not go quite far enough for some commentators. For Beck and colleagues, the PIR fell short in identifying the causes of these risk- communication failures, and it did not – and perhaps could not – "explain why government decision makers stubbornly maintained a specific and highly controversial scientific position in the light of contrary evidence" (Beck et al. 2005, pp. 402–403).

The James Report (1997): Towards a Food Standards Agency

Another important document that speaks to the changing approach to risk communication is the James Report, which proposed a new institution to deal exclusively with the safety of food – the Food Standards Agency (FSA). A more detailed discussion of the make-up and structure of the FSA is presented in the later section, which addresses institutional transformations. It is important to discuss here the context and form of the report, which is relevant to these transformations.

When the Labor Party was still the official opposition in the UK, one of its campaign issues was a proposal for a new structure to govern food safety (Food Standards Agency [FSA] 2007). This proposal proved to be especially relevant in the wake of the March 20, 1996, announcement on the transmissibility of BSE, and even more relevant to the two conflicting roles of the MAFF, which were to be highlighted as a major source of the BSE/nvCJD problem. While still in opposition, the Labor Party asked Professor Philip James to propose a structure and function for the Food Standards Agency. James was a Professor at the Rowett Research Institute in Aberdeen, where he specialized in food and public health policy. He officially presented his report to the new Prime Minister, Tony Blair, shortly after Labor came to power in early May 1997.

Although the structure and role of the FSA would play a central role in future risk communication and management strategies concerning food (and ergo, BSE and beef), the most notable quality of the James Report was its open and consultative form. As soon as the report was given to the Prime Minister, it was "immediately published for public consultation" (FSA 2007). For the next 2 months, the documents were scrutinized by "representatives from the consumer sector, public health medicine, local government, veterinary services, scientific research, all sectors of the food production and distribution industries and a significant number of private individuals." There were over 600 responses to the report (FSA 2007).

This opening-up of the mechanisms by which new institutional structures would eventually manage the food risk is significant. Although this was not a risk-communication strategy per se, it did signal that risk communication in the future would be characterized by consultation and transparency, two features that were sorely lacking in the previous era. In November 1999, the Foods Standards bill, which was based largely on the recommendations of the James Report, received Royal Assent and ultimately became the Food Standards Act. The Food Standards Agency began operations in April 2000 (FSA 2007).

New Actors in Risk Communication

The introduction to this section on risk communication pays attention to the widening of the institutions and actors involved with, and interested in, the communication and management of BSE risks during this period. Jasanoff (1997) has highlighted the fact that the government ministries and scientific committees had created a void in reliable and trustworthy advice concerning beef when their transmissibility announcement of March 1996 explicitly contradicted all of their previous risk communications that had assured the British public that domestic beef products were safe. Jasanoff goes on to point out how new actors in the risk-communication sphere emerged to fill the public need for credible information. Among other things, she shows how segments of the food industry attempted to mollify the concerns and fears about beef products with "detailed fact sheets that not only vouched for the quality of their beef and beef products, but also explained the precautions being taken to monitor production" (Jasanoff 1997, p. 224). One such "fact-sheet" from Sainsbury – one of Britain's top four supermarket chains – reads as follows: "brand products containing beef are sourced from a select group of *approved* suppliers who are *visited regularly* by *qualified* food technologists … [and that] an additional control is provided by our long standing policy of not using MRM (mechanically recovered meat) in any Sainsbury brand product"(Jasanoff 1997, p. 224, emphasis added). Similarly, the fast food superpower McDonald's had some of its franchises in the UK post "signs promising that they would not serve hamburgers until they could establish a secure supply-line from Argentina" (Jasanoff 1997, p. 224).

These policies by segments of the food industry were not likely to meet the threshold that was now being set for informational transparency, or openness of risk contingencies in the case of government-issued risk communications. As Jasanoff (1997) pointed out, these statements of reassurance were "offered under private auspices and without the constraints of legal and political accountability" (p. 224). What is important to note here is not the content or effectiveness of the food industry's risk-communication policies, but rather the diversification or involuntary outsourcing of locales for risk communication. This phenomenon, characterized by supermarkets and food retailers actively promoting "healthy options", openly displaying fat and salt content, and generally taking a more active role in communicating food risks, seems to have been an ever-present feature of the BSE saga.

The Role of the Media in Risk Communication

The food industry was not the only non-governmental actor engaging in risk communication during the BSE/nvCJD affair. Crucially, the media did more than communicate risk; its presence, and the public concern it could provoke, determined how much uncertainty the MAFF and the scientific committees were willing to express. As this was one of the central findings of the Phillips Inquiry, the following quote from Forbes is worth repeating:

> Throughout the BSE story, the approach to the communication of risk was shaped by a consuming fear of provoking an irrational public scare. This applied not merely to the government, but to advisory committees, to those responsible for the safety of medicines, to Chief Medical Officers and to the Meat and Livestock Commission. (Phillips et al. 2000, Vol. 1, para. 1294). This is a crucial finding. All the expert, policy-maker and political advice was leaning in the same direction (Forbes 2004, p. 349).

David Miller, who has studied the role of the media in framing BSE, went so far as to argue that "calculations about the media – at least – influence the way in which things are said and can – at worst – engage scientists, advisors and politicians in misinformation" (Miller 1999, p. 1247). There was, however, no hiding from the media when the announcement of the transmissibility of BSE to humans was made on March 20, 1996. Miller has shown how this announcement led to a steep spike in reporting on BSE.

Miller went on to argue that this media reporting had a huge "influence [on] public belief and behaviour about risk. In the case of BSE, consumption of beef and beef products dropped dramatically in both 1990 and 1996 (by around 28% for household purchases) in 1990 [at the time BSE transmissibility to felines was announced] and 40% (for the market as a whole) in 1996 in the immediate aftermath [of the announcement of BSE transmissibility to humans]" (Miller 1999, p. 1249).

Miller did not suggest that the presence of the media was responsible for the policy of concealment adopted by the MAFF regarding BSE, nor did he argue that the media's only role in risk communication is fear-mongering. On the contrary, he demonstrated that the media's presence and involvement with risk communication could in fact lead to pressure for – or involvement in – more effective risk-management strategies. For instance, in the process of filming for British television a journalism program entitled *World in Action*, Stephen Dealler described how producing this piece of reporting "involved carrying out a number of tests which would not otherwise have been done" (Dealler 1996, as cited in Miller 1999, p. 1248). Similarly, the Institute of Environmental Health Officers (IEHO) used the media to bring about pressure onto the MAFF concerning abattoir practices. After expressing its concerns to the MAFF on numerous occasions, the IEHO finally turned to the print media, which reported on the practices in question. It was only after such pressure by the media, which occurred years after the original IEHO complaint to the MAFF, that action was taken to remedy the situation (Miller 1999, p. 1248).

Clearly, the media plays a large role in risk communication; whether that role is in the communication of risks itself, or in influencing how policymakers and regulators conduct their business. Portraying the media as the source of the risk-communication problems is probably as incorrect as portraying them as the "savior" or the solution to these problems. Blaming the media is a common tactic at times of crisis, when government agencies and scientists face criticism of their approaches to risk communication. Rather than making normative judgments about the media's role in risk communication and management strategies, Miller argued that what we should understand is that:

> when BSE declined on the media agenda (in June 1990, for example) the policy process could move back towards the closed and secretive model often preferred by policy makers and politicians re-emerging only fitfully in the years before the fully-fledged re-entry to the public agenda in March 1996. A key conclusion, therefore, is that there is no such thing as a general policy process or a general model of agricultural policy, since the policy process can and does change depending on the varying relationships between the public, the media, interest groups and policy actors. (Miller 1999, pp. 1249–1250).

Successes and Failures of the Strategic Management and Communication of the BSE Risk in the UK from March 1996 to March/April 2000

The risk-management and communication strategies fundamentally changed after the transmissibility of BSE was announced in March 1996. For Period 1, it was stated as a preliminary conclusion that there was much procrastination by the principal committees involved in the early BSE consultations and by the relevant policymakers. Further, it was concluded that these advisory committees had too close a management relationship with both the industry and the government, resulting in the marketing of beef becoming the top priority. The BSE issue had been framed in a way that understated the BSE risk and therefore discouraged serious public debate about it. Moreover, the British science establishment sought to sustain a position of non-transmissibility, which was not, from today's perspective, driven by an evidence-based approach. Although these activities undermined effective risk management and communication, this judgment is made after the fact. Did the MAFF and the other relevant regulators and policymakers continue with the same approach to the management and communication of the BSE risk? Did they learn any lessons from the early stages of the saga?

The following section explores some of the successes and failures of the period from March 1996 to March 2000. We shall see that many of the failed strategies that characterized the earlier period were not repeated. But new problems arose that involved new and different actors and required new and different policy measures; some of these could even be called mild "successes." One of the more important of these new policies was to embrace the human-disease model of risk, which raised transmissibility concerns that the animal-disease model did not.

The Cattle Passport Order: Tracking Policies as a Model for the Management of Other Risks

Section on Period 1 made reference to a phenomenon of legacy and "path dependency" that the Southwood Working Party had on the subsequent approaches and understandings of the BSE risk. It would appear that tracking policies (which along with the Date-Based Export Scheme were instrumental in reassuring the EC of the relative safety of British beef and thus causing a partial lifting of the total beef ban) have now secured a certain legacy in British approaches to risk management. Not only were such tracking policies instrumental in somewhat containing the outbreak of foot-and-mouth disease in the UK in August 2007, but a similar tracking system is being developed by the Department for Environment Food and Rural Affairs (DEFRA) to deal with the problem of avian flu that is currently affecting the poultry industry.[3] However, because a tagging system such as the one developed for the beef industry is impossible, given the gigantic number of chickens, the DEFRA is proposing to track the relevant industry players (i.e., farms, slaughter houses, transporters) so that the network of the poultry industry can be understood and the potential spread of the disease modeled.

A policy crisis can often hasten change, and the development and implementation of effective tracking and modeling systems may be one of the few positive outcomes to emerge from the BSE saga. Although these tagging and modeling techniques do not address the warehouse- and manufacturing-like conditions that make these systems necessary, and they will never really model levels of uncertainty effectively (i.e., taking account of the unknown unknowns), they are nevertheless schemes that are being re-used in other contemporary risk situations and ones that appear to fit the agribusiness context that Britain finds itself in.

Was the EC Ban Successful in Reassuring and Protecting the Public?

> Of course the veterinary authorities in these countries were well-aware that this [ban on all beef products from the UK] would achieve very little in term of protecting the public. But that was not the point of adopting national bans on imports from the UK; the point was to reassure the public. An irrational worry might, it was hoped, be mitigated by such a response even if it were, strictly, itself irrational. (Packer 2006, pp. 165–166).

During this period – and even before and after – the health risk to humans and animals from BSE was relatively, if not largely, limited to the UK. By 1997, the vast majority (99.5%) of the 179,087 cases of BSE in the European Union were located within Britain's borders (Beck et al. 2005, p. 399). A comparison between the number of cases reported outside the UK (Table 8.1) and inside the UK (Table 8.2) should make this clear (Fig. 8.1).

Further, as illustrated below in Table 8.2, there have been to date only six cases of nvCJD in continental Europe, and only four in the Republic of Ireland. France is

[3] Personal Interview, by Andrew Webster, (August 19, 2007).

Table 8.1 Number of reported cases of Bovine Spongiform Encephalopathy (BSE) in farmed cattle worldwide

	–1987	1988	1989	1990	1991	1992	1993	1994	1995	1996	1997	1998	1999	2000	2001	2002	2003	2004	2005	2006	2007
Austria			0	0	0	0	0	0	0	0	0	0	0	0	1	0	0	0	2	2	
Belgium			0	0	0	0	0	0	0	0	1	6	3	9	46	38	15	11	2	2	
Canada			0	0	0	0	1	0	0	0	0	0	0	0	0	0	2	1	1	5	2
Denmark			0	0	0	1	0	0	0	0	0	0	0	1	6	3	2	1	1	0	
France			0	0	5	0	1	4	3	12	6	18	31	161	274	239	137	54	31	8	
Germany			0	0	0	1	0	3	0	0	2	0	0	7	125	106	54	65	32	16	4
Ireland			15	14	17	18	16	19	16	73	80	83	91	149	246	333	183	126	69	41	13
Italy			0	0	0	0	0	2	0	0	0	0	0	0	48	38	29	7	8	7	
Japan			0	0	0	0	0	0	0	0	0	0	0	0	3	2	4	5	7	10	2
Netherlands			0	0	0	0	0	0	0	0	2	2	2	2	20	24	19	6	3	2	
Portugal			0	1	1	1	3	12	15	31	30	127	159	149	110	86	133	92	46	33	
Spain			0	0	0	0	0	0	0	0	0	0	0	2	82	127	167	137	98	68	
Switzerland			0	2	8	15	29	64	68	45	38	14	50	33	42	24	21	3	3	5	
USA			0	0	0	0	0	0	0	0	0	0	0	0	0	0	0	0	1	1	
UK	446	2,514	7,228	14,407	25,359	37,280	35,090	24,438	14,562	8,149	4,393	3,235	2,301	1,443	1,202	1,144	611	343	225	114	24

Source: World Organization for Animal Health, OIE (2007).

Table 8.2 Worldwide cases of variant Creutzfeldt-Jakob disease as of July 2007

Country	Total number of primary cases	Total number of secondary cases: blood transfusion	Cumulative residence in the UK > 6 months during 1980–1996
UK	163 (5)	3 (0)	166
France	22 (2)	–	1
Republic of Ireland	4 (1)	–	2
Italy	1 (0)	–	0
USA	3 (0)	–	2
Canada	1 (0)	–	1
Saudi Arabia	1 (1)	–	0
Japan	1 (0)	–	0
Netherlands	(0)	–	0
Portugal	2 (1)	–	0
Spain	1 (0)	–	0

Note: Number of patients alive in parentheses.
Source: National Creutzfeldt-Jakob Disease Surveillance Unit (2007).

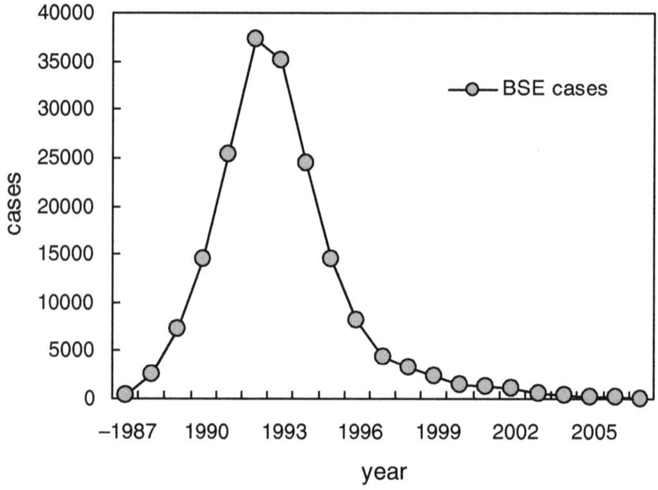

Fig. 8.1 Number of reported cases of Bovine Spongiform Encephalopathy (BSE) in the United Kingdom (Source: World Organization for Animal Health, 2007) Note. Cases are shown by year of restriction

a notable exception, which probably explains their reluctance to lift the beef ban in favor of the DBES. Despite the large number of French cases, the UK accounts for 84% of nvCJD cases in the EU. In other words, BSE and nvCJD are realities that are faced primarily by the British.

Given the fact that BSE and nvCJD have largely been limited to British producers and consumers, were the EU ban, and the national bans that illegally preceded it, irrational? Were the measures taken to reassure the public and protect public health effective? The latter question can be answered by consulting the data. If one accepts that nvCJD is linked to BSE and the rarity of nvCJD throughout Europe (France notably excluded), it follows that the public was relatively well protected from these diseases. Whether or not this "protection" is the result of the blanket ban on British beef is difficult to ascertain. It would perhaps be spurious to argue that the ban was solely responsible for preventing the spread of BSE and nvCJD, but to say that it was ineffective in preventing the spread is probably just as wide of the mark.

The question of whether or not the ban helped to reassure the various publics across the UK and Europe is entirely another matter. Richard Packer, Permanent Secretary of the MAFF from 1993 to 2000, is probably correct to assert that the EC ban on British beef created domestic problems for the Major government, as certain beef products were still being sold in the UK. Perhaps a more appropriate question is whether the choice to keep beef products on the domestic market was rational. Clearly, the British public became confused when they saw that their government was issuing one set of policies on beef consumption, while the EU and the EC were making conflicting statements about the safety of British beef. This could not have been reassuring. When one examines the literature on how "publics" go about conceptualizing risk, a number of important points come to light. In her study of the public and effective risk communication, Lynn Frewer concluded that "when the public want information about a risk, they prefer a clear message regarding risks and associated uncertainties, including the nature and extent of disagreements between different experts"(Frewer 2004; Frewer et al. 2004; Frewer & Fischer 2005). Such clarity was evidently not accorded to British citizens, but to lay the blame for the confusion at the door of the EC, as Packer does, is simply misleading.

Did the national bans and the EC ban reassure the European public? Packer (2006) asserts that "continental consumers did not find this reassuring. They seem to have taken the view that if the import ban was justified in 1996 then it ought to have been imposed much earlier, when agreed the risk had been greater" (p. 166). Again, this question about reassurance is difficult to answer. Although this particular chapter focuses on the UK and its risk-management and communication strategies, that perhaps misses the point. Questions about the effectiveness of these strategies in reassuring the public at a time of crisis assume a very narrow understanding of "risk," coupled with a belief that government management and communication policies (whether domestic or European) are the only, or at least the principal, vehicle through which such reassurance can be achieved. Alison Shaw (2004), whose study explicitly explored lay accounts of food safety and BSE, has argued that there is a need to acknowledge the British public's complex and multilayered understanding of food issues. Rather than conceptualizing BSE as "risky," Shaw has argued that "within the themes [of food safety], a variety of inter-connected concepts emerged in relation to risk, including danger, chance, safety, vulnerability, trust, responsibility, blame, control and choice. This adds empirical weight to social science work that, in seeking to define risk, has shown how risk forms the tip of an 'iceberg' of related concepts" (Alaszewski et al. 1998, as cited in Shaw 2004, p. 168). Therefore, it is perhaps not possible for **any** risk management or communication strategy to **reassure** "the public" about the "risks" related to BSE, as the concept does not adequately address the realities the public experiences. At the same time, it is important not to deny the role that risk communication can and should play in addressing the uncertainties within science and thereby making science more accountable to the public, or more "socially robust" (Nowotny et al. 2001). The issue is clearly not so much about reassurance as about accountability and transparency.

Further, as Shaw also makes plain, the public perception of risks (if we want to continue using this term in light of Shaw's argument), and the degree to which it feels reassured that the risks are not serious, are not solely the product of the government's management and communication strategies. To believe otherwise is not only too narrow, but it may contain a certain level of hubris on the part of ministers, regulators, and other officials. Shaw (2004) was in line with the prevailing understanding of risk within the social sciences when she stated that "food risks are constructed and negotiated by social actors within the varying relationships and social contexts of their everyday lives"; they are not defined simply by the policies of the MAFF and the EC. She went on to show how sociologists have come to alternatively conceptualized risk. By drawing on the work of well-established theorists such as Brian Wynne in the public understanding of science, Shaw (2004) rightfully stated that "risks are constructed from the 'situated knowledge' of lay actors (Wynne 1996, 2005), where risk is usually dealt with at the level of the local, the private and the everyday, rather than the more universal notions of risk in 'expert' and academic discourses" (Shaw 2004, p. 168). In other words, to expect the British and other Europeans all to have the same understanding of risk, or to understand it the same way that scientific advisors or policymakers do, neglects the contexts in which people live, as well as the "alternative rationalities" they apply to deal with the risk (Lupton 1999).

Therefore, although Richard Packer might have believed that both the public and EC were acting irrationally in issuing a total beef ban, such a belief falsely assumes a single (and cognitivist) "rationality" in how people make sense of risk. Simply because the perception of the BSE risk held by the public and the EC is (deemed to be) "irrational" from Packer's perspective, and maybe even from the perspective of members of the SVC or CVL, that does not de-legitimize the policy course that was followed, nor does it mean that the European public would have felt any more reassured about the safety of their food had the total British beef ban **not** been issued.

Period 3 (March/April 2000 to the Present): Institutional Change

At the height of the BSE crisis, when Britain was reeling from the transmissibility announcement, Shelia Jasanoff made the following statement regarding what she felt underlay the problems associated with governments, their policies in addressing the BSE risks, and the public:

> My own assessment is that British society has changed in profound ways that call for new forms of engagement between citizens and their government, and that institutions which may have been robust enough in their time will have to reconsider some of their fundamental assumptions in order to catch up with the altered state of things. (Jasanoff 1997, p. 226).

This profound change that Jasanoff believed to be occurring in the 1990s bears witness to institutional transformations, as well as to serious modifications in the state's approach to risk management and communication in the early part of the new millennium. Although substantive BSE policies were continually being introduced during this period, this part is largely concerned with the institutional changes as well as some of the novel approaches to risk management and communication that were – and continue to be – undertaken in the UK. Just as the BSE/nvCJD problem has "quieted down" during this period (particularly from 2003 onwards), so too have the risk-management and communication strategies receded.

Institutional Transformations

The Dissolution of the MAFF

It is no real surprise that in the summer of 2001 the MAFF was replaced by the Department for Environment, Food and Rural Affairs (DEFRA) (van Zwanenberg & Millstone 2003, p. 27). It is surprising, however, that it took 3 years to establish the Food Standards Agency (FSA), which took over the regulation of food (and related substances) from the MAFF in 2000. Many commentators have blamed the MAFF's poor communication of the BSE risk on its two conflicting roles of promoting the interests of British agribusiness while at the same time supposedly being responsible for public health as it pertains to food. These dual roles proved to be unworkable, and as Rothstein argues, "the regime [set up to deal with BSE] was poorly linked-up. Horizontal relations between the MAFF and its relatively weak policy partner – the Department of Health – were poor, whilst vertical relations between central government policy making and local government enforcement were virtually non-existent" (Rothstein 2003a, p. 4).

Furthermore, a phenomenon of institutional attenuation of risks, processes that serve to diminish inspectors' perceptions or awareness of a risk (management policy) was discerned (Rothstein 2003b). It was hoped that this dissolution of the MAFF would usher in a new era in the management and communication of food and animal risks. It was also hoped that some lessons had been learned during the past 15 years and that BSE would remain fixed on the radar screens of these new institutions.

The Food Standards Agency, and the Department for Environment, Food and Rural Affairs

As described, the establishment of the FSA as an institutional reform had been on the Labor Party's agenda for quite some time. With its landslide electoral victory in 1997, Labor was able to move ahead with this proposal. The conflicting roles that had plagued the MAFF (and perhaps had been responsible for the risk-communication strategy of concealment it employed in the early days of the BSE saga) were resolved by giving the FSA **no** role in the promotion of British agriculture and farming (Rothstein 2003a, p. 4). Further, the FSA was to be a non-ministerial agency, thus affording it a level of independence that had not been afforded the MAFF. It was hoped that this independence meant that the FSA would be less involved than the MAFF had been in direct political interventions. The problem of institutional fragmentation that Rothstein cited was specifically a problem for the MAFF, the DoH, and the other levels of government dealing with BSE. It was hoped that this problem could be resolved through the "consolidation of food safety standard-setting responsibilities within the FSA and by giving the agency extended monitoring powers over enforcement activities by local government and other agencies" (Rothstein 2003a, p. 5).

The MAFF had received much criticism for its failure to balance its obligations as food regulator and agriculture promoter. Therefore, the FSA was structured differently. It had a 12-member "stakeholder-style" board of directors that provided a mixture of expertise, experience, and consumer interests, which it was hoped would "prevent regulatory capture" (Rothstein 2003a, p. 5). Ultimately, it was also hoped that this diverse board, ministerial independence, and the consolidation of food responsibilities would make the FSA transparent, help to improve its risk assessments and decision making, and, perhaps most importantly, enhance the public's confidence in their food regulatory bodies.

With the FSA able to concentrate on the safety and regulation of food, and food alone, the DEFRA took over the other regulatory responsibilities that had been under the remit of the MAFF. This is not to say that the DEFRA has no interest in food issues, but rather that the responsibilities of the DEFRA and the FSA are quite different. It is clear from an examination of its website (see excerpt below) that the DEFRA's organizational mandate is much more market-orientated than the FSA's:

> The UK food and drink manufacturing and retailing industries contain many world leaders, but the need to secure competitive improvement has never been more important. DEFRA as sponsor department for these industries continues to work closely with individual sectors and key businesses to foster and promote greater competitiveness and to remove obstacles to growth. The *Food and Drink* pages on our site have been divided into the following subject areas:

Help with marketing – describes the incentives which DEFRA offers to improve marketing at home and abroad.

Help with exporting/importing – covers exports and trade in various products and services, including Services to Exporters and other export promotion work of DEFRA's Agri-food Exports Division. Also *Illegal imports*.

Eggs and poultry – provides general guidance and reference information for the eggs and poultry sectors, including regulatory and market information, and links to other poultry-related organisations.

Milk and milk products – provides information on milk production and marketing policy.

Beef labelling – provides guidance and information for the compulsory EU beef labelling system.

Competition – information on competition in the grocery retailing sector.

Regional and local foods – provides information on the regional and local food and drink sector as well as guidance on European Union regulations for the protection of food names on a geographical or traditional basis.

Food sponsorship – information for the Food & Drink Manufacturing Industry (including alcoholic drinks).

Organic production – information on the organic sector, including certification and imports.

Innovation, technology transfer and research grants – Information on Defra programmes, available on the DEFRA Sustainable Farming and Food Science website (DEFRA 2007b).

The DEFRA is a large organization with a variety of interests, ranging from climate change to wildlife and the countryside. With respect to the management and communication of BSE issues, the DEFRA claims to be the "authoritative source for BSE."[4] It provides information on legislation concerning BSE and TSEs, as well as up-to-date news on the BSE issue. It has produced guidelines for monitoring and eradicating TSEs, and it is a source for regulations concerning almost all farming practices and animal protection issues.

It should also be noted that the EC/EU continues to play a central role in the creation of farming regulations and legislation – and therefore of risk-management strategies – for Britain and the other EU member states. For instance, the document EC 999/2001 and its amendments "set out the requirements for the monitoring, control and eradication of TSEs as well as the controls on feeding stuffs and specified risk material (SRM) in relation to TSEs" (DEFRA, BSE Legislation). These regulations apply to all member states, including the UK, and enforcement power is provided by domestic laws.

The Spongiform Encephalopathy Advisory Committee

In light of the transformations that have taken place since 2000, it is perhaps worth noting some of the continuity that exists in the institutions dealing with

[4]See DEFRA website: http://www.defra.gov.uk/animalh/bse/index.html

BSE risk management and communication strategies. The Spongiform Encephalopathy Advisory Committee (SEAC), which was centrally involved in giving advice on risk management before and during the transmissibility stage, remains intact. Given that there are now three primary government bodies concerned with BSE, it is apt that members of the SEAC are jointly appointed by the Ministers and sponsored by the DEFRA, and the DoH, and the FSA (SEAC 2007).

The SEAC's role continues to be that of providing independent, expert scientific advice on all TSEs. The remit that the SEAC currently displays on its website is indicative of some of the previous critiques that scientific advisory committees faced in the wake of the BSE crisis. Note the emphasis on the recognition of uncertainty, as well as independence, in the following excerpts outlining the SEAC's functions: (1) Provide independent scientific advice on food safety and public and animal health issues relating to TSEs, while taking account of the remits of other bodies with related responsibilities. (2) Provide a scientifically based assessment of the risk that TSEs pose for public and animal health and food safety, while taking appropriate account of scientific uncertainty and assumptions in formulating advice. This advice conveys the nature and extent of the scientific uncertainty (SEAC 2007).

Policy Measures

As mentioned in the overview of institutions, many policies have been formulated in the UK to manage the BSE risk and to uphold the European legislation on TSEs, feeding stuffs, and specified risk materials (SRMs). Many of these pieces of legislation have been instrumental in the continued – albeit still only partial – lifting of the European beef ban that took effect in December 2000 with the Commission adopting a regulation that required "the UK to ensure that any meat from animals aged over 30 months can only be released for human consumption in the Community or third countries if tested negative for BSE. The Regulation also laid out the rules for the purchase for destruction schemes in other Member States" (DEFRA, Q&A). It is perhaps surprising to note that the total ban on British beef was not lifted until May 2006. When EU Regulation 657/2006 came into force, it amended "Annexes III and XI of the Regulation and repeal[ed] Decisions 98/256/EC, 98/351/EC and 1999/514/EC to allow exports of boneless meat and meat products from UK bovines, born after 31 July 1996, slaughtered after 15 June 2005; [as well as] exports of bone-in meat from UK bovines, born after 31 July 1996, [and] slaughtered after the Regulation came into force" (DEFRA 2007a).

The details of these specific regulations and pieces of legislation (as well as those that led up to them and made them possible) are available through the DEFRA website.

Novel Forms of Risk Management and Communication

FSA Stakeholder Decision Making with BSE in 2002

Since its inception, the FSA has been dedicated to altering the processes by which risk management and communication takes place with regards to food. According to their website, as well as the detailed case study undertaken by Rothstein (2003a), the FSA strives to espouse three principles: "putting consumers' interests first", "openness and accessibility", and "independence." We have already detailed some of the institutional structures built into the FSA that it was hoped would ensure that these principles are upheld.

Relatively early in its history, the FSA faced a risk-management problem concerning BSE in sheep. Examination of the existing controls for this risk in sheep was driven in part by the significant uncertainty about the levels of infectivity within the herds, as it is difficult to distinguish BSE from scrapie. Moreover, in 2001 the French food authorities came to the conclusion that "processing intestines from scrapie-affected sheep into sausage casings could leave some residual infectivity and had recommended a ban on intestine" (Rothstein 2003a, p. 6). In reexamining the controls on food products from sheep, the FSA did not handle the issue internally nor did act unilaterally. In an attempt to conform with its founding principles of "putting consumers' interests first", "openness and accessibility", and "independence," the FSA held a stakeholder meeting in December of 2001. There were over 100 stakeholders in attendance, from which a subgroup was hand-picked by the FSA to deliberate on the risk in more detail during 2002 (Rothstein 2003a, p. 6). Specifically, the subgroup was asked to make recommendations concerning the safety of natural sausage casings made from sheep intestine, as these had undergone scrutiny in France.

One of the central issues in this case of stakeholder risk-management deliberation is the fact that the select group was presented with two conflicting risk assessments: one from a group at Imperial College and the other from a private scientific consulting agency – DNV Consulting. The Imperial group argued that converting the intestine into casings would reduce the infection rate 10-fold, but DNV claimed it would be closer to 100-fold. The difference is important, because if the reduction was 10-fold, the sausage casings could account for up to a third of the total exposure to BSE infection, whereas if it was 100-fold, the casings could account for only about 9% of the total potential exposure. The stakeholder group, and ultimately the FSA, took the safe route and recommended a precautionary ban in 2002 (Rothstein 2003a, p. 7). As with all bans, this one had to conform to the EC's regulatory framework, namely the Scientific Steering Committee (SSC). Given the hard-line approach that the EC had previously taken on the BSE risk, it is somewhat surprising that the SSC waffled. After reviewing the evidence (including some provided by the sausage casing manufacturers) they rejected the ban, justifying their decision by the results of their conservative risk assessment. However, they subsequently proposed a ban themselves, although it applied only to the part of sheep intestine that could be the most infective (Rothstein 2003a, p. 7; 2004). Rothstein proceeded to investigate how closely the FSA's risk-management in practice squared with their founding principles, which were supposed to be the cornerstone of a new era and a new approach to the management and communication of the BSE risk.

Was the ban on sausage casings necessary to put the "interests of consumers first"? Were the stakeholders' deliberations indeed "open and accessible"? Was the FSA, in fact, acting "independently"?

According to the FSA, "if there is uncertainty we [the FSA] shall take a precautionary approach" when it comes to consumers' health as it pertains to food (FSA 2001, p. 23). However, the FSA "does not elaborate on how to apply the principle of proportionality where the uncertainties of precautionary action make the calculation of cost and benefits difficult. Nor does the FSA explicitly make clear where the public interest lies in balancing paternalistic protection against choice" (Rothstein 2003a, p. 8; 2004). Perhaps a more important question is why the FSA, if it was genuinely concerned about protecting consumers against the risk of BSE in sheep, singled out sausage casings.

> If, however, it was assumed that there was a significant reduction in infectivity by processing intestines into casings, as assumed by the DNV study, then the greatest risk to consumers would not be from natural casings but from infective materials such as lymph nodes that are endemically present in sheep meats. On that basis, the DNV calculated that potential infectivity in a leg of lamb could be five times greater than in a 250 g meal of sausages (FSA 2002). In that case, the most precautionary measure would be to ban normal carcass meat such as in joints and chops, not sausage casings. (Rothstein 2003a, p. 9).

In the face of these conflicting risk assessments, what could be judged as "the precautionary approach" to "putting consumers interests first" became complicated and anything but straightforward. Rothstein went on to argue that sausage casings were "singled out" for economic reasons. A ban on mature sheep would cost £115 m, whereas the ban on casings would cost only £6.5 m. Rothstein continued: "... from the FSA's viewpoint, therefore, not only was the recommendation [of the casing ban] precautionary but also it was easy to implement and proportionate" (Rothstein 2003a, p. 10). The FSA did not want to be tarred with the same brush as the other organizations and ministries in the UK, which historically had done nothing when faced with a possible BSE risk. The FSA felt strongly that some kind of action was needed, but did they want to decimate the sheep industry with a near-blanket ban? According to Rothstein (2003, 2004), "the casing recommendation was not so much a precautionary ban, but rather more a sacrificial lamb" (pp. 10–11).

Rothstein went on to question the degree to which the stakeholder meetings about sausage casings were indeed "open and accessible". He described in detail how the core stakeholder groups were "overwhelmed with FSA and other state-related personnel or advisors – some of them very senior (FSA 2002: Annex 1 [Annex E])" (Rothstein 2003a, p. 12). The select stakeholder group comprised the Chief Executive, the Board Chairman and Deputy Chair of the FSA, two scientists who were serving on government advisory committees, a representative from the Meat and Livestock Commission (a non-departmental public body), a representative from the Welsh Assembly (government), a farmers' representative, and two consumer representatives. The group did not include key stakeholders, such as retailers, the abattoir industry, or the natural-sausage-casing industry. It thus is debatable how "open and accessible" the deliberations were, and whether and how the agenda was, and has been set relatively pre-defined and closed.

Rothstein claimed that there was a potential conflict of interest within the FSA stakeholder (deliberation) group, and this jeopardized the group's "independence." Because the group was chaired by the Chairman of FSA Board, it could have been embarrassing if the FSA Board would have chosen to go ahead with the casing ban had the stakeholders decision not recommended it. Rothstein also argued that the FSA can never be truly independent, because its work is so closely tied to EC/EU policy making.

Whether or not the FSA was able to maintain its founding principles during the stakeholders' deliberations on the BSE risks of sausage casings would make an interesting case study, and we examine this in more detail in the "successes and failures" section to follow. Even if it is true – as Rothstein asserted – that the FSA was unable to adhere to these principles, in 2003/4 the EC banned the use of sheep ileum (i.e., the portion of the intestine at high risk for infection). This risk-management strategy might be considered moot, as the UK, European, North American and International Sausage Casing Association had already put in effect best-practice recommendations that included the routine removal of ileum before manufacturing (Bradley 2002).

The FSA nevertheless continued to "engage publics" and attempted to uphold its founding principles through such events as the consensus conference on genetically modified foods in 2003.

Successes and Failures in the Strategic Management and Communication of the BSE Risk in the UK from 2000 to the Present

One of the most notable characteristics of British policy making has been the tendency to form expert and advisory committees when the government and regulators are presented with potentially dangerous situations (Jasanoff 2005). This approach was not abandoned during the most recent period, but we do see a move towards more independent forms of review, as well as risk assessments that are kept at arm's length from the government; a good example is the FSA. In reviewing the case as a whole, we consider below some of the key changes.

Risk Communications

From the time BSE was discovered, through to the creation and proceedings of the SWP, the MAFF remained very quiet about the potential risks associated with the new cattle disease, BSE. According to van Zwanenberg and Millstone (2003), this approach amounted to a "strategy of concealment" (p. 34). Although in the summer of 1987 various farming and trade newsletters were reporting on a "mystery disease" in cattle, it was not until October 25, 1987, that the *Sunday Telegraph* published the first report about BSE in a national newspaper (Phillips et al. 2000, Vol. 16, chap. 1). It is important to note that the newsletters and the *Telegraph* article were **not** based on news releases from the MAFF, but rather on investigative reporting. Once the SWP had been convened in 1988, "[t]hose [MAFF] officials also persuaded scientific expert advisers not to mention the possible risks in public or in print, and deliberately chose not to warn importers that the material was contaminated" (van Zwanenberg & Millstone 2003, p. 34). At this stage, there simply was **no** risk-communication strategy, or, in other words, the strategy was **not to communicate** with the public about BSE. Indeed, regarding this early stage of the BSE saga, it has been observed that more information was obtained more easily from the British Veterinary Association than from the MAFF (Miller & Reilly 1995). The MAFF claimed that at that stage there was no scientific evidence indicating that BSE was transmissible to humans, and that to suggest otherwise would be alarmist, unscientific, and detrimental to the British beef industry.

This (risk-communication) strategy of concealment lasted until the SWP made its report about BSE concerns central, and its potential subsequent risks (especially, the potential risk of its transmission to humans) notified. Millstone and van Zwanenberg (2001, p. 107) argued that the content of this report, as a form of risk communication, was shaped by what they called "upstream framing assumptions." Drawing on the work of Jasanoff and Wynne (1998), they argued that the risk policy-making needs to be seen in the context of various socio-economic, political and ethical factors. They contended that these factors work to shape risk policy and communications by acting as a set of "social framing assumptions," as illustrated in Fig. 8.2.

It is clear that the MAFF exerted pressure on the SWP to produce results that would not alarm the British public and not lead to large regulatory expenditures.

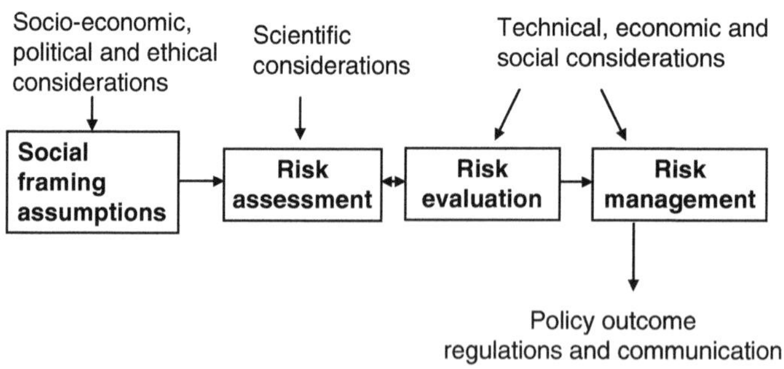

Fig. 8.2 The risk model and framing assumptions (revised from Millstone & van Zwanenberg 2001)

These "upstream framing assumptions" worked in combination with "downstream judgements about the acceptability of uncertain risks in exchange for the economic and political benefits of sustaining (at least temporarily) the UK meat and dairy trade." Millstone and van Zwanenberg (2001) argued that these two concerns of the MAFF played a key role in framing the SWP's subsequent communications. This framing led the SWP Report to stress the importance of acting only on risks that were known to be certain and could be addressed in practice. Raising awareness about potential risks, to which nothing could be done, would only scare the public and threaten the British beef industry.

The Rearrangement of Institutions

As so much of the blame for the mismanagement and miscommunication of the BSE risk was laid at the doorstep of the MAFF, it is natural to point to its breakup as one of the successes in this era. Much of the problem related to miscommunication, which in turn reflected a rather cynical judgment about what the public could in fact understand (Miller & Reilly 1995, p. 321). At the same time, the public had to digest a diverse assortment of claims and counterclaims, so it is perhaps not surprising that misunderstandings occurred. Here we see a classic case of the government mismanaging uncertainty while at the same time presuming a misplaced certainty on their own part about the potential risk of BSE (Myers & Raffensperger 2006, p. 110).

The Phillips Inquiry Report concluded that the risk-communication policy – which can be characterized as one of concealment and secrecy – was the major failure in the BSE saga, trumping the problems surrounding the government's risk-management policies. Consequently, a more open and transparent approach (to reviewing risk communication and in risk communication itself) was taken in the BSE Inquiry led by Lord Phillips and for the James Report, which led to the establishment of the Food Standards Agency. This new approach was welcome, not only because it could be linked to the lessons learned from the failures of the MAFF, but also because it was a sign that government institutions – now under the leadership of the Labor party – may be "catching-up" to the profound changes that Jasanoff (1997, p. 226) referred to as meeting the need for "new forms of engagement between citizens and their government."

It is also true that following the dissolution of the MAFF, the interests of the farming and agribusiness communities could no longer be so closely intertwined with those of the public regarding food safety. The creation of the Department for Environment, Food and Rural Affairs to manage the commercial interests of British farmers, as well as the creation of the Food Standards Agency to look after the public interest as it pertains to food, was a positive step forward.

According to van Zwanenberg and Millstone, who have traditionally not shied away from making strong criticisms of how risks are dealt with at the institutional level in Britain, "the FSA recently and boldly recommended that all its expert advisory committees should conduct their business in open sessions. It stipulated that unorthodox and contrary scientific views should be considered, and that advisory committees should always provide a clear audit trail showing how and why they reached their decisions, where differences of opinion had arisen, and which assumptions and uncertainties were inherent in their conclusions" (van Zwanenberg & Millstone 2003, p. 35). To these analysts, these recommendations signaled a significant change from the previous era and represented institutional transformations that could rectify some of the earlier failed approaches to the management and communication of risk – especially with regards to BSE. As meticulous as the analysts have been in their approach to risk management, they also have reservations concerning the – now current – institutional arrangements: "... separating industrial sponsorship from responsibilities for regulatory policy-making was essential but not sufficient" (van Zwanenberg & Millstone 2003, p. 35).

Of course, we must recognize that risk management and communication strategies are not the jurisdiction of one single government (national) body. Harmonization of the risk communiqués between the British government and other European governments is essential so that the public's confusion on the risk matter at hand does not turn to distrust and disenfranchisement with the risk-management policies. Furthermore, the monopoly on risk communication has been broken up and is now distributed among industry sectors, consumer organizations, and the media playing ever important roles.

This redistribution suggests that a more complex and layered understanding of how the public conceptualizes risk is necessary if risk-management and communication strategies are ever to become effective. The traditional notion of "irrational public responses" to risk, as well as to communication and management strategies, is inconsistent with social science research that has led to a more nuanced and multifaceted understanding of how risk is constructed and negotiated by people in their everyday lives. What policymakers understand to be "rational" communication and management strategies may neglect these processes and thus fail to reassure the public. As Miller (1999) has argued:

> Much of the process of deliberation by and conflict between scientists and policy makers is informed by assumptions about the social world and how it works (how many farmers will comply? how will the public react? will the media interpret it in a particular way? what will ministers or civil servants accept?) This is an area where social science can help to understand both the policy process as it is and how it could be. Yet, social science perspectives tend to play little part in these debates. (Miller 1999, p. 1249).

Stakeholder Engagement: A Way Forward in Risk Management and Communication?

In his assessment of the FSA's risk-management strategies concerning sausage casings, Rothstein (2003) made a good point about the problems it encountered, or perhaps even produced, in attempting to put its principles of consumer interests, openness, and independence into practice. His assertion about the futility of singling out sausage casings (while omitting the rest of the lamb industry), and his point that this ultimately amounted a sacrifice so that the FSA could be seen to be taking action, are particularly strong. He went on to argue, however, that the precautionary approach is only as good as the sense/degree of uncertainty one is trying to manage:

> Indeed, whilst precautionary action is often advocated in a situation of scientific uncertainty, it is often difficult to judge the precautionary nature of proposed action precisely because of the presence of scientific uncertainty. The sheep and BSE case was one such classic case, in which alternative and conflicting policy options could be equally represented as precautionary and proportionate, depending on the framing of risk management questions and choice of evidence. (Rothstein 2003a, p. 14).

The central questions appear to be which risk assessment to act on and what is the (relative) value of gaining scientific certainty for the development of risk-management strategies. Are these institutional problems specific to the FSA ("institutional factors have prevented the agency from fully fulfilling its remit in relation to each of its three guiding principles" (Rothstein 2003a, p. 14)), or – as we claim – is it, unfortunately, primarily a problem of science in the broad sense and its application to risk management?

Rothstein remains critical of the FSA's attempt to uphold its three central principles, putting consumers' interests first, openness and accessibility, and independence. He stated that these principles may "have rhetorical purchase, but meeting such objectives in practice can be more difficult" (Rothstein 2003a, p. 18). This is true, almost to the point of being banal. Further questions should be asked regarding the attribution of normative value to the achievement of scientific certainty, as well as about the degree to which a scientific consensus is desirable in the context of risk management and communication.

Should the "new" approach of the FSA to risk management and communication be labeled a success or a failure? As mentioned above, our view is that the processes that the FSA now have in place for risk management and communication are a great improvement over those of the previous MAFF era. Risk management is complex, and it will almost certainly upset some actors – such as the sausage casing industry. In light of that, caution should be exercised in condemning the relatively young FSA. Nevertheless, Rothstein's observations are at times apt. Rather than rendering a judgment of "success versus failure," it would perhaps be more prudent to point out that Rothstein's research has very effectively demonstrated the value of social scientists' examination of the institutions responsible for risk management and communication. By scrutinizing these processes and institutions, and holding them accountable to their self-professed founding principles, social scientists make a constructive contribution to risk management and communication per se. If there is to be a "success" here, it is this ongoing, and nearly real-time, ballasting of the institutions responsible for these very important risk strategies. Although more work remains to be done by social scientists, some comfort can perhaps be taken in the fact that reconstructions of the risk-management decision-making processes – such as that undertaken by Rothstein (2003) – were indeed "partly reliant on the FSA's open procedures, such as the open Board Meetings" (p. 15).

Conclusion

Through this detailed discussion of the BSE case within the UK, we have shown that there is a range of factors that have influenced the management of risk and its communication to the public since the first case of BSE was reported over 20 years ago. Many of the public policy issues raised by the case relate to how and through what mechanisms government agencies can secure a social license for what they do, and as such, are seen to act in a socially legitimate way. In the earlier stages of the BSE saga, the agencies were less accountable and incorporated within an established set of interests that, paradoxically, worked against the interests of both the state and the industry over the medium term. The public demands for greater openness made during this more recent period were driven not only by criticisms of the government's handling of the various crises and the growing anxieties over vCJD, but also by broader changes in the relationship between expert authority and lay audiences.

The lessons to be learned from this case are many, but perhaps the most important relates to the need to be much better prepared to embrace uncertainty and what Collingridge (1980) described many years ago (before the original BSE case) as making policy "under conditions of ignorance." That is, risk management is socially (not just technically) rational only when it is grounded in a preparedness to monitor developments over time, to be open to voices (from the public and elsewhere) that are often ignored, and to provide for decision-making structures that retain a sense of the implications of actions. Such an approach could lead to a much more socially robust form of science policy-making (Nowotny et al. 2001), because the focus is extended beyond a limited definition of technology and technical management to one that includes the social processes themselves through which technologies are defined, managed and used.

The BSE crisis was in that sense a crisis that resulted from a failure to incorporate the specific bio-agricultural elements of farming within the broad social technology of the food production system. A consideration of the context within which policy is made – the importance of the food production system as a key contributor to the spread of BSE – could ensure that a more optimal form of risk management and communication is put in place. Rather than simply attempting to maintain the status quo (which is often short-term), it would be much better to formulate policies within a longer and more future-oriented time frame, but policies that nevertheless are based on a rigorous risk assessment of current policies and practices. Governance for food safety should be certainly discussed from this viewpoint (McKee et al. 1996).

References

Beck, M., Darinka, A., & Gordon, D. (2005) Public administration, science, and risk assessment: a case study of the U.K. bovine spongiform encephalopathy crisis. *Public Administration Review, 65*(4), 396–408.

Bradley, R. (2002). *Report on the safety of sheep intestine and natural casings derived therefrom in regard to risks from animal TSE and BSE in particular*. Report prepared for the TSE/BSE Ad Hoc Group of the Scientific Steering Committee. Brussels: European Commission, May 7, 2002.

Brown, D. (2001, June 19). The 'recipe for disaster' that killed 80 and left a £5bn bill. *The Daily Telegraph*. Retrieved, August 17, 2007, from http://www.telegraph.co.uk/news/uknews/1371964/The-recipe-for-disaster-that-killed-80-and-left-a-5bn-bill.html
BSE Inquiry. (2000). *BSE Inquiry homepage*. Retrieved August 19, 2007, from http://www.bseinquiry.gov.uk/index.htm
Collingridge, D. (1980). *The social control of technology*. London: Pinter.
Dealler, S. (1996). Bovine spongiform encephalopathy. Disease is due to pressure on farming industry. *British Medical Journal, 313*(7050):171.
Department for Environment, Food, and Rural Affairs. (2004). *Bovine spongiform encephalopathy: Chronology of events (as at 20 July 2004)*. Retrieved August 15, 2007, from http://www.defra.gov.uk/ANIMALH/bse/publications/chronol.pdf
Department for Environment, Food, and Rural Affairs. (2007a). *EU export rules (Q&A, Section 3)*. Retrieved August 16, 2007, from http://www.defra.gov.uk/animalh/bse/general/qa/section8.html#q4
Department for Environment, Food, and Rural Affairs. (2007b). *Food and drink homepage*. Retrieved August 20, 2007, from http://www.defra.gov.uk/foodrin/index.htm
Department of Trade and Industry (2000). *Review of risk procedures used by the Government's Advisory Committees dealing with food safety*. London, Department of Trade and Industry. Retrieved, December 10, 2007, from http://archive.food.gov.uk/pdf_files/acrisk.pdf
Food Standard Agency. (2002). *BSE and sheep: report of the core stakeholder group*. London: Food Standard Agency.
Food Standards Agency. (2001). *Strategic plan 2001–2006: Putting consumers first*. London: Food Standards Agency.
Food Standards Agency. (2007). Origins of the Food Standards Agency. Retrieved, August 20, 2007, from www.foodstandards.gov.uk/aboutus/how_we_work/originfsa
Forbes, I. (2004). Making a crisis out of a drama: the political analysis of BSE policy-making in the UK. *Political Studies, 52*(2), 342–357.
Frewer, L.J. (2004). The public and effective risk communication. *Toxicology Letters, 149*, 391–397.
Frewer, L. J., & Fischer, A. R. H. (2005). Consumer perceptions of risks from food. In H. L. M. Lelieveld, M. A. Mostert & J. T. Holah (Eds.), *Handbook of hygiene control in the food industry* (pp. 103–119). Cambridge, UK: Woodhead publishing.
Frewer, L.J., Lassen, B., Kettlitz, J., Scholderer, V., Beekman, K., & Berdal, G. (2004). Societal aspects of genetically modified foods. *Food and Chemical Toxicology, 42*(7), 1181–1193.
House of Commons Records. (1997). *Agriculture, Fisheries and Food: BSE: Cattle cull, and BSE*. Accessed, August 15, 2007, from http://www.publications.parliament.uk/pa/cm199697/cmhansrd/vo970115/text/70115w12.htm
Jacob, M., & Hellstroem, T. (2000). Policy understanding of science, public trust and the BSE-CJD crisis. *Journal of Hazardous Materials, 78*, 303–317.
Jasanoff, S. (1997). Civilization and madness: The great BSE scar of 1996. *Public Understanding of Science, 6*(3), 221–232.
Jasanoff, S. (2005). *Designs on nature: science and democracy in Europe and the United States*. Princeton, NJ: Princeton University Press.
Jasanoff, S., & Wynne, B. (1998). Science knowledge and decision making. In S. Rayner & E. Malone (Eds.), *Human choice and climate change* (pp. 1–112). Columbus, OH: Battelle.
Kewell, B., Beck, M., & Asenova, D. (2007, March 13). *Defensive science and the making of the BSE crisis*. Paper presented at the Science and Technology Studies Unit Seminar Series. York, UK: The York Management School, University of York.
Lupton, D. (1999). *Risk: key ideas*. London: Routledge.
Maxwell, R. J. (1999). The British Government's handling of risk: some reflections on the BSE/CJD crisis. In P. Bennett & K. Calman (Eds.), *Risk communication and public health* (pp. 95–107). New York: Oxford University Press.
McKee, M., Lang, T., & Roberts, J. A. (1996). Deregulating health: policy lessons from the BSE affair. *Journal of the Royal Society of Medicine, 89*, 424–426.
Miller, D. (1999). Risk, science and policy: definitional struggles, information management, the media and BSE. *Social Science & Medicine, 49*(9), 1239–1255.

Miller, D., & Reilly, J. (1995). Making an issue of food safety. In D. Maurer & J. Sobal (Eds.), *Eating agendas: food nutrition as social problems* (pp. 305–336). New York: Aldine de Gruyter.

Millstone, E. & van Zwanenberg, P. (2001). Politics of expert advice: lessons from the early history of the BSE sage. *Science and Public Policy, 28*(2), 99–112.

Ministry of Agriculture, Fisheries and Food & Department of Health. (1989). *Interim report of the Consultative Committee on Research into Spongiform Encephalopathies (the 'Tyrell Report').* London: Ministry of Agriculture, Fisheries and Food & Department of Health.

Moran, M. (2001). The rise of the regulatory state in Britain. *Parliamentary Affairs, 54*(1), 19–34.

Myers, N. J., & Raffensperger, C. (Eds.) (2006). *Precautionary tools for reshaping environmental policy.* Cambridge, MA: MIT.

National Creutzfeldt-Jakob Disease Surveillance Unit. (2007). Variant Creutzfeldt-Jakob disease current data. Retrieved, August 3, 2007, from http://www.cjd.ed.ac.uk/vcjdworld.htm

Nowotny, H., Scott, P., & Gibbons, M. (2001). Re-thinking science: knowledge and the public in an age of uncertainty. Cambridge, UK: Polity.

Packer, R. (2006). *The politics of BSE.* London: Palgrave Macmillan.

Phillips, N., Bridgeman, J., & Ferguson-Smith, M. (2000). *The BSE inquiry: return to an Order of the Honourable the House of Commons dated October 2000 for the report, evidence and supporting papers of the Inquiry into the emergence and identification of Bovine Spongiform Encephalopathy (BSE) and variant Creutzfeldt-Jakob Disease (vCJD) and the action taken in response to it up to 20 March 1996.* Norfolk, UK: The Stationary Office. Retrieved August 17, 2007, from http://www.bseinquiry.gov.uk/report/index.htm

Rothstein, H. (2003a). Precautionary ban or sacrificial lambs? Participative risk regulation and reform of the UK food safety regime. *London School of Economics Discussion Paper, 15*, 1–23. Retrieved (last access), March 12, 2009, from http://www.lse.ac.uk/collections/CARR/pdf/DPs/Disspaper15.pdf

Rothstein, H. F. (2003b). Neglected risk regulation: the institutional attenuation phenomenon. *Health, Risk & Society, 5*(1), 85–103.

Rothstein, H. F. (2004). Precautionary bans or sacrificial lambs? Participative risk regulation and the reform of the UK food safety regime. *Public Administration, 82*(4), 857–881.

Shaw, A. (2004). Discourses of risk in lay accounts of microbiological safety and BSE: a qualitative interview study. *Health, Risk & Society, 6*(2), 151–171.

Southwood Working Party. (1989). *Report of the Working Party on Bovine Spongiform Encephalopathy*, London: Ministry of Agriculture, Fisheries and Food.

Spongiform Encephalopathy Advisory Committee. (1995). *Transmissible spongiform encephalopathies: a summary of present knowledge and research.* London: Her Majesty's Stationary Office.

Spongiform Encephalopathy Advisory Committee. (2007). *SEAC homepage*. Accessed, August 1, 2007, from http://www.seac.gov.uk/committee/about.htm

Treece, D., Pisano, G., & Pisano, A. (1997). Dynamic capabilities and strategic management. *Strategic Management Journal, 18*(7), 509–533.

United States Department of Agriculture, Animal and Plant Health Inspection Service. (2001). *Bovine Spongiform Encephalopathy: an overview*. Washington, DC: United States Department of Agriculture. Retrieved (last access), March 12, 2009, from http://www.aphis.usda.gov/publications/animal_health/content/printable_version/BSEbrochure12-2006.pdf

Van Zwanenberg, P. & Millstone, E. (2003). BSE: a paradigm of policy failure. *The Political Quarterly, 74*(1), 27–37.

Vogel, D. (2001). *The New politics of risk regulation in Europe.* London: Centre for Analysis if Risk Regulation. Retrieved (last access), March 12, 2009, from http://www.lse.ac.uk/collections/CARR/pdf/DPs/Disspaper3.pdf

World Organization for Animal Health. (2007). *Bovine Spongiform Encephalopathy (BSE): geographical distribution of countries that reported BSE confirmed cases since 1989.* Paris, France: OIE. Retrieved (last access), March 12, 2009, from http://www.oie.int/eng/info/en_esb.htm

Wynne, B. (1996). May the sheep safely graze? A reflexive view of the expert-lay knowledge divide. In S. Lash, B. Szerszynski & B. Wynne (Eds.), *Risk, environment and modernity: towards a new ecology* (pp. 44–83). London: Sage.

Wynne, B. (2005). Reflexing complexity: post-genomic knowledge and reductionist returns in public science. *Theory, Culture & Society, 22*(5), 67–94.

Chapter 9
Governing Uncertain Threats: Lessons from the Mad Cow Saga in France

Pierre-Benoit Joly and Hajime Sato

Introduction

The mad cow crisis has often been described as emblematic of the "risk society," for at least three reasons. First, the dangers to human health from beef consumption were not the result of fate; rather, they were "manufactured" (Giddens 1994, 1997) and a byproduct of modernization. Second, science was at the core of the crisis. Although the source of the problem was technology, scientific knowledge was instrumental in solving it by providing surveillance tools, a better understanding of the causes of the disease, early tests, and so forth. Thus, the mad cow crisis perfectly illustrates Beck's (1992) thesis of reflexive scientization. Finally, this risk appeared to be the product of the institutions and economic participants that had concealed the actual nature of the disease for a long time. Consequently, the mad cow problem was ripe for high-profile publicity and politicization. Indeed, as stated by Kasperson et al. (2001), the mad cow disease case typifies a special class of hazards – ones that trigger intense media coverage, strong public concern, and a great deal of institutional attention, and produce large-scale secondary or higher consequences.

In addition, the mad cow crisis is used to exemplify the precautionary principle "better safe than sorry," – in other words, it is better to act at the beginning, when the risk is still hypothetical –, as evidenced by the report to the French Prime Minister on precautionary measures (Kourilsky & Viney 2000) and other reports from Europe (Harremoës et al. 2001).

A look back at this crisis sheds light on some features of these "new risks" that deserve closer scrutiny. Although science and technology have played an important role in the mad cow crisis, it is necessary to examine the relation between knowledge

P.-B. Joly
UR 1216 TSV (Transformations Sociales et Politiques liées au Vivant),
French National Institute for Agricultural Research, France

H. Sato (✉)
Department of Public Health, Graduate School of Medicine, The University of Tokyo, Japan
e-mail: hsato-tky@umin.net; hsato@post.harvard.edu

H. Sato (ed.), *Management of Health Risks from Environment and Food*,
Alliance for Global Sustainability Book Series 16,
DOI 10.1007/978-90-481-3028-3_9,

and action in closer detail. In doing so, the following basic ideas emerge: (1) the early conjectures on the cause of mad cow disease were surprisingly on-target, (2) the impact of the disease on human health remained uncertain until 1996, (3) the hypothesis of the transmissibility of mad cow disease, technically labeled Bovine Spongiform Encephalopathy (BSE), to humans was systematically dismissed by government officials, (4) the decisions made by the government in the early stages of the crisis were accordingly quite relevant, and (5) these decisions were not well implemented, because of inadequate enforcements and a lack of compliance.

In this context, the problem is not early warning, but how socio-technical systems may increase their ability for early listening – that is, the ability to take advantage of new warning signals. One key element, as shown in various analyses, is that prior to 1996 mad cow disease was defined exclusively as an "animal health problem"; any attempt to define it as a human health problem was dismissed. The hypothesized transmission of BSE to humans did not enter the picture, even though there were numerous signs that such transmission is possible. This transmissibility to humans by definition had tremendous implications, not the least of which was the problem of compliance with the measures that were supposed to address it.[1]

Thus, this paper includes a discussion of problem definition, using the classical theories of public problems and socio-cognitive frames of reference. This approach leads to the explanations of the BSE saga, more balanced in terms of the factors affecting the course of events, such as interests, institutional factors, and the production/circulation of knowledge. Erik Millstone is right when he points to the importance of government strategies of reassurance aimed at protecting the beef industry (Millstone & van Zwanenberg 2001). This, in turn, leads to a focus on risk communication, or how to convey uncertain risks to the public and promote transparency. However, an analysis that only looks at interests is not sufficient; it is also necessary to analyze the construction of socio-cognitive frames. Although related to interests, these frames cannot be reduced to that dimension; and characterized by specific dynamics.

Furthermore, risk is always multi-faceted, incorporating the environmental, sanitary, economic, and political issues (Borraz et al. 2006). Strong politicization is obvious in the second mad cow crisis of October 2000. The high political profile was mainly triggered by a boycott of beef by French consumers, resulting in a 40% drop in consumption. The boycott induced a major economic crisis in the beef industry. Labeled a "collective psychosis" at the time, this crisis undoubtedly was partly determined by the previous one. The failure to publicly acknowledge that BSE may be a risk to human health provoked a trauma in public officials and adversely affected interactions between politicians, policymakers, and scientific experts.[2]

In this chapter, we use the above considerations to analyze what we call the "politics of risks and uncertainty," in which: (1) the risk is transformed into a political stake, which was influenced by the recent history of sanitation scandals and the

[1] This lack of compliance does not appear to stem from uncertainty as such – as stated by Setbon, Raude, Fischler, and Flahaukt (2005) – but from the problem definition.

[2] Indeed, this is reflected in the transformation of the organization of scientific expertise, which began in March 1996 with the establishment of the statutorily independent Dormont Committee and later with the creation of the French Agency for the Sanitary Security of Food (AFSSA).

political competition induced by cohabitation; (2) a sense of urgency appeared as a result of political competition and the need to take symbolic action to stop the economic crisis, which followed from the collective memory of the mad cow saga; in other words, a "scapegoat policy"[3]; and (3) no consideration was given to the costs and benefits of the meat-and-bone meal (MBM) ban, an example of backwards thinking: act and then justify.

In discussing these various dimensions of the mad cow saga, we concentrate in this paper on three series of events. The first section covers events from the early days of BSE to the 1996 crisis. The second section focuses on the "beef war," which set France against the European Union and served as the first test since the 1996 crisis of the new relationship between scientists and policymakers. In the third section, the discussion turns to an analysis of the 2000 crisis and its consequences – a crisis that has been called a phantom risk.

Science and Public Actions in the Mad Cow Saga: From Early Signals to Early Listening (1989–1998)[4]

The UK BSE and the French MBM Industry: Alerts Regarding Animal Feedstuffs Prior to 1990

The first case in the UK of what later was to be identified as BSE was recorded by veterinary pathologists at the Central Veterinary Laboratory (CVL) of the Ministry of Agriculture, Fisheries and Food in 1985. The brains showed characteristics similar, but not identical, to scrapie in sheep. In December 1986, the head of the pathology section sent a note commenting on the brains of the animals examined: "… the lesions observed have similarities to spongiform encephalopathy of other species and in particular scrapie of sheep" (Parker 2006). In the same note, Parker mentioned two major concerns that would apply if the disease turned out to be "bovine scrapie": possible effects on exports and human health. At the time, this new cattle disease aroused very little interest in French authorities. It was generally considered to be a UK problem, and the threat to French cattle was considered remote.

Initial Management of the Problem: A Need to Act, but No Urgency

At first, only a few people in France were interested in this new disease. Marc Savey,[5] then a Professor at the National Veterinary School at Maisons-Alfort, was instrumental in creating early interest in the issue. Since the end of the 1970s, Savey had been studying

[3] See the UK Philips Report pertaining to food risks (Phillips et al. 2000).

[4] This section draws on a report prepared for the European research project "BASES," coordinated by Pierre-Benoit Joly (Joly et al. 1999).

[5] Interview by P. B. Joly, the *Centre National d'Etudes Vétérinaires et Alimentaires* (CNEVA), Maisons-Alfort, January 1999.

slow viruses (viruses with long latencies before the presentation of symptoms). In 1979, he published a paper outlining his proposal to use scrapie as a model for Creutzfeldt-Jakob disease (CJD) (Savey & Espinasse 1979). He subsequently tried to get funding for this program, but the French government funding agencies were not receptive to his requests. Although he had not been involved in the research on transmissible spongiform encephalopathies (TSEs), he had frequent contact with Dr. Court and Dr. Dormont of the *Centre de Recherches des Armées* and the *Commissariat à l'Energie Atomique* (CEA, Commission on Atomic Energy), who were studying the transmission of TSEs.[6] Savey was also involved in veterinary research throughout Europe and had visited the laboratories of Kimberlin and Barlow.

When the first paper on this new cattle disease was published in the *Veterinary Record* (Wells et al. 1987), Savey was eager to apply this new information to his own research. As an expert working for the *Direction Générale de l'Alimentation* (DGAL, Division of Food Affairs), *Ministère de l'Agriculture* (Ministry of Agriculture), he sent out a notice about this new problem. In June 1988, he contributed to the production of an internal memo that summarized the available information on BSE,[7] which was circulated by the *Sous-Direction de la Santé et de la Protection Animales* (SDSPA, Veterinary Services section) of the DGAL. The memo was intended to warn the ministry about the potential threat of BSE, which at the time was still unknown in France.

At the end of 1989, Savey published a paper in collaboration with E. Maillot, the head of the DGAL, that accurately reflected the perception of the BSE problem in France at that time (Savey et al. 1989). The paper began by summarizing the BSE problem in the UK, pointing out that: (1) although sheep and cattle have been living for a long time in close proximity, the BSE problem is quite recent and, therefore, the most plausible explanation for its emergence is transmission through MBM from them; (2) the acceleration in the number of cases in the UK is spectacular, increasing from four in December 1986 to 8,100 by the end of November 1989; and (3) the similar clinical and anatomic-pathologic characteristics of CJD and BSE, which suggest a risk of BSE transmission to humans through the consumption of beef (Savey et al. 1989).

According to Savey et al. (1989), these features explain the importance of decisions made by the UK government in 1988 and 1989. Referring to the recommendations of the Southwood Report (Southwood 1989), they emphasized the risk to human health, and the need to prevent the human consumption of meat and milk from cattle with clinically significant BSE symptoms as well as specified bovine offal[8] (SBO) from all cattle older than 6 months. They also observed that the decisions taken in the UK increased the danger to French cattle, as MBM containing the banned SBO was itself banned from cattle feed in UK.

[6]Françoise Cathala, of the *Institut national de la santé et de la recherche médicale* (INSERM), played an important role in the emergence of research on the infectious nature of encephalopathy in France in the 1960s.

[7]*Note de service*, DGAL/SDSPA/N88-8114 (June 20, 1988).

[8]Offals: brain and central nervous system products.

In this early period, the French government took only a few measures to mitigate the BSE risk: (1) on August 16, 1989, the SDSPA issued an *avis* (notice) forbidding French firms from using MBM from the UK as ruminant feed, but they were still allowed to use it for other purposes (see the next section for a further discussion); (2) starting March 15, 1990, European Commission (EC) Decision 89/469 of July 28, 1989 (banning UK bovine carcasses in which the nervous and lymphatic system was intact), and EC Decision 90/59 of February 7, 1990 (banning UK bovine imports aged more than 6 months), were enforced through an *avis aux importateurs*[9] (notice to importers).

Savey et al. (1989, p. 489) also warned that BSE cases would appear in France as early as 1991, either through animals imported from the UK between 1982 and 1988 or through French animals that had been fed with UK MBM between June 1988 and August 1989. They stressed "to the administration and to the media" the need to implement an efficient national surveillance system to identify infected cattle. The reporting of BSE was made compulsory on June 12, 1990.

The Lack of Control of the Importing of MBM from the UK

As a result of the UK ban of July 18, 1988, on the use of MBM for ruminants, MBM prices fell in the market; consequently, the French animal-feed producers greatly increased their use of these meals. As a result, the government banned the use of MBM for cattle feed (Notice of the Ministry of Agriculture, August 16, 1989). However, as the regulation notice was ambiguous and only called for a partial ban, enforcement was very weak.

Starting in September 1989, the President of the Association of Producers of MBM, the *Syndicat des Protéines et des Corps Gras Animaux* (SPCGA, the Union of Animal Protein and Body Fat), alerted the Ministry of Agriculture that this *avis aux importateurs* (notice to importers) was very inefficient.[10] At a meeting on BSE held at the Ministry of Agriculture on September 26, 1989, the President of the SPCGA presented data on the importation of MBM from the UK and Ireland (see Table 1).

Table 9.1 French imports of MBM, 1988–1989

	From UK	From Ireland	Total
January/July 1988	4,569	1,275	5,844
August/December 1988	5,711	4,449	10,170
January/July 1989	10,794	17,970	28,764

In tons.

[9] An *avis* is not compulsory but intended to inform the participants of the government's policy.

[10] Testimony by J. P. Lugan before the Parliamentary Mission on September 18, 1996, and published later (Guilhem & Mattei 1997). Of course, this kind of report has to be interpreted with caution, as such a mission does not have the same power of inquiry as a tribunal. Therefore, it is fairly obvious that persons testifying before the mission might present a rather biased viewpoint, because the procedure did not force them to give a statement based on facts and evidence.

As a direct result of the ban and the resulting price decline, the importing of the suspected meal had increased fivefold within 1 year. This caused the President of the SPCGA to demand a stronger ban, criticizing the inefficiency of the *avis*: "If only one case of BSE is officially registered in France, 100,000 tons of meat meal will be banned from exportation."

The meeting concluded with agreement on the need to ban MBM. As a result, on December 15, 1989, the ministry decided to extend the *avis aux importateurs* to the Republic of Ireland. In a letter dated January 4, 1990, the President of the SPCGA warned the Minister of Agriculture, Nallet, that this latest decision would also be ineffective. He emphasized the potential risks – not only for the meat industry, but also for human health. He also pointed out that the media might make much of the situation: "Some key newspapers have already seized upon the subject and not only the so-called specialized press. They have evoked concerns about consumer safety."

One example of how some newspapers covered the BSE problem at the time comes from the French magazine *Sciences et Vie*. The following excerpt requires no further comment:

> BRITISH PALMED OFF THEIR INFECTED MEAL ON US:... On 13/8/89, a decree indeed banned importation of MBM for ruminants from the UK: but it was too late: The British, who suspected at this time the infection of their meal, had exported to France 8,574 tons within 6 months (from January to July 89). The traces of poison were lost in the 450,000 tons consumed in French farms each year (Torny 1997, p. 223).

In a second letter to the minister, dated January 11, 1990, the President of the SPCGA summarized the situation in the UK, pointing out that since November 14, 1989, SBO had been banned from human food. Therefore, it naturally was added to MBM, thus increasing the risk. In a third letter to the minister, dated February 7, 1990, the President provided the latest information on French imports of MBM. In November 1989, a total of 7,654 tons had been imported – 2,953 from the UK and 470 from Ireland. From January to November the total was 41,101 tons, compared to 12,942 tons in the same period in 1988. The President received a reply from the Minister, dated February 20, 1990, which made reference to the *avis aux importateurs*; this was followed by a letter from the Chief of International Trade and Health Regulations, dated March 23, 1990, which specified that it was necessary to wait to appreciate the effects of the *avis aux importateurs*, at least until there was a European decision about MBM.

On November 27, 1989, the two animal-feed trade unions, *Syndicat National des Industries de la Nutrition Animale* (SNIA, National Union of Animal Feed Industries) and *Syndicat des Coopératives d'Aliments Composés* (SYNCOPAC, Union of Composite Feed Cooperatives), issued notices to their members. So as not to contribute to the propagation of the BSE agent and to preserve the health of French cattle, the unions recommended that their members: (1) ban the use of imported or French MBM in animal feed formulas, and (2) not use the same equipment to produce ruminant formula that they use to produce animal-feed formula containing MBM.

Despite the silence of the EC matter and the lack of expertise at the outset,[11] the *Comité Interministériel et Interprofessionnel de l'Alimentation Animale* (CIIAA, Inter-ministerial and Inter-professional Animal Alimentation Committee) was asked on April 5, 1990, to look into the question. The committee gave its recommendations on June 26, 1990. This led to the Decree of July 24, 1990, which banned the use of any MBM for cattle feed (not only made in the UK and Ireland, but also made in France).[12]

The 2001 Parliamentary Mission on MBM subsequently revealed that this 1990 ban, unlike the 1988 UK ban, applied only to cattle and not to all ruminants. The French ban was a response to pressure exerted by the animal-feed trade unions on the Ministry of Agriculture (Sauvadet & Vergnier 2001). The SYNCOPAC strongly opposed a ban, fearing that the pressure of public opinion would lead to a global ban of MBM for all animal feed. It argued that such a ban would destabilize the entire rendering industry. The weak nature of this decision resulted from an ambiguous situation in which the DGAL, a division of the Ministry of Agriculture, was responsible for both food safety and the competitiveness of the food industry. This led to compromises that were not in the best interest of the public in terms of health risks. The situation also created internal conflicts of interest, which in 1999 led to reform of the DGAL under Minister Glavany.

"Mad Cats" and the First Embargo of the 1990s

In spring 1990, BSE emerged as a public issue. The subsequent change in the management of BSE in France was strongly linked to the public announcement on May 10 that "mad cats" had been identified in the UK. According to Nallet, the evidence of oral transmission of the disease to carnivores was a major influence on how the BSE risk was perceived by the public. It was also the first piece of evidence ever that mad cow disease could cross species barriers (Council of the European Communities 1990).

In the UK, this event triggered a massive reaction. Hundreds of local authorities banned British beef in school meals; in the week following the announcement, domestic beef sales in the UK fell by about 25% (van Zwanenberg & Millstone 1999). Nallet's Cabinet determined that the situation had to be monitored closely. Several meetings were organized and experts were consulted.

[11] In a January 1999 interview, Savey recalled that, at that time, he was not aware of the importance of the problem of the continuing importation of UK MBM. Although the President of the SPCGA had reliable information on the matter, he might be suspected of trying to protect the SPCGA membership rather than the public interest.

[12] *Arrêté du* 24/7/90 (Decree of July 24, 1990) published in the *Journal Officiel de la République Française (JORF)*, on August 11, 1990. Although it does not explicitly refer to the precautionary principle, the decree is clearly in the spirit of the principle, in that it refers to the potential risk of transmission that would be created by the ban.

In late May, Nallet proposed a set of measures to the French Prime Minister that included: (1) a ban on all types of UK bovine products, (2) the establishment of an eradication program at the national level, along with subsidies to farmers (Decree of July 24, 1990), and (3) better enforcement of the regulations regarding UK MBM in cattle feed. These decisions were announced in the French Parliament on May 30, 1990.

By this time the French mass media had also begun to report on the threat of BSE in France. The Savey prediction, published earlier in the *Technical Report from the Veterinarian Academy*, was reproduced on February 23, 1990, in the magazine *L'Express* (Savey et al. 1989). On May 23, 1990, an announcement from the *Agence France Presse* (AFP, French Press Agency) also carried the prediction. Later, it was quoted by a deputy in the form of a question to the minister, to which the minister replied that no case of BSE had occurred so far in France, but it was necessary to remain very vigilant. He also recalled the measures taken up to that time (*avis aux importateurs* and EC Decision 89/469).

The ban provoked very little reaction in the French media, except in the newspaper *Le Monde*, which was rather critical in arguing that such a decision might be motivated by commercial interests. At any rate, the ban seemed to attain one of its secondary goals: the media would no longer report on the problem at length and public interest would wane.

The European Crisis

Other European countries, such Austria and Germany, also adopted a ban on UK beef. An extraordinary Council of Ministers meeting took place in Brussels on June 6 and 7, 1990. It was convened by R. McSharry, the Commissioner for Agriculture. France's unilateral decisions were severely criticized. The UK Minister of Agriculture argued that such decisions had been taken to protect national markets and not public health. The French measures were also attacked by the Scientific Veterinary Committee (SVC) as being unjustified and an arrogation.[13] Because the Council of Ministers had decided to go beyond the recommendations of the SVC and adopt a set of compulsory measures to be implemented in the UK (Decision 90/261, June 8, 1990[14]), France lifted its ban, a measure that conformed to the various decisions made previously.

[13]Nallet estimated that the French decision was a breakthrough in the relationship between the Agricultural Council and the SVC: "Up to 1990, BSE was the responsibility of the Veterinary Committee which is quite autonomous and whose conclusions are usually adopted in 'point A' of the Ministry Council, without any discussion."

[14]This Council Directive obliges the UK to identify the cattle and restricts the exporting of UK bovine meat (adding restrictions to Commission Decision 90/200, which was adopted on April 9, 1990, before the crisis). It also requires the examination of the MBM production equipment in all member states and further research on BSE. However, the measure had little effect.

Public Decisions in the First Stages of the Crisis

The French government subsequently made a set of important decisions. Overall, these decisions were the direct consequence of the measures, announced by Nallet on May 30, which were intended to remedy the shortcomings of the French regulatory system.

Decree 90/478, issued on June 12, 1990,[15] required the reporting of BSE and represented the implementation of EC Decision 90/200, which had been issued on April 9, 1990. A decree of December 3, 1990, went even further, mandating that the DGAL create the French epidemic-surveillance network (ESN) on BSE. The *Centre National d'Etudes Vétérinaires et Alimentaires* (CNEVA, National Center of Veterinary and Food Studies) Bovine Diseases Laboratory in Lyon was given the task of coordinating the activities of the various laboratories involved in implementing the national plan. This permanent network was nationwide and its mandate included all bovine breeds. Its goal was the systematic surveillance of cattle displaying unnatural neurological signs, and its work led to the identification of the first case of BSE in France on February 28, 1991. However, the official announcement came as no surprise, both because it had been widely predicted and because the system appeared capable of dealing with the problem.

The implementation of hygiene measures, including the isolation of suspected herds, increased. The moment a case was confirmed, the animal was slaughtered. The slaughter of the rest of the herd was not made compulsory until 1994; however, the whole herd was usually slaughtered even before then, provided that the farmer received compensation.[16] According to Savey,[17] in most cases, any herd that included an animal with suspected BSE was slaughtered. However, with the exception of SBO, the remainder of the slaughtered herd was not removed from the human food chain.[18]

The regulation of MBM was also reinforced. A decree issued on July 24, 1990, and modified on September 26, 1990, banned the use of MBM for cattle feed. This decision was reached 18 months after the UK MBM ban was first put in place, but 4 years before the EU ban took effect. Although this decision may appear timely, it proved to be rather ineffective because of the difficulties involved in implementing a partial ban and regulating MBM from foreign sources. In addition, these measures, which were taken to protect animal health, were irrelevant to human health. Although all cows identified as

[15] Published in the *JORF* on June 13, 1990.

[16] The December 4 (second) decree details the financial part of the ESN.

[17] Interview, January 1999.

[18] This information was subsequently revealed by *Le Canard Enchaîné* in 1996 but not reported in other journals. Information on the methods for detecting mad cow disease before 1994 remains a little fuzzy, as the accounts are not always the same. The chronology established by the DGAL in 1996 recalls that Decree 90/478 implied that, if a case of BSE is confirmed, the whole herd must be slaughtered and all SBO incinerated. However, this was not obligatory before the *Arrêté* of July 27, 1994; at this earlier stage, it is probable that all BSE animals were destroyed, but other animals were marked and kept isolated on the farm. When the latter were slaughtered, their SBO were incinerated. (For a different account, see Sauvadet & Vergnier 2001, p. 88.)

having mad cow disease were slaughtered and destroyed, the use of French SBO was not banned until 1996. Indeed, after the June 1990 crisis, the species barrier continued to be cited as the reason why the disease would not spread.

Lessons from the Early Days

As of July 1990, 18 months after the publication of the Southwood Report in the UK and 2 years after the UK MBM ban for ruminants, the French authorities had made two key decisions: (1) the compulsory reporting of BSE (Decree 90/478, June 12, 1990) and the establishment of an epidemic-surveillance network (Decrees 3 and 4, December 1990), and (2) a ban on the use of MBM for cattle feed (July 24, 1990).

Both decisions were important for the protection of animal health and, indirectly, for human health. They were formulated despite limited knowledge about the new disease. However, the knowledge increased with the identification of similarities between BSE and TSEs. In France, the small and marginal group of scientists who were working on BSE/TSE played a key role. Although uncertainty was high and it was not possible to do a comprehensive risk assessment, the information they had obtained allowed them to design preliminary measures that could mitigate the risk.

The restriction of the MBM use for cattle feed was decided on by the DGAL. As noted above, the DGAL had two possibly contradictory objectives, food safety and the competitiveness of food industry.

Although the UK had banned MBM domestically, it did not forbid its export to other EU member states. In addition, the information procedures of the European Committee and the governments of the member states were found to be inappropriate. Therefore, as revealed in the Philips Report and by the French Parliamentary Mission, the UK took steps to prevent alarming other member states and thus protected its own market (Sauvadet & Vergnier 2001). This lack of coordination contributed to the first cases of BSE in French cattle. It was later estimated that 25 cows were infected.[19] These first cases could have easily been prevented.

During this first period, the European Committee, influenced by the UK government, was reluctant to make decisions restricting the free circulation of goods. Therefore, protection decisions were generally made in response to the member states' unilateral decisions.

From January 1991 to October 1995: A Long "Silent Period" … with a Number of Key Decisions

After the first crisis, which prompted the implementation of preventive measures, France entered what might be called a long period of silence. This is not to say that nothing happened, but what did happen was not unexpected. The numbers of identified

[19] To be compared with more than 1,000 infected cows in France until December 2007.

cases of mad cow gradually increased. Paradoxically, these identifications simply confirmed the experts' predictions. Adequate systems seemed to be in place to deal with all these problems – "everything was under control." Indeed, the government paid little attention to this problem; they were preoccupied with the transformation of EC decisions into French regulations.

One important point is that policymakers who had experienced such a crisis before, and thus knew all-to-well the uncertainties and difficulties involved, were replaced. Nallet left the Ministry of Agriculture to be replaced in October 1990 by Mermaz, who remained Minister of Agriculture until a new government under Balladur was installed in 1993; Puech then became Minister. After the presidential elections of June 1995, yet a new government was formed: Juppé became Prime Minister and Vasseur became Minister of Agriculture (the "minister of the crisis"). The terms of the ministers and their respective cabinets were thus rather short.

To some extent, similar events occurred at higher levels of the administration. Guthmann, who had headed the DGAL during the first crisis, was replaced by Guerin in April 1994. Indeed, the period between the first crisis and the second crisis involved not just a passage of years, but also changes in the political and administrative personnel in charge of dealing with the problem.

Another important feature of the 1991 to 1995 period was the attempt to improve French scientific expertise on TSEs, under the initiative of the Minister of Research.

The Difficulty in Mobilizing the Research Community to Study TSEs

In a letter dated April 6, 1992, French Minister of Research Curien asked Dormont, Head of the Laboratory of Experimental Neuropathology and Neurology at the CEA, to prepare a report on TSEs. The letter cited the increase in BSE cases in the UK and the potential risk of transmission to humans. Curien asked Dormont to report on the most up-to-date knowledge, and proposed recommendations that included the development of research and the prevention of risks.[20]

The Dormont Report (Dormont 1992), a 117-page document prepared by a group of French experts,[21] drew on an extensive literature review. This bibliographic study pointed out that the available knowledge about TSEs and related diseases raised more questions than they answered; uncertainty was the rule. As of 1992, the nature

[20] The preparation of this report was first informally requested of Marc Savey, reflecting the concern over BSE. Note that in 1991, two important conferences were organized: a consultation on BSE organized in Paris from October 30 to 31, 1991, by the *Office International d'Epizooties* (OIE, International Office of Animal Health), and a WHO consultation on public health issues related to animal and human spongiform encephalopathy, held in Geneva from November 12 to 14, 1991 (World Health Organization, 1991).

[21] Note that only one veterinarian and no researchers from the *Institut National de la Recherche Agronomique* (INRA, National Institute for Agricultural Research) participated in this first working group. This composition shows the weak mobilization of agronomic and veterinary researchers on TSE problems at that time.

of the contaminating agent was not known and the uncertainty was increasing. New TSEs appeared in species in which they had not previously been observed. The discovery of the unsettling reality of inter-human transmission was the result of progress in surgical and therapeutic techniques.[22] A new approach to the management of the microbiological security of medical products, biomaterials, and cosmetics was mandated. A genetic susceptibility to BSE in some humans was strongly suggested by the data but had not yet been confirmed. The incubation period following peripheral (e.g., oral) infection was so long that the consequences of BSE on human health could not be determined with certainty for 10 to 20 years.

The report outlined the weaknesses of the French research system. Following a detailed examination of the problem, it noted that only 9 full-time and 14 part-time researchers were contributing to the generation of knowledge about TSEs in France, and this knowledge was being disseminated to only nine research institutes (Dormont 1992). It called for a major reinforcement of the French research system, noting the importance of the potential risks and the lack of knowledge about how to deal with them. An optimal, long-term research effort would require the creation of an inter-ministerial organ to oversee the development of a strategic research program and the proposal of measures deemed necessary to deal with the risk of infection (Dormont 1992).

However, the report did not go into much detail on the risks associated with beef consumption. It stated that such consumption created "a potential risk" and that, thus far, no experiment had shown that contamination was the result of the ingestion of muscle.[23] However, as BSE was quite new and little was known about it, the report stated that it was necessary to take the risk seriously and to assess it.

How Was the Dormont Report Received?

According to Dormont,[24] Curien was very receptive and asked Dormont to work on a project undertaken by the *Groupement d'Intérêt Scientifique* (GIS, Group of Scientific Interest). Dormont's idea was to set up a funding agency similar to the

[22] The growth hormone produced by the extraction of hypophysis was a cause of the transmission of CJD in the late 1980s (The first case in France occurred in 1989 and in the US in 1985). By late 1998 it had caused 54 cases of CJD in France, the most exposed country. The working group, largely composed of specialists in medical research and practice, were very keen to focus on such important problems.

[23] Even at that time, research on experimental animals that had orally ingested brain and then had been tested for scrapie showed that the quantity of brain tissue ingested was very important (Broxmeyer 2005). The authors of the report estimated that inter-species transmission through oral ingestion would require 106 times more substance than through intra-cerebral injection. The possibility of a change in the infectious properties of the tissue was not considered. Further experiments showed that BSE had been transmitted to sheep through oral ingestion of 0.5 g of infected brain (Foster et al. 1996). This result changed completely the evaluation of the BSE transmission risk to humans. As Dormont commented, "… these data demonstrate that we are now completely changing the frame [for the evaluation of risk – note from the editor] that had been elaborated in the last few years" (*Le Monde*, May 6, 1996).

[24] Interview, Grenoble, February 1999.

Association Nationale de Recherche contre le SIDA (ANRS, the National Association of Research against AIDS). However, public research organizations strongly opposed this idea. When the government changed after the 1993 elections, the GIS idea was dropped altogether. Instead, Fillon, the new Minister of Research, proposed a program on prions to be undertaken as part of a larger program on the life sciences launched in 1995. The budget devoted to such activities was not large (about 10 million Francs), but it did allow research across 23 research laboratories, including the *Centre national de la recherche scientifique* (CNRS, National Center for Scientific Research), the *Institut national de la santé et de la recherche médicale* (INSERM, National Institute for Health and Medical Research), the *Institut National de la Recherche Agronomique* (INRA, National Institute for Agricultural Research), the CEA, and various universities. In general, the financed projects were basic research.[25] In 1996, the Parliamentary Mission highlighted the disconnect between Curien's personal involvement and the negative reactions of the French research establishment (Guilhem & Mattei 1997).

The heavily solicited opinions of the chief officials of the INRA's Department of Animal Pathology may help in understanding how the Dormont Report was received by certain segments of the research community. In a letter addressed to the Director General (October 1992), the officials explained the position taken by the INRA and highlighted three dimensions of the problem:

(1) *Science*: they pointed out that, in the US and, to a greater extent, the UK, large research teams were working on these problems, which created a problem of scientific competition. (2) *Animal health*: the actual rate of BSE in France was very low (five confirmed cases); thus, the proposed strategy was to expand the resources of the CNEVA so they would be better able to eradicate BSE, thereby enable its classification as an exotic disease; this in turn would allow the removal of the non-tariff barriers against the export of French cattle – in other words, the BSE project was not a priority for the INRA. And, (3) *Human health*: the Dormont Report focused on the impact of BSE on human health; the few laboratories conducting basic research on scrapie had to be funded, and establishing a network or working group on scrapie was greatly encouraged.

Retrospectively, the account given by Grosclaude, the INRA's Scientific Director of Animal Productions, is consistent with this position. He stated that the INRA is responsible for dealing with certain diseases and their risks to human health: "... We concentrate on researching bacteria, which represent greater and more frequent risks for consumers."[26] This choice is also explained by the competitive disadvantage of French research laboratories compared to those of other European countries, particularly the UK.

---

[25] Note that of the 23 selected projects, only 8 were directly related to TSEs (of which only one was conducted at the INRA). It is interesting to compare this lack of mobilization to what took place after March 1996. In 1997, the French TSE Program sponsored 50 projects, all directly related to TSEs.

[26] Hearing of the Parliamentary Mission, July 9, 1996 (as cited in Guilhem & Mattei 1997).

The Dormont Report outlined the importance of the cognitive breakthrough, the lack of knowledge, the high degree of uncertainty, etc. It was implied that if a zoonose ever developed, the situation would be very difficult to manage, because effective risk-management tools did not exist, due to the lack of knowledge. Therefore, although BSE infection in humans was only a remote possibility, it was important, in fact urgent, to undertake a serious research effort.

The INRA's position was based primarily on the available data, which stressed the very small number of BSE cases in France. The concept of an "exotic disease" was introduced to signify that BSE was mainly a UK problem. Since the UK research was very good, "the subsidiary principle" might apply, and, thus, the INRA followed its own priorities. Its position illustrated how public research organizations normally deal with such problems. Tambourin, who was then the Scientific Director of the Department of Life Sciences in the CNRS, remembers that the CNRS was less concerned with CJD, which affects only a few people, than with cancer or AIDS. Also, the better knowledge of the disease did not seem to facilitate progress in the associated basic research. Many people recognized that although the CNRS was monitoring the situation, it was not necessarily anticipating problems.[27] Despite a significant, multi-year financial investment in Court and Dormont's laboratory and following the preparation of a report for the Minister, Dormont (the CEA) was instructed by the Chief of the Department of Life Sciences in 1993 to abandon its research on TSEs. Of course, the research did not stop.[28]

The working group set up by Dormont to prepare the report continued to meet informally, mainly to exchange scientific and epidemiological information. These exchanges were focused on establishing a future epidemiological network to study CJD.[29]

The Actions of the Ministry of Health

In this so-called silent period, the Ministry of Health made several important decisions related to TSEs. It focused its activities on its traditional area of competence: In June 1992, the Ministry of Health removed from the market 19 medical products containing bovine tissues; on July 3, 1992, it issued a decree prohibiting the execution of the "issue of magisterial preparations based on bovine-origin tissues." On July 22, 1992, this decision was extended to homeopathic products, and on July 31, 1992, a decree was issued prohibiting the use of SBO and other bovine or ovine

[27] Tambourin refers to the 1984 *Rapport de Conjoncture et de Prospective de l'INSERM* (Report of the Situation and Prospective of INSERM), recalling that it did not say a word about AIDS, despite the presence of Montagnier on the editorial board. Interestingly, a 1994 report from the INSERM on therapeutic risks cites the TSE problem and notes: "Even though rare, this pathology is all the more troubling as the infectious agent is not identified but is potentially present in our food". However, this report had no impact.

[28] Interview Grenoble, February 1999.

[29] Interview Grenoble, February 1999.

tissues in jarred baby food.[30] This decree was initially in effect for just 1 year, extended on July 29, 1993, but not further extended in 1994.

These decisions concerning medical products were closely linked to a group of physicians working at the *Direction Générale des Médicaments, Ministère de la Santé* (*Direction de la pharmacie et du medicament*), French Ministry of Health, who had been involved in the AIDS and contaminated blood crises. Since the early 1990s, they had been undertaking a review of all pharmaceutical products, looking for any possible sources of contamination – including contamination from TSEs. This group forwarded its results to the EC, which prepared a note for all its members.[31] This summary eventually served as the basis for a decision to remove certain medicinal products from the shelves in various member states, including France.

The transmission of CJD through the use of human growth hormone (HGH) became obvious in 1992. At the request of the Ministry of Health, Dormont prepared a report on the problem. From 1989 to 1992, 13 cases of CJD contamination by HGH were confirmed. According to Dormont, this was a "strong signal" that non-conventional transmissible agents were causing infections.

The epidemic-surveillance network for CJD was established in 1991 at the initiative of the INSERM's Alperovitch Unit. The network was initially funded by the Ministry of Health, the INSERM, and two EC programs (Biomed 1 and 2). The network's mission was to report as soon as possible on the possible impact of BSE on human CJD.

However, the head of the Surveillance Division at the Department of Health received a strong message in a paper by Savey and Baron (1994), the content of which differed greatly from what was generally being said about the risk of BSE to human health.[32] Commenting on the possibility of transmission of BSE to humans, the authors pointed out that one way to verify this possibility was through the surveillance of CJD; if the number of cases increased, it meant that this hypothesis should most likely be taken seriously. They added that the UK data seemed to confirm this point. An extract from Savey and Baron's paper, which contains the notice sent to the head of the Surveillance Unit at the Ministry of Health,[33] reads as follows:

[30] The July 31, 1992, joint decree of the *Direction Générale de l'Alimentation, Ministère de l'Agriculture* (DGAL, Division of Food Affairs) and the *Direction Générale de la Santé* (DGS, Directorate of Health) implemented a recommendation of the *Commission Interministérielle d'Etude des Aliments destinés à une Alimentation Particulière* (CEDAP, Inter-ministry Commission on the Study of Foods for a Particular Alimentation).

[31] *Note for guidance for minimizing the risk of transmitting agents causing BSE via medicinal products (III/3281/91)*, prepared for the European Community's Committee for Proprietary Medicinal Products (Brussels, June 1991).

[32] He notes an important difference with the way these problems were discussed at a 1993 session of the *Comité Supérieur d'Hygiène Publique, Ministère de la Santé* (CSHP, Higher Committee on Public Health, Ministry of Health). In this session, the committee did conclude that the situation presented no specific dangers and, thus, did not require further action.

[33] Ibid.

> This phenomenon seems to occur in the United Kingdom, because the annual number of CJD cases identified has gone from 32 (period 1990/1991) to 48 (period 1992/1993). It is nevertheless evident that the United Kingdom is the country where the risks of an increase in the cases of CJD related to BSE are the greatest. It is important to note that this situation is unique to the world since in other countries, and in particular France, the disease remains very sporadic and public health protective measures were taken much more quickly, considering the United Kingdom's experience (Savey & Baron 1994).

The head of the Surveillance Division then examined papers published by the Surveillance Unit of Edinburgh, which confirmed his concerns. He described the problem at an inter-ministerial meeting and was given instructions by both the Minister of Agriculture and the Minister of Health, which eventually led the EC to reinforce its preventive measures in 1994.[34]

The Actions of the Ministry of Agriculture

In the next period (1994 to the crisis), the DGAL was given back the task of managing the BSE risk. This decision resulted in part from pressure by the *Direction Générale de la Santé* (DGS, Directorate of Health) (February 23, 1994, Decree about the processing of mechanically recovered meat), but more so as a result of the enforcement of the 1994 EC decisions. After a period of silence on the part of the EC – which was affected by the UK's influence on the Agriculture Committee and the Scientific Veterinary Committee (SVC) of the DGS – Germany requested a new set of measures to mitigate the BSE risk. The following measures were adopted:

1. The banning of MBM from animal feed for all ruminants (EC Decision 94/381 June 27, 1994). As the first European regulation on the use of MBM, this decision was stated in a decree by the DGAL and the *Direction Générale de la Consommation et de la Répression des Fraudes* (DGCCRF, Directorate of Competition, Consumer Affairs and Fraud Control, Economic Ministry) on December 20, 1994.
2. Compulsory heat treatment of ruminant waste in order to inactivate the BSE agent in the production of MBM (EC Decision 94/382 June 27, 1994). The standard heat treatment is to be done at 133°C at 3 bar for 20 minutes.[35] However

[34] The French demand coincided with one from Germany, thereby reinforcing the pressure on the EC and the UK.

[35] Directive 90/667 of November 27, 1990, on the treatment of animal waste was supposed to be enforced in all member states within 30 days of its publication. Note that, recently, several of these states (France, Benelux, Germany, Spain, Sweden, Finland, Italy, Portugal, Denmark, and Greece) have been subjected to infringement proceedings to Directive 90/667/EEC, as well as Decisions 94/381/EC and 96/449/EC. The proceedings followed the European inspections in these countries. In the French *Arrêté* of December 30, 1991, this treatment was made obligatory for both high- and low-risk animal waste. If the shortcomings persist, the commission claims that it will pursue the proceedings further (Commission of the European Communities 1998).

based on research financed by the EC in 1991, the SVC concluded that the inactivation of the BSE agent by the standard heat treatment should not be taken for granted, which explains why EC 94/381 came into force at this time.

3. An export ban on meat from the UK if it came from bovines 6 months or older (EC 94/474 July 27, 1994). This decision was stated in a DGAL decree issued on September 9, 1994.
4. EC Decision 94/794 (December 14, 1994). This decision authorized the export of meat from animals born after January 1, 1992, or coming from farms certified as free of BSE for 6 years. It was issued in a DGAL decree on June 1, 1995.

Animal (Cattle) Health in France During the Period

By June 1993, only six proven cases of BSE had emerged. Indeed, the occurrence of the disease continued to be very sporadic. The French researchers in charge of the epidemic-surveillance network concluded that this number of cases was not incompatible with the hypothesis that the disease had existed in sporadic form for many years, with very low incidence – as with the incidence of CJD in humans (Savey 1993). However, this did not mean that the risk to human health was over. In another paper, these researchers simultaneously insisted on the need to monitor CJD because of the risk of transmission of BSE to human beings (Savey & Baron 1994). However, the risk to public health remained a UK problem.

In the beginning of May 1995, 118 suspected clinical cases had been fully investigated, leading to the confirmation of 12 cases. All were dairy cows; nine were from the western part of France, and six of these were from two departments – Finistère and Côtes d'Armor (Coudert et al. 1995).

However, the work did not go smoothly. The CNEVA managers, who at the time were in charge of the coordination of the epidemic-surveillance network, complained of a "lack of acceptance" of their work by farmers. They added that "the motivation of veterinaries diminishes over time," which could explain "a probable under-reporting of suspected cases" (Coudert et al. 1995).

According to the estimates of the 2001 Parliamentary Mission on MBM (Sauvadet & Vergnier 2001), 250 cattle were contaminated between 1993 and 1994. This spread of the disease occurred despite the banning of MBM for cattle (French Decision of July 24, 1990) and all ruminants (European Decision on June 27, 1994). As some factories prepared foodstuffs for pigs and chicken – for which MBM was still allowed – and for ruminants, traces of MBM still contaminated some cattle feed. This effect, later named cross-contamination, was fully identified in July 1995, when it became obvious that cattle born after the MBM ban (they were called BABs: "born after the ban"), were infected. The French authorities learned little from the experience in the UK, where such an effect had been identified 2 years earlier.

Until 1996, no measures were taken specifically designed to mitigate human risks. This was a difference from the UK, where SBOs were banned from human food in 1989. The specific ban of SBO in baby food was more a product of risk ownership (as the responsibility for baby food was shared by the Ministry of Health and the Ministry of Agriculture) than the result of a comprehensive risk assessment, which was never performed. In fact, neither ministry performed a risk assessment, but the Ministry of Health was more inclined to adopt precautionary measures.

Lessons from the Silent Period, 1991–1995

Between 1991 and 1995, BSE was a "normal" problem handled by the government, using its standard procedures. The important decisions were made by the EC. Other decisions were made by the Directorate-General for Agriculture (DG Agriculture: DG VI), which exclusively provided the grounds for action. These decisions were implemented by the Ministry of Agriculture. The Ministry of Health failed to influence the dossier, because it was not qualified to act on food safety (except for baby food).

At the European level, the issue was framed primarily as a problem of animal health and a threat to the economic fortunes of the British cattle industry. This is why the EC acted rather slowly and was reluctant to take measures that could seriously harm the cattle industry. During this period, meetings about BSE were even prohibited by the internal note of the DG Agriculture "Stop any meeting on BSE" (Ortega 1997). Thus, no country could be benefited greatly from the experience of the other(s), either bilaterally or through the EU. Consequently, the French management of MBM did not take into account the cross-contamination problem, which, as noted above, had already been observed in the UK.

Finally, this period was characterized by a failed attempt to launch a research program and articulate the already available and/or compiling research results and risk assessments. Research remained disconnected from the official/governmental expertise, which, according to the French technocratic tradition, resided in the administration.

The First Mad Cow Crisis in France, 1996

The Early Signals

In October 1995, some very important information was reported in *The Lancet*: two new cases of atypical CJD were detected in the UK. The people affected were less than 20 years old and had not been infected through the use of HGH.

In France, the scientific experts did not react to information from either Savey, who had tried to draw attention to the transmission of BSE to humans for many years, from Dormont, who also was very interested in the problem, or from Alperovitch (1998), the epidemiologist in charge of the French epidemic-surveillance network on CJD. Alperovitch was collaborating with Will, who co-authored the paper reporting the 10 new variant Creutzfeldt-Jakob disease (nvCJD) cases, which was published in *The Lancet* on April 6, 1996 (Will et al. 1996).

Meanwhile, because of strong concerns about food safety, four of Germany's states (*Länder*) decided to ban UK beef. In addition, the French administration began to pay close attention to the problem. Exchanges of information between the Cabinet of the Agriculture Ministry and the French Embassy in London were very frequent, thereby demonstrating genuine concern at the highest levels.

The Early Days of the Crisis

On March 8, the CJD Surveillance Unit informed the SEAC that 10 cases of what appeared to be nvCJD had been discovered. Strangely, this information did not leak out and thus was unknown to other European experts. This happened despite all the links that had been in existence since the early 1990s between committees, colloquia, and even research institutions/laboratories engaged in collaborative research.

A few days before March 20, rumors about the discovery began to spread. For some time, a colloquium on TSEs had been planned for Paris. It began on March 19, 1996, in the presence of Gajdusek, who first described the prion disease *Kuru*. The experts expected this meeting to reveal detailed information about what was going on in the UK. Instead, the UK experts left the conference on the morning of March 20 without revealing anything. That afternoon, Dorell announced to the House of Commons that 10 new cases of nvCJD had been diagnosed and that the possibility of transmission of BSE to human beings could not be excluded. On the morning of March 21, Vasseur, the French Minister of Agriculture, announced a unilateral ban on UK beef,[36] a decision reached overnight by his cabinet.[37]

An exceptional meeting of the European Agricultural Ministers was hastily arranged, and on March 27, 1996, a Europe-wide ban on the export from the UK of bovine products, including those destined for use in cosmetic, medicinal or pharmaceutical products, was adopted (EC Decision 96/239, signed by Fischler).

French Prime Minister Juppé decided to set up a *cellule ministérielle de suivi* (follow-up department), which he chaired himself; its members included State Health Secretary Gaymard, Agricultural Minister Vasseur, Minister of Finances and Foreign Trade Galland, and State Secretary for Research d'Aubert. This "departmental unit" met twice between March 21 and April 10. The General Directors of key ministries met once a week during this crisis period.[38]

---

[36] *Arrêté du* 21/3/96 (Decree of March 21, 1996) published in the *JORF*, on March 22, 1996. It was later confirmed by the EC Decision and by the judgment of the European Court of Justice (May 1998).

[37] The extreme reaction of the minister may be explained, inter alia, by certain biographical facts. His Cabinet director was well aware of all the problems associated with meat production and consumption, as he was previously Director of the *Office Interprofessionnel de la Viande et du Lait* (OFIVAL, National Inter-Professional Office of Meat, Breeding and Poultry Farming).

[38] The *Directeur Général de la Santé* (DGS, Director General of Health), Ministry of Health; the *Directeur Général de l'Alimentation* (DGAL, Director General of Food Affairs), Ministry of Agriculture; and the *Directeur Général de la Consommation, de la Concurrence et de la Répression des Fraudes* (DGCCRF, Director General of Consumption, Competition and Repression of Fraud), Ministry of Economy and Finances.

French consumers reacted promptly, which made the situation difficult. One week after Dorell's announcement, beef sales dropped 50% on the Rungis national market. Information on the risk of nvCJD from beef consumption, as well as the revelation of supposedly hidden practices in the cattle industry (feeding animal meal to animals, adding cattle brain to hamburgers, etc.) created a sense of confusion and diminished public trust in such key actors as the government, scientific experts, and the beef industry. The interpretation of these events by the French public was colored by the blood scandal, which was still high on the public's agenda (Hermitte 1996; Dodier 2003).

Soon thereafter, one key point became obvious: France lacked scientific expertise on TSEs. It subsequently became necessary to set up an expert committee, which, in a sense, was a return to the procedure recommended in the Dormont Report in 1992.

Establishment of an Expert Committee to Manage the Crisis

On April 17, the Inter-ministerial Committee, consisting of experts on TSEs and prions, was officially established. The importance of this event was marked by the presence of three ministers (Agriculture, Health, and Research). These ministers gave the committee a dual mandate: (1) collect and analyze the information on the current state of knowledge about BSE and provide guidance for making decisions regarding human and animal health; and (2) propose a national research program aimed at characterizing the factors responsible for these animal and human diseases and how they are transmitted, analyzing their physiopathology, designing diagnostic tools, and exploring potential pharmacological tools.

This second mandate is particularly important because, as the threat of BSE transmission to humans became increasingly real, the decision makers began to realize that they did not have even the minimum amount of information and knowledge necessary to control the problem – a situation that arose from the failure to launch appropriate new research projects in the 1990s. This well-documented point was one of the State Secretary for Research d'Aubert's key messages during his address to the 1996 Parliamentary Mission.[39]

The committee's composition and organization were determined by policy makers. The Prime Minister's Cabinet defined the general rules. One of his concerns was that such a committee should not be suspected of being influenced by either administrative or economic factors. Indeed, its key function was to propose measures to protect human health. To restore the public's trust, the committee had to be seen as credible and impartial (Joly et al. 1998). This was especially important, as beef consumption in France had been decreasing sharply. Each of the three ministries had proposed eight experts who in turn had been nominated by

[39] Hearings of the Parliamentary Mission, July 9, 1996.

their respective research institutes. However, it was agreed from the outset that the Ministry of Health would appoint the chairman and that the committee members would all be scientists, not members of either the administration or the economic sector.

The first act of the committee was to set an agenda. Its first conclusion on the problem of nvCJD was stated as follows: "In this context of uncertainty, the precautionary principle means that, in all decisions surrounding animal and human health, the BSE agent has to be considered as being transmissible to human beings." This first statement was important because it reflected agreement on the importance of the precautionary principle, which proved instrumental to the successful completion of the committee's work. It was not easy to obtain a consensus among veterinarians, medical doctors, and scientists from the fields (structural, cellular, or molecular biology), all of whom had from the outset different opinions about the ongoing crisis. Thus, committing this first piece of advice to paper was key to the committee's collective learning process (Dormont 1998). The statement was also important because it was circulated; it was sent to the government on May 9 and soon reached the public through a leak to the news magazine *Le Monde*:

> MAD COW: French experts warned the government about the risks of transmission ... That conclusion was meant to be kept secret (*Le Monde*, June 8, 1996).

After this "scoop," the committee members' opinions were generally made public – at least during the "hot period" of the crisis – although transparency was not a formal rule.

One tour de force of the Dormont Committee was that, despite great uncertainty, controversy, and conflicts of power among the various research organizations, a common position was established in the scientific community. Therefore, during this period of crisis – and with the help of the media – scientists reinforced their importance in the decision-making process, despite the gaps in their knowledge. The scientific information that emerged in the months following March all seemed to confirm their claim of authority, thereby reinforcing the committee's central position.

The committee also organized several thematic seminars to which all French researchers potentially interested in the various aspects of the study of TSEs were invited. They represented a wide range of disciplines, such as molecular biology and epidemiology (including the sociology of risks). These seminars were important in structuring the call for a tendering of a national research program, which was published later in June 1997.

The Expert Committee and Government Bodies

The committee essentially agreed that it would only respond to questions posed by government bodies. It also decided to produce several comprehensive reports to provide the state with scientific knowledge (Dubot 1996).

Soon after the *Le Monde* revelation, the government announced that the recommendations of the Dormont Committee should be applied in a systematic way.[40] Before the committee was established, however, the administration/government had to manage this project for itself. Consequently, the following three periods during which the committee's role qualitatively changed can be distinguished:

1. From March 20 until the end of May, the committee was divided up into working groups and had not yet produced precise recommendations. Public decisions were made on the basis of the experience of the administration and by taking into account the recommendations of the WHO.[41] The government decided that SBO from cattle born after July 31, 1991, had to be excluded from the human food chain and incinerated (Decree of April 12, 1996). This decision implicitly acknowledged that the July 1990 partial ban of MBM had not been immediately effective.
2. From June to November 1996, the government implemented the recommendations of the Dormont Committee, including an adaptation of its earlier decision on SBO. On June 28, 1996, it decided to ban all SBO from cattle older than 6 months and from sheep and goats older than 12 months. This measure followed the Dormont Committee's recommendation. The committee did not believe that the spread of BSE had been stopped by the implementation of the decisions made before 1996 and, furthermore, that the risk of "ovine BSE" could not be excluded.
3. Although the committee continued to play an important role in designing research on TSEs, after November 1996 its role as provider of expertise for public decision making progressively diminished. For instance, it was not consulted in the first stages of the negotiations that resulted in lifting the ban on British beef, which was becoming one of the key elements of the beef war.

Let us have a look at the committee's main recommendations during the second period above. On May 23, 1996, the committee answered a series of questions it had received on April 26 from the administration. First, it restated its fundamental position: Despite the high degree of uncertainty, precautionary decisions must be made as if the possibility of transmission of BSE to humans had been confirmed. Another observation provoked intense media attention: "Another hypothesis must be considered: sheep BSE may be hidden by natural scrapie. This is possible because sheep may have been fed with MBM containing BSE." Because of such a risk, the epidemic surveillance of sheep, which had always been neglected in the past,[42] now had to be enforced. On June 14, a ministerial decree added scrapie to

[40] The DGAL produced a chronology, available on its website, that presents a detailed explanation of how each decision resulted from a committee recommendation.

[41] The committee of WHO experts met in Geneva on April 2 to 3, 1996. It recommended banning SBO from human food. The DGAL published a circular completed on April 9, stipulating that all SBO from animals that may have been fed with MBM must be banned. This partial measure is another illustration of the conflicts of interest at the DGAL.

[42] Actually, the surveillance for scrapie had been implemented in the Atlantic Pyrénées district in 1995, through the initiative of the Farmers' Veterinary Association, because, with some 100 herds infected, this district had been severely affected.

the list of reportable diseases and created the epidemic-surveillance network for scrapie (Decree 96–528).

The experts also raised the question of inactivation of the prion: The papers of Taylor (Taylor et al. 1994; Taylor 1996) suggest that the BSE agent may have a different sensibility to treatments than scrapie: chemical treatment with soda 1N and 2N and heat treatment under pressure, at 134/138°C for 18 minutes, only led to a partial inactivation. This observation explains the committee's lack of confidence in the EC regulations, which are based precisely on the assumption that such treatments guarantee the quality of MBM. On June 9, the complete set of recommendations was made public by the French government in a press conference organized to demonstrate its transparency in managing the crisis.

On June 27, the committee sent out a second report. Again dealing with MBM, it stressed the presence of significant uncertainty and recalled that only a few of the many questions raised by the government could be answered, because of the lack of scientific data. Due to the problems surrounding the inactivation of the BSE agent, the following set of measures concerning the upstream selection of the raw materials used for preparing MBM was proposed: (1) animals not fit for human consumption should be excluded from the MBM circuit and be incinerated, (2) the central nervous system of *all ruminants* should be removed, and (3) a strict separation of the ruminant animal feed devices should be implemented. It further stated: "This set of measures must be enforced and their control must be rigorously enforced. These are the conditions which would allow pigs, chicken and fish to be fed with MBM containing ruminant tissues."

This strong concern about the implementation of policy measures is an interesting feature of the Dormont Committee. In its deliberation, this committee focused on the "real risks," thereby taking into full consideration the concrete and practical conditions (namely, the real state of implementation) of the measures proposed (Joly 2007). On various occasions, the committee took the initiative of visiting slaughterhouses and feed or rendering plants to get a clear picture of how risk management was really practiced.

The recommendations on MBM was the most controversial issue in the committee's deliberations. For example, some members favored a total ban on MBM in animal feed (Dormont Interview in Grenoble, February 1999). These recommendations were important because of their economic impact,[43] but more so because they put France in an isolated position. The EU and France still had different assessments of safety in the production and use of MBM. The decree of September 10, 1996, banned imports of MBM by one European country from another if the requirements of the June 28 decree were not met. The EC subsequently required France to enforce the EC's rules on heat treatment (EC decision 96/449).[44] As such actions

[43]Their implementation provoked a crisis in the rendering industry because of the sharp increase in costs. The volume of "fallen stock" to be incinerated was about 450,000 tons a year, which represents a total cost of about 700 million Francs. It was necessary to change the status of this industry through the passage of a law in December 1996. The law defined rendering as a public service and imposed a tax on meat to finance the activity.

[44]The French government considered that upstream selection was more efficient, taking into account the doubts about the efficiency of the heat treatment of MBM. The latest evidence seems to confirm this position. However, the government was obliged by an *Arrêté* dated February 6, 1998,

were published in full by *Le Monde*, they received good media coverage. Prime Minister Juppé, accompanied by all the ministers concerned, announced these recommendations, stating that the government would enforce them all. A decree was published on June 28.

From a Sanitation Crisis to a Political Crisis

Several hypotheses may be formulated to explain this first crisis:

1. In France as in the UK, the public announcement of the possible transmission of BSE to humans was obviously the first factor that provoked a decrease of beef consumption and triggered the crisis. Because of the great uncertainty about the nature of the infectious agent, estimates of the number of potential fatalities were very high – up to 150,000 in the worst-case scenario.
2. The first reaction of the French authorities, which occurred on March 20, 1990, was to close the borders to protect French consumers. This is a common action in French public policy (cf. the first ban in May 1990). Meat professionals also discussed the origins of French beef. In May 1996, in an effort to reassure French consumers, the Center for Information on Meat (CIV) launched a communication campaign and introduced a new kind of beef labeled French Bovine Meat (VBF).[45] When information on the importation of MBM from the UK was published in *Nature* and *Le Monde* on June 13, the policy of maintaining current levels of beef consumption by distinguishing British beef from French beef was abandoned, as it appeared to the public that French cows had been exposed to the same infectious agent as British cows. The spate of BSE cases just after March 20 confirmed the public's suspicions.
3. The scope of the contaminated blood affair surely played an important role as well. In the blood case and thereafter, the political risks appeared to be very high and the attention paid to the problem by the media was significant. *Le Monde*, in particular, played a key role, following the early stage of the crisis day-by-day. On several occasions, the agreement between *Le Monde* and *Nature* allowed the newspaper to publish key information even before the public authorities had been informed.
4. The political responsibility for the crisis appeared to be great, especially in the UK, where the government first allowed BSE to spread in an effort to protect economic interests at the expense of public health. In its early months, the crisis was fed by a number of factors, including the so-called revelations (actually, all the information had been previously published), the role of the expert committees in Brussels, and the absence of control of MBM imported from the UK.

---

to implement EC Decision 96/449, which had been rendered on July 18, 1996. Nowadays, these standards are obligatory for all factories producing MBM.

[45]CIV is under the control of the "*filière* (processes)" of meat production and receives subsidies from the Ministry of Agriculture.

5. Given the many uncertainties, this crisis left policymakers with few benchmarks. Who were the victims or potential victims? The cows? The farmers? The tripe butchers? The rendering industry? The general public? Who was guilty? Thatcher? The neo-liberals? The EC? The rendering industry? Given the growing presence of prions and the expansion of the network of victims and the guilty, managing the crisis was becoming very difficult. This can be seen by following the communications in this phase. For instance, see the different elements in the detailed chronology of President Chirac's activities.

The expansion of the health and hygiene crisis led to political intervention. On June 18, the National Assembly decided to set up a parliamentary enquiry presided over by Evelyne Guilhem. The parliamentary mission was meant to shed light on past political responsibility. Interestingly, after some very significant work, the results were quite modest and no blame was assigned. The main problems were attributed to misbehavior by the UK and Europe. The Parliament's conclusion was an attempt to reassure the public: "… It is obviously necessary to expect the rise of new cases of BSE and Creutzfeldt-Jakob disease in the future and public opinion must be informed, but except for a dramatic, scientific turn of events, *the risks of contagion seem henceforth under control*" (Guilhem & Mattei 1997).

MBM, Cows and BSE: Data and Hypotheses by the End of 1998

By the end of 1998, France had 51 BSE cases (in a national herd of 4.476 million dairy cows).[46] Thanks to British veterinary research, as well as the scientific data accumulated on TSEs since Gajdusek's pioneering research, the key role of MBM in the transmission of BSE was identified in 1987. Vertical and horizontal types of transmission were also discussed as complementary hypotheses. This led to three main factors being considered as causing the spread of the disease: (1) the importation of live animals from the UK, (2) the importation of MBM from the UK, and (3) the extent of MBM use in cattle feed.

In France, the extensive use of MBM was recent. It evolved during the 1970s and 1980s, increasing from 140,000 tons in 1970 to 440,000 tons in 1987. Moreover, MBM was fed much more often to pigs and chicken than to cows. How much MBM was used as feed for dairy cows remained unknown because of the high variability in the amounts of animal feed produced by the factories, which are very reactive to price changes. However, it is estimated that before the ban, MBM was used to feed between 1% and 3% of the dairy cows in France versus 10% in the UK (Enjalbert 1996).[47]

[46]Compared with the UK's 2.496 million dairy cows, Germany's 5.193 million, Portugal's 0.364 million (in 1995), and Italy's 2.113 million (in 1995). Eurostat 97 (DG Agriculture website).

[47]Note that "estimated average use" means that the level of exposure to the infectious agent varied greatly. This surely explains the concentration of cases in large dairy herds in the western part of France.

Table 9.2 Trends of beef consumption in France: the two crises

	Beginning	3 Months later	6 Months later	10 Months later	Cumulative 10 months
1996 (from April on)	–20%	–18%	–6%	–5%	–10%
2000 (from October on)	–36%	–22.5%	–11%	–8%	–18%

Source: Wolfer (2004).

In addition, France had a much less extensive breeding system than the UK.[48] Although the procedures used by the rendering plants for processing MBM are not very well known, it seems that in the 1980s, as in the UK, many factories in France suppressed hexane treatment to produce fatty meals that were in high demand because of the relatively high cost of energy in producing animal rations. The relatively high exposure to MBM and bovine imports from the UK in the western regions of France, where a little less than half of French dairy cattle are raised, may explain the high proportion of BSE cases in this region (see Table 2).

The partial ban of MBM and the improper implementation of measures to control it are probably responsible for its continued use in cattle feed, whether this is done intentionally or accidentally. This conclusion is indicated by the high proportion of BABs, which has raised strong concerns in France. In a statement transmitted to the ministers on February 23, 1999, the Dormont Committee suggested that the ongoing use of MBM in cattle foodstuffs after the ban (between 1990 and 1996) was the most likely reason for the high proportion of BAB cases, and it considered cross-contamination to be the major explanation for it. In a March 10, 1999, press conference, the Minister of Agriculture also cited this explanation, acknowledging that BSE would not be eradicated before 2001. However, he refused to comment on the origin of the MBM, pointing out that its use might not have been intentional and that the heterogeneity of regulations within the EU was a matter of genuine concern (*Le Monde*, March 12, 1999). Although the importation of MBM from the UK had been limited since 1990, its importation throughout the period was still fairly significant.

Lessons from the 1996 Mad Cow Crisis

The magnitude of the 1996 mad cow crisis in France may surprise many, as the situation differed significantly from that in the UK. At the end of 1996, as stated in the French Parliamentary Report, 26 BSE cases were identified in France versus 170,000 in the UK, and only 1 case of nvCJD in France versus 14 in the UK (Guilhem & Mattei 1997). Consequently, the threat to human health due to exposure to the BSE agent was much lower in France. Furthermore, unlike their counterparts in the UK, the French authorities could not be accused of concealing the threat to human health. Despite these differences, the magnitude of the crisis was

[48]In France, 29.3% of dairy cows are in herds numbering more than 50 cows, compared to 84% in the UK, 36.2% in Germany, 41.1% in Ireland, 50% in Italy, and 16.4% in Portugal (DG Agriculture: Eurostat 1995 website).

similar in terms of economic damage and political impact. Consumers decreased beef consumption, waiting for reassuring safety measures (Latouche et al. 1998).

All this might mean that the UK was responsible for transmitting to the other EU member states not only the infectious agent, but also the main casualties of the crisis, namely, public trust in government institutions and the markets. Indeed, the nature of mad cow disease contributed to the amplification of the risk: the infectious agent was not known, the exposure to the danger was not voluntary, nvCJD led to a dreadful death, and its spread was potentially catastrophic. However, the French context also played a role in the dynamics of the crisis, through a combination of distrust in public institutions left over from the blood scandal and the weakening of the technocratic model of decision making.

Regarding this context, it is important to note that the government created a new actor to manage the crisis: an expert committee that was independent of the ministerial departments and economic interests and whose members were chosen for their competencies. The members also adopted the principle of transparency, even though that was not clearly required when the committee was formed. During the crisis, the government claimed that it followed the recommendations of the committee, which, if true, meant that its actions conformed to the precautionary principle. Here again we observe the re-nationalization of the mad cow problem; the primary decisions were initiated by the member states and then adopted by EU institutions, which retook the initiative after the situation normalized.

During the crisis, three main decisions were made: (1) an embargo on all UK beef products, (2) the exclusion of SBO for all human consumption, and (3) a new set of measures preventing contamination of MBM by potentially infected material, which amounted to a ban on ruminants.

The Post-crisis Period (1997–1998)

In addition to the short-term decisions it evoked about the crisis itself, the mad cow crisis stimulated long-term structural reforms in the French system of public health. As this crisis occurred in the same time frame as a series of sanitation crises (contaminated blood, asbestos, Chernobyl, etc.), many actors were ready to initiate or accept institutional changes. Such changes followed two complementary lines: (1) the acknowledgement of sanitation security as a legitimate area for state intervention, and (2) new principles of public management that entailed the delegation of risk assessment and/or risk management to independent public authorities (Benamouzig & Besançon 2005; Borraz 2007).

With the passage of Law 98–535,[49] a new sanitation security system was established to cover all aspects of human health (Besançon 2007). It consisted of four agencies. The *Institut de Veille Sanitaire* (IVS, Health Watch Institute) replaced

[49]Law 98-535, enacted July 1, 1998, is related to the intensification of the health watch and the control of the sanitation of products intended for humans.

the *Réseau National de Santé Publique* (RNSP, National Public Health Network) and was responsible for the surveillance and permanent observation of the French population. The *Agence Française de Sécurité Sanitaire des Produit de Santé* (AFSSPS, French Agency for the Sanitary Security of Health Products) was responsible for the regulation of all therapeutic or cosmetic products. The others are the *Agence Française de Sécurité Sanitaire de l'Environnement* (AFSSE, French Agency for the Sanitary Security of the Environment) and the *Agence Française de Sécurité Sanitaire des Aliments* (AFSSA, French Agency for Food Safety), created by Decree 99–242 on March 26, 1999.

The AFSSA was placed under the authority of the Ministers in charge of Health, Agriculture and Consumption. It employed 700 people, most of whom had previously worked at the CNEVA.[50] It integrated several expert committees, such as the relevant sections of the *Comité Supérieur d'Hygiène Publique, Ministère de la Santé* (CSHP, Higher Committee on Public Health, Ministry of Health), and satellite centers.[51] Its first general director came from the Ministry of Health.[52]

The AFSSA was given several mandates. The first was *risk assessment* of all food or feed products destined for humans or animals, animal diseases, veterinary or agrochemical products, GMOs, etc. The second was to provide *expertise and technical assistance* to the government in the preparation and the implementation of national, European, and international regulations. This second mandate also included consulting on public services programs, proposing new priorities, formulating recommendations, and asking for specific controls or inspections. The AFSSA was also to implement its own research program and seek assistance from other public research institutes as needed. Furthermore, the AFSSA could, on its own initiative, propose any measures necessary to improve sanitation security. In addition, consumer associations were given the right to call for an AFSSA assessment on specific points. The opinions produced by the AFSSA are made public, a policy intended to guarantee the agency's transparency.

The AFSSA embodied a basic principle quite new to the French system: the separation of risk assessment and risk management.[53] This separation was intended

---

[50]The decision to transfer CNEVA from the Ministry of Agriculture to the AFSSA was added to the law by an amendment of the *Sénat*.

[51]*Commission de Technologie Aimentaire* (CTA, Food Technology Commission); *Commission Interministérielle d'Etude des Produits Destinés à l'Alimentation Particulière* (CEDAP, Interministry Commission on the Study of Products Destined for Particular Alimentation); *Centre National d'Etudes et de Recommandations sur la Nutrition et l'Alimentation* (CNERNA, National Centre of Studies and Recommendations on Nutrition and Alimentation); and *Observatoire des Consommations Alimentaires* (OCA, Observatory of Food Consumption).

[52]Martin Hirsch, who was Director of the Cabinet of the Minister of Health, Bernard Kouchner. Hirsch published a book on the mad cow crisis: *L'affolante Histoire de la Vache Folle* (The terrifying history of mad cow), cited, in which he writes that this crisis shows the need for a major change in the food safety system (Hirsch & Duneton 1996).

[53]This standard distinction has been elaborated in the joint expert consultation of the FAO and the WHO (Geneva, March 13–17, 1995). It has now been integrated into the text of the Codex Alimentarius.

to guarantee the objectivity of risk evaluations and the use of all relevant data. Formerly, neither of these separate functions had been precisely defined. There were also important changes involving risk management. The DGAL devoted all its time to this task, which entailed the transfer of the *Service des Politiques Industrielles et Agro-Alimentaires* (Industrial and Agro-Alimentary Policy Service) to the Division of Production and Trade under the Ministry of Agriculture.

In short, the mad cow crisis has indeed been instrumental in the transformation of the food safety system, even though this transformation was initiated as the result of a new sanitation security policy.

Acting Under Conditions of Uncertainty: A First Look Back

A detailed analysis of this first period allows us to draw key lessons about how the public behaves under conditions of uncertainty. Although there were many unknowns, the knowledge available in the early days of the crisis proved – in retrospect – to have been sufficient to allow the development of measures that would efficiently mitigate the BSE risk. The deficiencies in the public management of the risk did not stem from the lack of scientific knowledge per se, but from its translation into action. The problem was caused primarily by three interrelated factors: the framing of the issue, the design of the policy measures, and the implementation of these measures.

First, prior to 1996, the problem was framed mainly (but not exclusively) as a matter of "animal health" and "British beef," a decision probably based on the fact that the Ministry of Agriculture was the main "risk owner." Many days transpired from the announcement of possible BSE transmission through MBM, to the UK decision to ban MBM from cattle, and then to the French decision to do likewise. This was a long time, even for protecting the health of animals. Second, the initial choice of low-cost, partial measures turned out to be inefficient. The 1990 partial ban on MBM left the possibility of cross-contamination wide open, and this subsequently proved to be a matter of genuine concern. Although the concept behind this decision (protection at a lower cost) was interesting, its practical implementation proved problematic.

A third issue is the linkage of analysis to risk management and policy enforcement. The implementation of a policy requires many different actors, such as the *Direction des Services Vétérinaires* (DSV, Management of Veterinary Services: local service in each department), DGCCRF inspectors, rendering-plant managers, animal feed companies, and farmers. Even if the decision is right and timely, it may prove difficult to implement if these various actors are not aware of the gravity of the problem and do not understand the rationale for the protective measures.

A fourth issue is the resistance to putting TSEs on the agendas of national/public research organizations. Despite the decision of the French Minister of Research Curien to commission a report on French TSE research, and despite the very clear conclusions of this report, the changes that took place before 1996 were minimal. A research program was set up in 1995, but it was not designed to produce scientific knowledge on key issues such as the inactivation of the BSE agent in MBM, the transmission of BSE to primates by oral ingestion, or the early detection of BSE. This limiting feature of the program's design, which resulted in substantial intellectual and organizational inertia at the intermediate levels, did not allow the minister to recruit and adequately motivate researchers.

The uncertainty created by this lack of scientific knowledge had a major impact on the 1996 crisis by contributing to the looming fears among actors in the society at large. It also had a major impact on the financial costs of the crisis, because the lack of knowledge resulted in unfocused and sometimes poorly defined precautionary measures. For instance, results from early BSE tests to identify asymptomatic animals became available only at the end of 2000.

Last but not least, we must consider the uniform pattern of communication in European countries. Governments' best efforts to reassure the public may entail dismissing worst-case scenarios, suppressing certain information, and framing the problem accordingly. In France, the threat to human health was said to be caused only by British beef; French beef was considered safe. The animal health problem was supposed to be "under control." Such reassurances are quite fragile when they are based on partial information and unknowns. When new information contradicts previous communications, the trust in risk management institutions may be destroyed. Therefore, the reassurances might pave the way for a future political crisis fed by the subsequent "revelation" of "hidden" information. Furthermore, such reassurances erode the efficiency of the corrective measures when those who are supposed to implement them feel that they do not have to do so because such measures were adopted only for political reasons.

To Lift or Not to Lift the Ban: The "Beef War"[54]

Following the March 1996 announcement of the possible transmission of BSE to humans, the EC followed the lead of some member states in banning the export of UK beef. Then in July 1999, following the advice of its Scientific Steering Committee (SSC), the EC decided to lift the ban. After 2 years of tough negotiations, the EC concluded that the measures proposed by the UK government in its Data Base Export Scheme (DBES) to be sufficient to guarantee the safety of UK beef for human consumption.[55]

---

[54] This section draws on the article by Joly and Barbier (2001), "Que faire des désaccords entre comités d'experts?: Les leçons de la guerre du bœuf."

[55] The first version of the DBES was submitted to the commission by the British government on October 2, 1997. The SSC at first gave a favorable notice under the conditions in place on December 9, 1997. The new version of the DBES received a favorable notice from the SSC on February 20, 1998. On November 4, 1998, the Permanent Veterinarian Committee (consisting of leading veterinarians of the member states) adopted a decision that placed an embargo on British bovine meat (eight votes to five – France, Italy, Spain, Germany, and Austria voting no). The commission made this decision on November 25, 1998 (Decision 98/692/EC). Decision 99/514/EC, rendered July 1999, fixed the starting date of the embargo as August 1, 1999.

According to French Law 98–535, the implementation of this measure required the French government to consult with the AFSSA. This event marks the beginning of a conflict between the AFSSA and the SSC, both of which were only recently established and thus lacked experience in handling such a difficult situation. Indeed, the so-called "beef war" was the first test of the new system for obtaining scientific advice after the 1996 mad cow crisis.

Sources of the Conflict

On July 1, 1999, before the AFSSA was officially consulted by the government, the *Comité Interministériel sur les ESST* (CIESST), the official name of the previously discussed Dormont Committee, published an Opinion on the progress in obtaining epidemiological data on BSE in the UK. It compared the data collected in 1996, 1997, and 1998 to the predictions of the Anderson and Donneley model. The discrepancies between the predictions and the observations could not be explained by the transmission paths so far considered (for example, MBM, banned since 1989, showed vertical transmission). The committee thus concluded that the situation could not be considered as completely under control and suggested that another explanation be sought.

Thus, when the government consulted the AFSSA, the committee stuck to its previous analysis. It worried about the increasingly slow decline in the number of BSE cases in the UK despite the measures taken. These data called into question the limited understanding of the possible contamination pathways. If the hypothesis of horizontal transmission was to be confirmed, the risks involved in exporting infected animals would have to be taken into account. The committee reminded the government that in the UK, only animals with clinical symptoms of BSE were slaughtered, not the whole herd as was the case in France. This observation is related to the fact that more than 50% of herds in the UK were impacted by BSE.

As the DBES did not exclude for export purposes animals from herds affected by BSE, the committee was skeptical about the guarantees offered by the scheme. It thus recommended that the horizontal transmission hypothesis be verified before any decision be made having public health implications. The DBES also proposed that beef introduced into the food chain be de-boned, de-nerved, and have the lymphatic ganglions removed. The committee did not consider these measures to be fully effective, as the process of infection in cattle was not well known. Last but not least, the reliability of the whole plan depended on the effectiveness of the identification and traceability system, which was brand new in the UK. The effectiveness of this new system could not be taken for granted; it had to be proven.

Having considered all these elements, the committee concluded that, given the state of the available scientific and epidemiological evidence, the risk of importing contaminated cattle or beef from the UK could not be considered totally under control. The committee also stated that uncertainty about the scope of BSE infection could be reduced very soon, thanks to ongoing research. The problems surrounding the transmission of BSE could also be solved on short notice because of the availability of biological tests providing improved detection capability. The expected information was thus considered necessary to reject hypotheses that cast doubt on the effectiveness of the DBES.

Martin Hirsh, General Director of AFSSA, accepted the Opinion of the committee and on September 30, 1999, expressed opposition to the decision to lift the ban.

The Position of European and Committee Experts

On October 14, the EC posed three precise questions to its Scientific Steering Committee (SSC): Does the French Opinion make reference to scientific or epidemiological information that would not have been taken into account in the previous SSC Opinions? Does such information require a new examination of the previous SSC Opinions regarding the scientific basis of the DBES? Does the SSC confirm that the DBES criteria, if respected, are satisfactory in term of the risks associated with the consumption of meat or meat products?

The ad hoc group on TSEs was captivated by these questions. This group included members of the SSC as well as recognized TSE specialists. After a two-day meeting, the group reached a consensus on the first question: The French notice did not provide new information unknown by the SSC during its previous deliberations. Regarding questions 2 and 3, the group declared that, first of all, a guarantee of zero risk cannot be given for any bovine product originating in any country where BSE is known to be currently active, not just the UK. However, there were two opposing positions. The first agreed with the previous SSC Opinions, stating that the residual risk was very low and the guarantees of the DBES were sufficient to justify the conclusion that the selected animals posed no risks. Besides, research programs aimed at showing that the AFSSA hypotheses were not confirmed could be envisaged. The second opposing position, on the other hand, stated that a new evaluation of the DBES – and the possible risks of (human) consumption of de-boned meat in all countries with a BSE problem – would be necessary within 12 to 24 months. This would allow for the consideration of new information, most notably the results of more transparent tests on the infectiousness of the tissues; the results of a follow-up on targeted animals that, although at risk, satisfy the criteria of the DBES; the results of a more detailed follow-up of the UK epidemiological data; the results of the validation attempts to confirm the post-mortem tests; and the results of the SSC working group on the slaughtering procedures. Given the need for this additional testing, the supporters of the second minority position suggested postponing the decision to lift the ban to a later date.

The SSC Opinion (October 28–29) considered the various arguments presented by the AFSSA. First, concerning the epidemiological data, it made clear that the AFSSA did not use the most appropriate data. The corrected data from 1998 fall within the 95% confidence interval for the prediction; thus, there was no reason to suppose a new way of infection. Concerning the infectiousness of the tissues, the SSC noted that the most recent results were encouraging, because – unlike those of the CIESST – they left no room for extrapolation from the spreading of the BSE agent among sheep. Also, the distribution of prions in the tissues seems to differ significantly by species. The SSC also observed that the available post-mortem tests were not reliable, but better alternatives could not be envisaged in the short term. Finally, the SSC recognized that the reliability of identification and traceability is of crucial importance. However, it considered this to be a matter involving control or risk management, not science. Thus, it is beyond the scope of risk assessment. In conclusion, the SSC stated that, if the measures proposed by the UK were meticulously applied, the risk to human health from UK beef exports would, in the worst case, be similar to that from such exports from other European countries.

Toward Maintaining the Ban

These differences of opinion between the two committees triggered political discussions aimed at reaching agreement. During a meeting of the French and UK authorities with the European Commissioner for Health and Consumption, a compromise was proposed to the effect that the embargo of UK beef exports could be lifted under the following five conditions: (1) entire herds must be slaughtered, not only the animals affected by BSE, (2) a series of tests must be conducted, (3) products containing any amount of beef must be traced, (4) animals and meat in the UK must be traced; and (5) meat must be labeled, and the label must include geographic origin.

After some exchanges between the committee and both member states, the French government again approached the AFSSA on November 23, 1999, for an opinion regarding the "conditions in which the import by France of meats and foodstuffs could be authorized resulting from cattle raised and slaughtered in the UK." In an Opinion dated December 6, 1999, the General Director of the Agency stated that the implementation of a test program would enable the risks to be reevaluated and more knowledge to be obtained at a faster rate, provided that the protocol was well conceived. However, he went on to state that nonetheless the test program would have no impact on the control of the risks. On a more positive note, he conceded that the additional measures concerning labeling and traceability should help mitigate the risks related to an imperfect application of the DBES. He concluded by stating that any decision had to take into account several factors.

First, the elements of the risk should be well considered, associated with uncertainty about the (distribution of) BSE infectiousness and (all) the manners of transmission. This concern is apparent if one follows the communications exchanged during this period, for instance, the detailed chronology of the communications involving President Chirac. Second, the measures taken to guarantee that the decisions taken will actually be respected may have no direct and immediate impact on these risk elements. Finally, it is necessary to be aware that these measures are reversible.

Based on this Opinion, the French government announced on December 8, 1999, the decision to maintain the embargo on English beef. The next day, the European

Committee attacked France in front of the European Court of Justice (ECJ) for noncompliance with the European treaties. On December 13, 2001, the ECJ condemned France for not having lifted the ban on UK beef as required by the European decisions. However, the court considered the arguments of the French government pertaining to the risks related to the lack of labeling and the traceability of UK cattle to be sound. Following a new AFSSA notice, France lifted the ban on October 2, 2002.

Lessons from the "Beef War"

This affair has been the subject of numerous commentaries. Godard (2001) carried out a comparative analysis of the AFSSA and SSC Opinions, which included the following observations: No significant difference existed between those Opinions, concerning the estimated and observed epidemiological facts. Differences come from a field that could be designated as non-scientific one: the viewpoints adopted by both sides regarding the hypotheses and the uncertainties were different; and the work of the experts and the form of the argumentation were responsive to different questions.

Concerning the first point, Godard argued that the French committee focused on residual uncertainties and (possible) hypotheses that had yet to be validated or invalidated. The SSC, on the other hand, considered the hypotheses to be plausible. Godard noted that the SSC's stance conformed more closely to Kourilsky and Viney's recommendations on the precautionary principle than did the AFSSA's (Kourilsky & Viney 2000). Concerning the questions, he opined that, forced to respond to an order, the AFSSA had to fix the question(s) to be answered: Is there a residual risk associated with the import of UK beef? On the other hand, the question the SSC had to answer was more precise: comparison of the risk from UK beef to the risk from beef from other countries. Finally, referring to a distinction introduced by Chevassus (2000), Godard observed that the European Committee estimated the "theoretical risks," whereas the French committee addressed the real risks, integrating into its analysis the possibility that the proposed policy/program would fail, but "… without adequately arranging for a scientific committee on the information needed to evaluate the level of measurement to be applied." (Godard 2001)

Rival Interpretations of the Precautionary Principle

In his analysis of this conflict, Godard makes a key methodological point: the need to make a comparative analysis of the risk related to the consumption of UK beef vis-à-vis that of products originating from other countries. The Opinion of the French committee appears to be seriously compromised by this blind spot.

Against this position, it may be argued that the French committee rejected a comparative analysis because they considered the data not to be comparable.

Even though the prevalence of BSE was declining in the UK, the change is incommensurable with that in other countries, including France, where in 1999 there were only 33 new cases compared to 3,000 in Britain.[56] Recall that because of this specific situation, the EC had created a specific category for the UK in its analysis of the "geographical risk of the BSE." However, the French experts adopted a rather narrow frame and failed to anticipate what could happen after autumn 1999 – namely, a continuing reduction in the number of BSE cases in the UK (from 1,200 cases in 2000), a remarkable increase in the number of cases in France (160 cases), and the reporting of cases in Germany, Spain, and Italy.

Nevertheless, the difference between these two Opinions is interesting because it illustrates the deep conceptual differences concerning the role of scientific evaluation in a context of uncertainty.

For Godard, the analysis of the CIESST leans on "theoretical suspicions." Calling upon hypotheses that might be valid implies an attempt to prove the null hypothesis (requiring a reversal of the proof, in this case, proving the absence of risk). On the contrary, the CIESST used the term *plausible hypothesis* as the basis for its Opinion. Even the experts could cite no experimental proof, although they proposed many hypotheses concerning the route(s) of transmission and the infectiousness of tissues. The French committee's examination of the epidemiological data, guided by its limited knowledge, led it to consider a third hypothesis about the contagion of the epizootic and/or the failure to control it.

> Since its creation, the CIESST repeatedly took into account in their reports of the hypotheses which seemed plausible, considering their knowledge at that time, when they could anticipate a risk for human health, even when it had not been shown ... While using the same initiative, in its notice of September 30th, 1999, the group of experts ... considered that ... the risk of UK meat from contaminated cattle was not completely under control. (Group of TSE experts at the AFSSA)

This context of great uncertainty also led the CIESST to take into account certain aspects of risk management within the framework of risk evaluation in France.

In this respect, the CIESST brought to light a key difference: in France, it was the practice to slaughter all cattle, whereas in the UK meat (and other parts) from apparently healthy cattle coming from herds with BSE cases were still consumed. Thus, the British policy increased the probability of consuming contaminated animals. The related question is whether the risk of contagion within a herd, which may not have been taken into account by the DBES, was stronger than the risk of the voluntary dissimulation of contaminated animals resulting from the shock of total slaughter. Certainly, as indicated by Godard, the committee did not have the expertise to undertake such an analysis.

The data needed for risk assessment, which in this specific case were quite complicated, were also needed for risk management. Hence, the boundary between evaluation and management in the application of expertise was blurred and could

[56]A total of 650 cases out of a million cattle more than 2 months old, compared with the French incidence rate of 1.5 to 2, and the rate in Northern Ireland of 10 to 15.

be reestablished only at the stage of writing the Opinion. Even if it is necessary to respect the separation between the roles of the experts who estimate the risks and the decision makers who manage them (and thus decide what level of risk is acceptable), the evaluation sometimes requires access to the management data.[57] This approach differs fundamentally from the doctrine of separation between risk evaluation and risk management that was established in Europe (cf. the Communication of the European Commission on the Precautionary Principle, COM 2000).

In contrast to Godard, who sees no formal contradiction between the Opinions of the two committees, we see a fundamental difference in the way risk was assessed. The CIESST did not focus on the quantitative aspect (i.e., the amount) of risk, but on the qualitative aspect of risk (e.g., the quality of the control), which was assessed with reference to the degree of uncertainty and the experience acquired in studying TSEs. This is why the CIESST stated that the French situation is not comparable to the British one. In this context, to proceed to a comparison of the BSE incidence levels can be misleading, because the nature of the data differs in the two cases. Unlike the SSC, which considered that, "if the measures proposed by the UK are accurately applied," the risks from UK beef were the same as those from French beef, the CIESST refused to conduct an analysis that would address this point. This position reflects their experience in dealing with public decision makers.

Thus far, we have explained the need for the CIESST to assess the effectiveness of the management measures designed in the DBES on a rational and/or cognitive basis. But there is also a legitimacy issue. Considering the fact that the UK exports had spread the disease to various European countries, the management measures had to be discussed openly so that the ban could be lifted in good faith. It was thus beneficial that, after this troublesome period, the CIESST openly asked questions that had been raised by various constituencies: Can we believe in the effectiveness of the current measures taken by the UK government? On what tangible elements can our confidence be based? From this point of view, the CIESST was only underlining a blind spot of the system: the opaqueness of the procedures used for risk management.

The Organizational and Political Dimensions of Risk Management

The public dispute between the two committees has been considered as undermining the scientific assessment and thus unnecessarily worrying consumers. To the contrary, we claim that the dispute helped produce an increase in knowledge, although the political translation of this knowledge remains problematic.

First, the dispute led to a more in-depth assessment of the available data and fostered an examination of critical aspects of the risk management. Moreover, it opened a new space for political negotiation. After the publication of the Opinion of the SSC, additional measures designed to mitigate the risks and improve the state

[57] In this view, we join Hermitte and Dormont (2000) when they suggest endowing expert committees with a power of inquiry, thereby guaranteeing them access to the relevant data for their analyses.

of knowledge were explored. How should a test protocol be designed and implemented? How should herds/cohorts be slaughtered? How should the reliability of the identification measures and the traceability of animals and products be measured? The ban/no ban dichotomy thus proved to be too limited. In this respect, the effect of the controversy was positive.

On the other hand, we can wonder why, after such an exploration, the political authorities stuck with the initial position of the expert committee. In this specific case, what was perceived at the time as a united front by the entire French society against the lifting of the ban seems to have had a determining effect. The dispute was thus a challenge for public decision makers, who could not hide behind the opinion of the scientists to justify lifting the ban. It thus may be feared that the identification of potential risks by the expert committees, and the publication of this information, systematically led to an excess in precautionary measures.

Mad Cow Disease as a Public Issue: The October 2000 Mad Cow Crisis and the Banning of MBM

October 2000 was marked by a second major mad cow crisis, with even more far-reaching effects on beef consumption than the first one (see Table 9.2).

This second crisis occurred at a time when public policymakers and professionals might have had the impression that the measures necessary to mitigate the risk of BSE in humans had already been implemented. All the human exposure to BSE had occurred before these policies were adopted, and so the residual risk appeared to be very low. As there was no longer a rational reason to worry, consumers' reactions were considered to be "irrational." Many observers used the label "collective psychosis." Indeed, this period is the key to understanding problems of communication and public reactions to phantom risks.

A Crisis Triggered by a Phantom Risk?[58]

As we now know, the origins of the 1996 crisis can be explained by the revelation of a serious threat that had remained hidden for economic reasons. On the contrary, the 2000 crisis came about for no apparent reason. None of the recently collected information about such topics as the propagation of the epizootic and the transmission of BSE to humans could explain what occurred after October 20. The impact of this crisis on beef consumption was huge. According to a poll conducted at the

[58]Phantom risks are risks the very existence of which is unproven and perhaps incapable of being proven. Nonetheless, they cause real problems at the interface between science and the law (Foster et al. 1999).

end of 2000, this second crisis was considered as the year's most important event by 68% of those surveyed. This percentage dwarfed that for the crash of the Concorde (32%) and for the war in the Middle East (CSA–La Croix poll).

The events that triggered the crisis are well known (Mer 2001; Barbier 2003). On October 20, one BSE-infected cow was identified at the SOVIBA slaughterhouse in Villers-Bocage. A butcher who owned the sick animal had separated that (BSE-infected) cow from the slaughtered herd several days earlier. The case might have gone unnoticed, but were actually disclosed and known to the public. Doubts and concerns set in. Could this be a hoax? Or more ominously, were the slaughtered cows that were introduced into the food chain also infected? The prompt responses of the protagonists (i.e., the prosecutor and veterinary services) aggravated those concerns, though they can also be construed as a sign of efficient control mechanisms. Most important from a practical standpoint, the Carrefour Supermarket chain decided to recall all ground beef from the SOVIBA slaughterhouse and widely publicized this decision.

The magnitude of this event is probably due to the growing anxiety in the public. On the one hand, since the summer of 2000 the newspapers had been reporting alarming information about the number of potential human victims of the nvCJD. The bovine active-surveillance program, started in June 2000, subsequently revealed a much higher number of infected animals than expected (see Fig. 9.1).[59] Such information was available because of the increased efficiency of the epidemic-surveillance measures and the policy of transparency adopted by the government.[60]

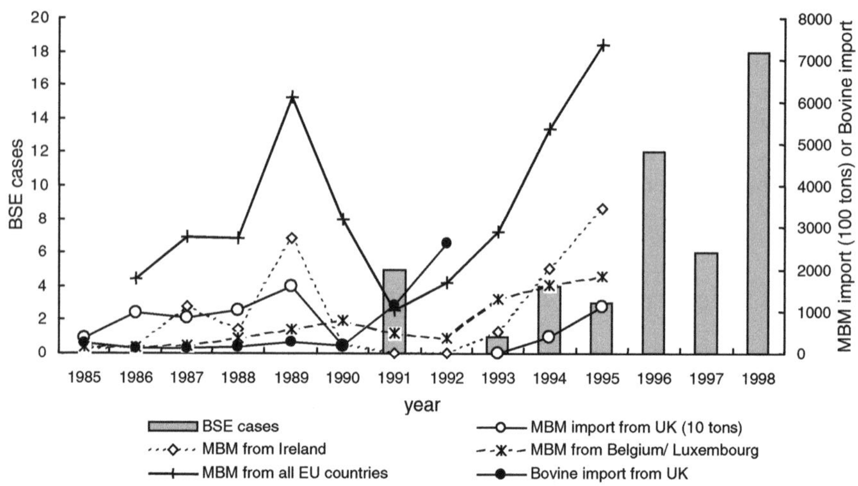

Fig. 9.1 Number of BSE cases and imported MBM and bovines in France

[59] See: http://vetolavie.chez-alice.fr/bse/details/graphfr/graphfr.htm

[60] The government decided to systematically publish epidemiological data on BSE in mid-2000, just a few months before the second crisis.

This transparency brought to light uncertainties, which in turn attracted public attention to the data and eventually induced a feeling of insecurity. Accordingly, the problem could no longer be considered specific to UK beef, as previously claimed.[61] Newspapers exploited the "frightening figures," including the "consumption of 1,200 contaminated animals each year in France" (*Le Figaro*, September 11, 2000) (Fig. 9.2).

Immediately after the SOVIBA event, images of a young man infected by nvCJD were shown on French television for the first time (*M6 TV Channel*, November 5, 2000). The mayors of many French cities consequently decided to ban beef from school meals:

> The psychosis overtakes the school lunch menus: Thousands of pupils deprived of beef. Bordeaux, Toulouse, Carcassonne, the Dunkerque, Strasbourg, Caen, Cherbourg, numerous municipalities announced the withdrawal of beef from the school lunch menus yesterday ... (*La Parisien*, November 7, 2007).

Fuelled by the circumstances of government cohabitation, and under intense political and media pressure, the government announced a total ban on MBM on November 15.

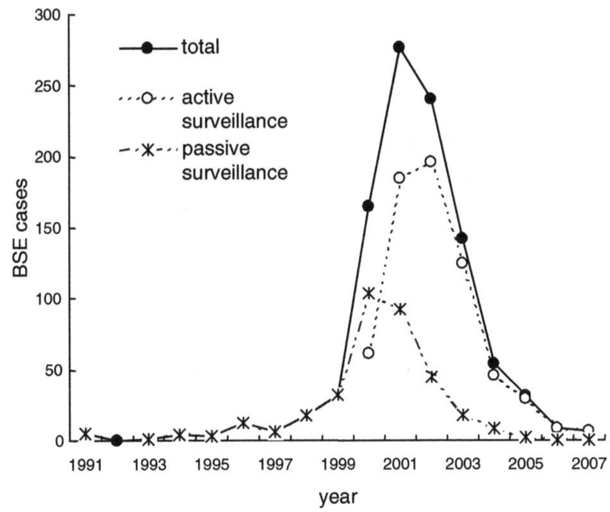

Fig. 9.2 Number of BSE cases in France

[61] Declared also in the governmental acts, because at the end of 1999 the French government was opposed to lifting the ban on British bovine meat.

In the public arena, many actors claimed that the crisis was a case of public hysteria. The M6 TV broadcast set the tone with its title "Mad cow, the great fear." The hysteria grew rapidly and dominated, foregoing the sober facts and explanations of the situation – both by the media and in political discourse. Hence, questions arose about individual perceptions of the food risks and the role these perceptions played in determining the dynamics of crisis.

Between Individual Perceptions and Crisis: The Notion of Collective Risk

It so happened that, completely by chance, the crisis occurred during a session of the *Etats Généraux de l'Alimentation* (EGA, Public Hearings on Food and Nutrition), organized by the Minister of Agriculture. The objectives of the EGA were to promote debate, dialogue, and information to clarify government policies related to food production and food risks. The formation of this policy was supposed to be based on the "real expectations and interests of the French" concerning the security and quality of food. The first step was to clarify their needs, especially for information.

The results of preliminary forums that brought together more than 500 people in five cities in September and October 2000, in addition to the results of large polls from the same period, confirmed the intuitions of the protagonists. The risk of disease was not an obsessive concern, just one of the many dimensions through which the French think about their relationship to food. The main results were published by the Ipsos in a document entitled "The French and the Quality and Safety of Food" (Ipsos 2000) (see Table 9.3).

This poll was backed up by other studies showing that dietary fears were a secondary concern among French consumers. In the same period, from October 16 to 31, 2000 – precisely when the second mad cow crisis unfolded – the *Institut de Protection et de Sûreté Nucléaire* (IPSN) conducted its annual poll on the perception of risks. In response to the question "In France, among the current problems, which is most worrisome to you?" only 3.3% of interviewees answered "food risks," far below the main preoccupations of violence, bankruptcy, unemployment, environmental degradation, poverty, and exclusion (IPSN 2000). Therefore, as shown by many studies, consumers' perceptions of risk, although qualitatively different from those of the experts,[62] are realistic.

[62] Nevertheless, as numerous studies have established, the perceptions of risk by consumers are not the same as those of experts. They are more qualitative and take into consideration criteria that are not explicitly considered in quantitative models of risk. Examples of these criteria include the voluntary nature of the exposure, level of awareness of the dangers, the number of people potentially concerned, and so forth. This difference between experts and the uninitiated has less to do with differences in cognitive capacity than with a difference of position: the expert considers the

Table 9.3 The French and the quality and safety of food

1. French cuisine: a source of national pride

   This first title highlights the French attachment to its culinary culture, which has both an essentially patrimonial aspect (the gastronomic tradition, the tradition of diversity, the agricultural tradition) and a social function (the importance of the meal as the event that seals family unity).

2. A French exception at risk: danger in the flavor

   Interviewees recognized that in the last 10 years the diversity of available food products has improved (77%), along with the modes of conservation (73%), the ease of meal preparation (71%), the hygiene of food production (66%), and the sanitation controls (64%). However, practicality and speed are detrimental to the taste of the products (32% of the French think that the taste has improved versus 30% who think it has deteriorated) and change how the meals are perceived (44% of the interviewees said that they do not want to abandon the meal and that they generally devote a good deal of time to it; 37% agree that its important, but they don't have much time to devote to it).

3. The perception of risks: fake food, bad food, and the disappearance of the meal

   The risks associated with the security of food sanitation are at the head of French preoccupations (45% of those interviewed), but a fifth (21%) are worried about the relative risks of an unbalanced diet and an identical proportion by the loss of our gastronomic idiosyncrasies; 11% mainly fear the degradation of the taste of food. When the interviewees speak of risk, it should be viewed as tied to their perception of "eating well," which gravitates toward two poles – namely, "healthy eating" (consumption of fresh products, research on nutritional balance, food variety, complimentary nutrients, reduction of the intake of fat and sugars) and, to a lesser degree, eating "safe, without risk" (without fertilizers, without colorants, without additives).

   This second dimension encompasses different dimensions: the risk of toxicity (expiration, rupture of the cold chain, etc.) and the fear of poisoning (27% of people fear genetically modified foods, 23% fear mad cow disease, 21% fear the use of chemical products or of hormones in food, etc.). This fear of poisoning is nourished by the fear of unscrupulous and corrupt industries or of the negligence of researchers and public leaders. These risks are less feared "for oneself" than for future generations. They nourish the fear of consumers who do not have an opportunity to change their opinions due to a lack of information.

Source: Ipsos Poll, October 2000.

In reality, there was indeed a crisis, and it caused a substantial proportion of French consumers to reduce their beef consumption. According to Setbon et al. (2005), most of the people interviewed did not consider such restraint to be a good way to avoid mad cow disease. However, most agreed that "reducing one's beef consumption is legitimate, since we don't have complete knowledge." It was also disclosed that changes in consumer behavior is positively correlated with how risky

"risk in itself" to be truly important, whereas for the consumer it's the "risk to him or herself." Such a difference in perception can lead to the distinction of two types of evaluation – that of a "theoretical risk," which results from a fundamental analysis and is not concerned with the contingencies involved with the implementation of preventative measures (errors, sabotage, etc.), and that of a "real risk," which integrates in its evaluation elements tied to effective practices (fraud, malfunctions, etc.) and actors, and to the inevitable deficiencies in institutional systems.

they perceived such consumption to be (Setbon et al. 2005). The level of risk perception, on the other hand, was strongly influenced by trust in government officials, with more than two thirds expressing "distrust" or "strong distrust." Outrage was also expressed, but to a lesser extent.[63]

Such results are compatible with the view that risk is a social relationship. It is not neutral to observe that the threat of mad cow disease is not the product of misfortune, but that it results from human activities, a consequence of the albeit unintentional, technological choices that transformed herbivorous animals into a carnivores. Likewise, changes in consumer behavior can be explained by the dynamics of a crisis (Laufer 1993). A crisis is a social state characterized by a general loss of signposts that affects most of the actors. The legitimacy of institutions that are supposed to provide for our safety can easily be questioned. It is understandable that there can be a discrepancy between a rarely expressed concern and a radical change of behavior. What counts is not only the risk itself, but also the way it was created and from where it originated (Irwin & Wynne 1996). Moreover, in a society full of hyper-choices, the cost of abstaining from the consumption of red meat is not necessarily very high.

The effects of individual perceptions are not a sufficient explanation of consumer behavior in a crisis. A further explanation must be sought in the nearly collective nature of a crisis. In this respect, the two mad cow crises are paradigmatic of the "society of risk," because they express the decline of tradition, the decline in the authority of science, and a crisis of confidence in public institutions. The incurred risk is even less acceptable because it was caused by the institutions charged with creating greater public well-being (Beck 1992).

Toward the Total Ban of MBM

On November 7 – 2 days after the M6 TV show – President Chirac gave a public address in which he urgently proposed a total ban on MBM:

> In this crisis, no imperative can be placed higher than the requirement of public health. No other consideration can influence public authorities' action. The trust of our fellow citizens in the safety of their food and that of their children must be restored. The very future of the bovine sector is completely subject to the response which will be made to this major imperative of public health. No objection, no technical constraint can be held in opposition to this imperative. Everything must be implemented in order to reach maximum safety. I already presented the objectives of the government 15 days ago. Without delay, we have to prohibit meat meal and start the systematic screening of the disease …(Wolfer 2004, p. 14).

Such a presidential address is exceptional. This initiative is related to the context of "cohabitation." President Chirac, from the political right wing, chose to act against

[63]This sense of outrage is related to reactions to the following statements: (1) "It is shocking that herbivores have been made carnivores," (2) "Lobbies work to prevent certain measures from being taken," (3) "Mad cow disease is linked to blind profit-seeking," and (4) "Profit is considered more important than public health."

the Socialist-Green government. Prime Minister Jospin was exploring at the time the possibility of a total ban of MBM. On October 31, 2000, he asked the AFSSA to give its advice on the ban of MBM. Chirac's address bypassed scientific expertise and implicitly criticized the Jospin government for hiding behind scientists and not being ready to assume its responsibilities.

Starting with this presidential address, the problem was strongly politicized. The result was the total ban on MBM announced by the Prime Minister on November 14. The European Commissioner David Byrne reacted promptly, stating that he understood the decision, as it was based on serious doubts about the effectiveness of the measures adopted to control MBM in 1996 in response to the first crisis. He asked other member states to check their own control systems. On November 28, the SSC proposed that measures be introduced gradually, based on the risk classification of each European country. It also stated that, if problems with the implementation of these gradual measures were to be confirmed, the result would be a total ban of MBM. On November 29, the European Committee announced a temporary ban of all MBM in all European countries. The ban was confirmed by the European Council of Agriculture on December 4.

This decision was criticized as having been made under pressure from public opinion and not taking into account the available information. Public opinion and the politicization of the issue are certainly important factors for the MBM ban. According to Wolfer, an economist at the INRA, the measures taken in 1996 were efficient enough and were responsible for the decline in BSE cases starting in 2001 (Wolfer 2004). However, the epidemiological curve reveals a problem with this view (see Fig. 1). The end of 2000 is too soon for one to really appreciate the impact of the 1996 measures. What the experts could observe beginning in November 2000 are the following: (1) the number of BSE cases was still increasing exponentially, and (2) two BSE cases were identified as *super NAIFs*[64] (born in August 1997 and May 1998); as these dates followed the second ban, they cast doubt on the ban's efficiency (Abrial et al. 2005).

In its report, the AFSSA focused on this key problem of implementing the control measures and accordingly supported the total ban of MBM:

> One of the major lessons of the experience is the difficulty of following-up, controlling and evaluating complex and fluctuating rules. The rules concerning products of animal origin that were authorized for animal feeding have changed several times a year during the last five years. … The frequency of these changes certainly has important consequences for the effectiveness of the promulgated measures, whatever may be the intent of those who have to conform to them and the intent of those who control them … This system led to a constant overlapping of transition periods, which facilitated a delay in their regulation, voluntarily

[64] French livestock and veterinary circles have been thrown into disarray by the first case of BSE in a cow born after the implementation of the more stringent feed safeguard measures in 1996. The cow in question, in which BSE was discovered in a systematic screening of cattle over 30 months old, was a cow of the Normande breed born in August 1997 in Seine-Maritime. The cow is being described by French officials as *'super-naif' (Ne Apres l'Interdiction des Farines animales)* i.e. born after the full ban on animal meal in mid-1996 (*Agra Europe*, April 20, 2001).

> or not, and among them were delays that had a sanitary impact. It is thus essential to have, in this domain and for this period, a simple enough measure, easy enough to implement. From now on, the system will rest on a generalized ban on the use of transformed animal proteins stemming from any species (except transformed animal proteins stemming from fish), cooking fats and fats stemming from the transformation of bones intended for the production of gelatin. This will allow the logic which has prevailed until now to be reversed ... (AFSSA 2001).

This second crisis and the related decisions were serious enough to trigger two parliamentary reports – one from the Senate (decision November 21, 2000) and one from the National Assembly (decision December 13, 2000). Both reports supported the government's decision. The report of the Senate, published in May 2001, took for granted that the spread of BSE had not been stopped by the partial bans on MBM. It suggested a definitive ban of MBM for all animals, coupled with a reshaping of the common agricultural policy, including a new plan to develop and utilize plant proteins (Deriot & Bizet 2001). The report of the National Assembly did not go as far; it simply suggested that the government continue the temporary ban.

The decision to ban MBM is interesting, as it shows the fragility of the new system that had been put in place after the 1996 crisis. As discussed earlier, this system was supposed to restore trust in government institutions and protect politicians from blame for the mismanagement. In the midst of the crisis, politicians breached the pact at the core of the new system. They bypassed scientific expertise and decided to take the lead themselves, marginalizing the AFSSA. Accordingly, they were trapped in an escalation toward a "zero risk" policy, whatever the cost. This was a far cry from any sound consideration of the relevant data. This process was at least partly a product of the political competition fostered by cohabitation.[65] However, it also reflects unresolved conflicts of legitimacy between the president, the government, and the new agencies; the boundaries between policy/polity and scientific expertise were shown to be unstable in such situations.

This problem was once again illustrated when the AFSSA published its advice on "ovine BSE" in February 2001. President Chirac criticized the agency at the opening of the *Salon de l'Agriculture* (Agricultural Show), stating that no new scientific evidence was available and that the AFSSA was misleading and alarming the public. But the AFSSA was acting legitimately according to the general principle of transparency, which was considered as a cornerstone of the new system. However, reverence for this principle was not shared by the *Elysée* which maintained that communication must remain under political control. For his part, President Chirac not only returned to the traditional regime of risk management, but he also reaffirmed his strong connection with the agricultural policy networks.

These conflicts are not only the result of power games. They also reflect two opposite ways of conceiving risk management. The government and the AFSSA

---

[65] In France, under the Fifth Constitution, it may happen that the President and the government come from different and opposed political parties. Such a situation existed from 1997 to 2002, with Jacques Chirac – right-wing – at the *Elysée* (President) and Lionel Jospin – Socialist Party – at the Matignon (Prime Minister).

were acting according to a new regime of risk management in which uncertainty has to be acknowledged. This regime also stipulates that it is necessary to act transparently and admit that part of the world we live in is out of control and that failure is endemic. The President was more than a cynical calculator, as he was described by Wolfer (2004); his behavior was also grounded by a functional and political need to maintain the myth of control and manageability, because that is what various constituencies and stakeholders seemed to demand. As stated by Power (2004), this contradiction is at the core of contemporary risk management by the government.

Lessons from the 2000 Crisis

The first and foremost lesson we can draw from this period is that the new system of risk management established after the first crisis in 1996 proved to be a straw man with respect to the second crisis. The new system failed to restore public trust in government institutions and the markets. But this failure is also related to another weakness of the government (risk management): its inability to stabilize the relationships among the key actors. The politicians could not resist taking the lead in risk communication. They thus broke the pact under which the new system was organized. Risk communication was at the core of the conflict, and it was difficult to deal with it in a sustainable way because transparency had different costs and benefits for different actors.

The second lesson concerns the parliamentary inquiries. As in the 1996 crisis, these inquiries were instrumental in resolving the conflicts and normalizing the crisis. A public inquiry is not meant to assign responsibility or determine truth. It is meant instead to reach a new national consensus through public hearings and discussion, as well as to serve as a vehicle through which the elected representatives of the French people can explore all facets of the problem. The 1996 crisis led to the transformation of the risk management system. The lesson of the 2000 crisis is that the system of production of animal products must be radically changed and agricultural policy adequately adapted.

Conclusion

As illustrated by this report, the French mad cow saga was instrumental both as a transformer of the politics of risk and as a social laboratory within which social actors interacted to better understand how to handle uncertain dangers. The saga not only stimulated academic research; there were also several parliamentary reports examining various aspects of the saga. It is through these debates in Parliament that the crisis was understood and the lessons drawn.

One of the most interesting features of the saga is the recurrence of crisis. The 2000 crisis took place despite major structural reforms that had been put in place immediately after the 1996 crisis. Perhaps the 1996 crisis can be explained as a reaction to a potential danger that had been concealed from the public to protect economic interests, but that is not the case for the 2000 crisis. This later crisis demonstrates that crises can occur for unknown reasons. In its early stages, public concern about mad cow disease was rather low, the exposure to risk was minimal, the controls were well implemented (as illustrated by the SOVIBA episode), and so forth. Furthermore, the new regime of risk management was in place; information was widely available, and the track record of the Dormont Committee, as well as the recent "beef war," demonstrated that the decisions were based on the precautionary principle.

Indeed, the 2000 crisis revealed that it was not enough to reduce the high vulnerability of French society. One plausible cause of the crisis is public concern about the possible policy implementation failures, which at that time could not easily be dismissed given the available scientific evidence. Along with this concern, it is certain that the relationships between risk management, society, and the polity caused the crisis in the first place and then aggravated it. This new regime proved to be fragile, as it was not easy for the chief policymakers to admit that they were powerless to do anything about the problem, that they were not in control of the situation, and that the public simply had to accept life in an uncertain world. Effective types of interactions between scientific experts and policymakers have yet to be devised. All this presents a new challenge in terms of risk communication. However, these difficulties are by no means a good reason to return to an antiquated regime.

References

Abrial, D., Calavas, D., Jarrige, N., & Ducrot, C. (2005). Spatial heterogeneity of the risk of BSE in France following the ban of meat and bone meal in cattle feed. *Preventive Veterinary Medicine, 67*(1), 69–82.

Agence Française de Sécurité Sanitaire des Aliments. (2001, April). *Avis de l'Agence française de sécurité sanitaire des aliments en date du 7 avril 2001: Réponse à la saisine du 31 octobre 2000 - Les risques sanitaires liés aux différents usages des farines et graisses d'origine animale et aux conditions de leur traitement et de leur élimination.* Paris: Author.

Alperovitch, A. (1998). Apports et limites de l'épidémiologie [Contributions and limits of the epidemiology]. *Annales des Mines: Responsabilité & Environnement, 9*, 74–79.

Barbier, M. (2003). La constitution de l'ESB comme problème public européen: Une interprétation à partir de l'étude de quelques configurations. *Revue Internationale de Politique Comparée 10*(2), 233–246.

Beck, U. (1992). *Risk society: towards a new modernity.* London, SAGE.

Benamouzig, D., & Besançon, J. (2005). Administrer un monde incertain: les nouvelles bureaucraties techniques. Le cas des agences sanitaires en France [Administering an uncertain world: New technical bureaucracies in French health agencies]. *Sociologie du Travail, 47*(3), 301–322.

Besançon, J. (2007). French agencies. In D. Benamouzig & O. Borraz (Eds.), *Food and pharmaceutical agencies in Europe: between bureaucracy and democracy* (Cahiers Risques Collectifs et Situations de Crise, no 7) (Section 2). Grenoble, France: MSH Alpes.

Borraz, O. (2007). Risk and public problems. *Journal of Risk Research, 10*(7), 941–957.

Borraz, O., Besançon, J., & Clergeau, C. (2006). Is it just about trust? The partial reform of French food safety regulation. In C. Ansell & D. Vogel (Eds.), *What's the beef? The contested governance of European food safety* (pp. 125–152). Cambridge, MA: MIT.

Broxmeyer, L. (2005). Is mad cow disease caused by a bacteria? *Medical Hypotheses, 63*, 731–739.

Chevassus, B. (2000). *L'analyse du risque alimentaire: quels principes, quels mode` les, quelles organisations pour demain?* Paper presented at OECD Edinburgh Conference on the Scientific and Health Aspects of Genetically Modified Foods, Edinburgh (February, 2000). UK, Edinburgh.

Commission of the European Communities. (1998). *First bi-annual BSE follow-up report: communication from the Commission to the Council, the European Parliament, the Economic and Social Committee and the Committee of the Regions* (COM (98) 282 final). Brussels, Belgium: Author.

Coudert, M., Belli, P., Savey, M., & Martel, J.-L. (1995). Le réseau national d'épidémiosurveillance de l'encéphalopathie spongiforme bovine [The national network of epidemic-surveillance of the bovine spongiform encephalopathy]. *Epidémiologie et Santé Animale, 27*, 59–67.

Council of the European Communities. (1990). *Press releases, January–June, 1990: Special council meeting – Agriculture*. Retrieved, November 12, 2008, from http://aei.pitt.edu/4429/01/000486_1.pdf

Deriot, G., & Bizet, M.J. (2001). *Rapport de la commission d'enquête (1) sur les conditions d'utilisation des farines animales dans l'alimentation des animaux d'élevage et les conséquences qui en résultent pour la santé des consommateurs, créée en vertu d'une résolution adoptée par le Sénat le 21 novembre 2000* (Sénat Rapport, no 321). Paris: Sénat.

Dodier, N. (2003). *Leçons politiques de l'épidémie de sida.* Paris: EHESS.

Dormont, D. (1992). *Les encéphalopathies subaigües spongiformes humaines et animales: description clinique et biologique, facteurs étiologiques, conséquences sur la santé publique, et axes de recherches développés en France (Rapport au Ministre de la recherche et de l'espace)* [Human and animal spongiform encephalopathy: Clinical and biological description, ethological factors, consequences for public health, and areas of research developed in France (Report to the Ministry of Research and Space)]. Fontenay aux Roses, France.

Dormont, D. (1998). Oral presentation. In G. Paillotin (Ed.), *European agricultural research in the 21st century: which innovations will contribute most to the quality of life, food and agriculture?* Berlin: Springer.

Dubot, J. (1996). Les encéphalopathies subaigües spongiformes transmissibles: Etat actuel des connaissances. [Transmissible Spongiform Encephalopathy: Current status of knowledge]. Paris: Institut national de la santé et de la recherche méd.

Enjalbert, F. (1996). Les farines de viande: intérêts dans l'alimentation des ruminants et réglementation. *Le Point Vétérinaire, 28*(179), 689–695.

Foster, J. D., Bruce, M., McConnell, I., Chree, A., & Fraser, H. (1996). Detection of BSE infectivity in brain and spleen of experimentally infected sheep. *The Veterinary Record*, 138(22), 546–548.

Foster, K. R., Bernstein, D. E., & Huber, P. W. (1999). *Phantom risk: scientific inference and the law*. Cambridge, MA: MIT.

Giddens, A. (1994). Living in a post-traditional society. In Beck, U., Giddens, A., Lash, S. (Eds.), *Reflexive modernisation: Politics, tradition and aesthetics in the modern social order*. Cambridge, UK: Polity.

Giddens, A. (1997). Risk society: the context of British politics. In J. Franklin (Ed.), *The politics of risk society* (pp. 23–34). Cambridge, UK: Polity.

Godard, O. (2001, February). Embargo or not embargo? *La Recherche, 339*, 50–55.

Guilhem, E., & Mattei, J.-F. (1997). *De la "vache folle" à la "vache émissaire"* [From "mad cow" to "emissary cow"] (Par la mission d'information commune sur l'ensemble des problèmes posés par le développement de l'épidémie d'encéphalopathie spongiforme bovine, Rapport d'information, no 3291). Paris: Assemblée Nationale.

Harremoës, P., Gee, D., MacGarvin, M., Stirling, A., Keys, J., Wynne, B., et al. (Eds.) (2001). *Late lessons from early warnings: the precautionary principle 1896–2000* (Environmental Issue Report No. 22). Copenhagen, Denmark: European Environmental Agency.

Hermitte, M.-A. (1996). *Le sang et le droit: Essai sur la transfusion sanguine.* Paris: Seuil.

Hermitte, M.-A., & Dormont, D. (2000). Propositions pour le principe de précaution à la lumière de l'affaire de la vache folle. In P. Kourilsky & G. Viney (Eds.), *Le principe de precaution* (pp. 341–386). Paris: Odile Jacob.

Hirsch, M., & Duneton, P. (1996). *L'affolante histoire de la vache folle* [The terrifying history of mad cow]. Paris: Balland.

Institut de Protection et de Sûreté Nucléaire. (2000). *Perception des risques et de la sécurité. Résultats du sondage d'octobre 2000*. Fontenay-Aux-Roses: Institut de Protection et de Sûreté Nucléaire.

Ipsos. (2000, October). *The French and the quality and safety of food.* Paris: Ipsos.

Irwin, A., & Wynne, B. (Eds.). (1996). *Misunderstanding science? The public reconstruction of science and technology.* Cambridge, UK: Cambridge University Press.

Joly, P.-B. (2007). Scientific expertise in public arenas: lessons from the French experience. *Journal of Risk Research, 10*(7), 905–924.

Joly, P.-B., & Barbier, M. (2001). Que faire des désaccords entre comités d'experts? Les leçons de la guerre du bœuf. *Risques, 47*, 87–94.

Joly, P.-B., Le Pape, Y., & Remy, E. (1998). Quand les scientifiques traquent les prions: Le fonctionnement d'un comité d'experts dans la crise de la vache folle [When the scientists pursue prions: the functioning of an expert committee in the crisis of the mad cow]. *Annales des Mines: Responsabilité & Environnement, 9*, 86–95.

Joly, P.-B., Le Pape, Y., Barbier, M., Estadès, J., Lemarié, J., & Marcant, O. (1999). *BSE and the France national action system* (Report for the BASES EU project). Grenoble, France: Institut National de la Recherche Agronomique.

Kasperson, R. E., Jhaveri, N., & Kasperson, J. X. (2001). Stigma and the social amplification of risk: toward a framework of analysis. In J. Flynn, P. Slovic, & H. Kunreuther (Eds.), *Risk, media and stigma: understanding public challenges to modern science and technology* (pp. 9–27). London, Earthscan.

Kourilsky, P., & Viney, G. (2000). *Le principe de précaution.* Paris: Odile Jacob/La Documentation Française.

Latouche, K., Rainelli, P., & Vermersch, D. (1998). Food safety issues and the BSE scare: some lessons from the French case. *Food Policy, 23*(5), 347–356.

Laufer, R. (1993). *L'entreprise face aux risques majeurs: a propos de l'incertitude des normes sociales.* Paris: L'Harmattan.

Mer, R. (2001). Vache folle: du rôle des médias en temps de crise ... *Le Courrier de l'Environnement de l'INRA, 43*, 79–92.

Millstone, E., & van Zwanenberg, P. (2001). Politics of expert advice: lessons from the early history of the BSE saga. *Science and Public Policy, 28*(2), 99–112.

Ortega, M. (1997*). Sur les allégations d'infraction ou de mauvaise administration dans l'application du droit communautaire en matière d'ESB, sans préjudice des compétences des juridictions communautaires et nationales.* Brussels, Belgium: European Parliament.

Parker, R. (2006). *The politics of BSE.* New York, Palgrave McMillan.

Phillips, N., Bridgeman, J., & Ferguson-Smith, M. (2000). *The BSE inquiry: return to an order of the Honourable the House of Commons dated October 2000 for the Report, evidence and supporting papers of the Inquiry into the emergence and identification of Bovine Spongiform Encephalopathy (BSE) and variant Creutzfeldt-Jakob Disease (vCJD) and the action taken in response to it up to 20 March 1996.* Norfolk, UK: The Stationary Office.

Power, M. (2004). *The risk management of everything: rethinking the politics of uncertainty.* London: Demos.

Sauvadet, F., & Vergnier, M (2001). *Rapport fait au nom de la Commission d'enquête sur le recours aux farines animales dans l'alimentation des animaux d'élevage, la lutte contre l'encéphalopathie spongiforme bovine et les enseignements de la crise en termes de pratiques agricoles et de santé publique* (Rapport, no 3138). Paris: Assemblée Nationale.

Savey, M. (1993). *BSE risk assessment and surveillance in France* (Proceedings of a consultation of the Directorate-General for Agriculture at the European Community (DG VI)). Brussels, Belgium: Commission of the European Communities.

Savey, M., & Baron, T. (1994). La BSE: un risque pour la santé publique? [BSE: a risk for public health?]. *Revue de Médecine Vétérinaire, 145*(11), 819–827.

Savey, M., & Espinasse, J. (1979). Scrapie in the sheep and laboratory animals: suitability of the sheep model for the study of slow virus diseases. *Veterinary Research Communications, 3*(1), 87–107.

Savey, M., Parodi, A.-L., & Maillot, E. (1989). L'encéphalopathie spongiforme bovine en Europe. *Bulletin de l'Académie Vétérinaire de France, 62*, 483–490.

Setbon, M., Raude, J., Fischler, C., & Flahaukt, A. (2005). Risk perception of the "mad cow disease" in France: determinants and consequences. *Risk Analysis, 25*(4), 813–826.

Southwood, R. (1989). *Report of the working party on bovine spongiform encephalopathy*. London, UK: Department of Health.

Taylor, D. (1996). Inactivation of the causal agents of Creutzfeldt–Jakob disease and other prion diseases. *Brain Pathology 6*(2), 197–198.

Taylor, D.M., Fraser, H., McConnel, I., Brown, D.A., Brown, K.L., Lawza, K.A., & Smith, G.R.A. (1994). Decontamination studies with the agents of bovine spongiform encephalopathy and scrapie. *Archives of Virology 139*(3–4), 313–326.

Torny, D. (1997). Surveiller et contenir. Trace des alertes aux prions. Alertes et propheties. In F. Chateauraynaud (Ed.), *Les risques collectifs entre vigilance, controverse et critique* (volume II, pp. 138–309). Paris: EHESS, Groupe de Sociologie Politique et Morale.

Van Zwanenberg, P., & Millstone, E. (1999). *BSE and the UK national action system* (Report for the BASES EU project). Brighton, UK: Science and Technology Policy Research, University of Sussex.

Wells, G. A., Scott, A. C., Johnson, C. T., Gunning, R. F., Hancock, R.D., Jeffrey, M., et al. (1987). A novel progressive spongiform encephalopathy in cattle. *The Veterinary Record, 121*(18), 419–420.

Will, R. G., Ironside, J. W., Zeidler, M., Cousens, S. N., Estibeiro, K., Alperovitch, A., et al. (1996). A new variant of Creutzfeldt-Jakob disease in the UK. *The Lancet, 347*(9006), 921–925.

Wolfer, B. (2004). *La crise des farines.* Ivry sur Seine, France: INRA/MONA.

Chapter 10
Policy and Politics of BSE in the United States

Rose Campbell and Hajime Sato

Rationale and Context for BSE Policy Discussions

Policy research is the study of how laws and regulations are formed, including the decision-making templates, the methods employed, and the motivations behind policy deliberations (Morgan & Henrion 1990). Policy analysis is the process of developing a regulatory decision to address a specific problem (e.g., reaction to an emerging public health threat) or to establish policy for a new phenomenon (e.g., genetics and food production). Those engaged in health policy research conduct case studies with the aim of identifying trends in analytical models employed by particular regulatory agencies. Policy development is dynamic, however; its a complex social process that often is played out in the public arena, which means no single model will prevail even within the same agency.

Extant research suggests multiple views of how policy parameters are defined. In contemporary American society, health policy is deliberated not only by regulatory bodies and independent analysts, but also by potentially affected parties, whose messages are managed and filtered to targeted publics (Morgan & Henrion 1990). One of the best contexts to study public influence on policy involves public health risk, such as food safety (Sobal & Maurer 1995). In the past few decades, irradiated food, Salmonella poisoning, *E.coli*, mad cow disease, genetically altered food, animal cloning, chemicals in food processing, and pesticide use in agriculture, to name a few, all have been addressed by regulatory agencies. Even in similar risk-level situations, policy outcomes have differed (for a review, see Morgan & Henrion 1990). One way of making sense of these differences is to study how scientific uncertainty is addressed in each case. Rarely is it the case that probability calculations and predictive modeling provide defining evidence pointing to one policy approach.

R. Campbell
Eugene S. Pulliam School of Journalism, Butler University, USA

H. Sato (✉)
Department of Public Health, Graduate School of Medicine, The University of Tokyo, Japan
e-mail: hsato-tky@umin.net; hsato@post.harvard.edu

H. Sato (ed.), *Management of Health Risks from Environment and Food*,
Abiance for Global Sustainablity Bookseries 16,
DOI 10.1007/978-90-481-3028-3_10, © Springer Science+Business Media B.V 2010

The following case study is designed to illuminate the complex policy deliberation process in the United States in the context of bovine spongiform encephalopathy (BSE). Specifically, the study identifies the different agencies and influential publics whose voices shaped US policy after a BSE crisis led to international restrictions on US beef. Arguments and strategies employed by these divergent publics to effect policy change or maintain the status quo will be a key focus.

A comprehensive list of sources and documents were consulted for the case study. These included official government documents, articles from national newspapers, reports issued by consumer advocacy groups, documents created by trade organizations representing the many stakeholders of the beef industry, research articles from academic and scientific sources, and reports from front groups representing coalitions whose finances are directly affected by policy outcomes.

Factors Influencing BSE Policy

Economics of Beef

Beef is the most important agricultural commodity in the US, representing approximately 20% of sales generated from farms (Hoppe et al. 2007). The National Agriculture Statistics Service of the United States Department of Agriculture (USDA) reported that in 2007, beef cattle production represented 31% of farm operations, more than any other agricultural product. The most current report on consumer spending on beef was $74.6 billion in 2007, which represents an increase of over $25 billion since 1999 (CattleFax 2008). Additionally, beef is the top-selling protein in the nation's restaurants, including full service and fast-food commercial operations (National Cattlemen's Beef Association 2008a).

Beef exports also represent a significant portion of farm sales, exceeding 10% of the farm value of cattle for a total of $3 billion (National Cattlemen's Beef Association 2008b). Any breakdown in consumer confidence for US beef would be devastating not only to farmers but all ag-related businesses, such as meat-packers, restaurants, and grocers. For these important reasons, diseases such as BSE represent a significant threat, should US cattle become infected.

Consumer Protection

Policy to prevent BSE from entering the food supply in the US was first initiated in response to the UK mad cow disease outbreak that became public in the 1980s. Several government agencies already existed and were activated to study and address the immediate concerns of preventing BSE and protecting cattle and consumers. The US began formulating policies based on existing science but also needed to avoid a public scare. Even if BSE risk were determined to be low, the US public would not be easily convinced, given the UK example and the government assurances that preceded human illness and death.

Several government organizations are involved in BSE policy and compliance. The agency with primary responsibility for policies regarding agricultural products is the United States Department of Agriculture (USDA). Within this agency, the Animal and Plant Health Inspection Service (APHIS) safeguards US agriculture from pests and diseases carried from abroad. The USDA's Food Safety and Inspection Service (FSIS) purportedly tests all live animals destined for human consumption. Another agency integrally involved in BSE-prevention policy is the Food and Drug Administration (FDA), which makes rules about the composition of animal feed. Finally, the Government Accountability Office (GAO) has been involved in reviewing policy enforcement (Scheier 2005).

Although BSE-prevention policies generally are welcomed by consumers and ag-related businesses alike, some proposed prevention measures are viewed as being too costly and time consuming for the risk reduction they would provide. Adding to the debate and driving policies in different directions include agricultural trade organizations such as the US Cattlemen's Beef Association, coalitions representing agriculture and agri-businesses such as the Meat Promotion Coalition, US manufacturers such as meat-packers, consumer advocacy groups such as Center for Science in the Public Interest (CSPI) and Organic Consumers Organization, and astroturf organizations (front groups for industry formed with the singular purpose of influencing policy) such as the Center for Consumer Freedom. Each of these organization types and their roles in the BSE policy debate are detailed below.

Scientific Uncertainty and Disagreement

Scientific debate about BSE continues regarding three key issues. First, scientists disagree about the foundational principles of the disease structure. Second, in spite of the UK case study linking BSE to variant Creutzfeldt-Jakob disease (vCJD), some scientific disagreement still exists about disease transference to humans. Finally, adding to the BSE discussion has been an increase in cases of chronic-wasting diseases that are affecting North American deer and elk. Although just a cousin of BSE, uncertainty remains regarding the ability of such diseases to jump to other species or infect humans through consumption of game meat. Because of the disease similarity with BSE, much attention is given to this rising problem in the deer and elk population.

A database search for articles in scientific journals using the terms "vCJD" and "BSE" (or spelled out) in their titles illustrates the divergent findings and foci, such as: "Are Prions the Real Cause of BSE and vCJD?" (Coghlan 2006); "The Same Prion Strain Causes vCJD and BSE" (Hill et al. 1997); "Latest Results Strengthen Link Between BSE and vCJD" (Bradbury 1996); and "The Transmission Dynamics of BSE and vCJD" (Ghani et al. 2002). Scientists continue to explore the relationship between BSE and vCJD. The public health threat, and whether that risk is serious enough to warrant new regulations, is at the center of US policy discussions regarding BSE prevention.

It is important to note that the debate about BSE disease structure and its link to vCJD is not limited to scientific journals; the controversy also has been reported in the mass media, where complex topics become accessible to laypeople, but in a limited context. An article from *The New York Times* illustrates the BSE cross-species debate:

> … mad cow disease is caused by an even more mysterious infectious agent, a misfolded protein called a prion. When animals eat the remains of their own species or other species, infectious prions sometimes pass among them. Prions can incubate for very long periods of time before they result in full-blown disease, with each animal or person experiencing the effects differently. But questions about which animal can infect which other animal in the animal world, including humans, continue to perplex researchers worldwide.Can sheep infect cows? Can cows infect sheep? What about goats and pigs? Or chickens? Can deer infect cattle living in the same pasture without cannibalism going on? Can deer infect people? Are some animals silent carriers of the disease, giving meat lovers of the world a false sense of security? Answers matter because regulations in United States intended to keep mad cow disease out of the country and to control related prion disorders like chronic wasting disease in deer and elk are based on fully understanding the biology of species barriers. (Blakeslee 2003)

As this excerpt reveals, the science of BSE is critical to policy and prevention. A goal is to develop tests that can diagnose BSE in cows that do not already exhibit outward symptoms of mad cow disease. This cannot be accomplished without greater understanding of the structure of BSE. Thus, studies of transmissible spongiform encephalopathies (TSEs) inform the BSE policy debate.

Some scientists support the evidence that a prion can, on its own, without influence from DNA or RNA, cause infectious diseases. Dr. Stanley Prusiner is a Nobel Prize-winning scientist whose research on prions has led to greater understanding of TSEs. His research, however, has been dismissed by a few prominent scientists who doubt that his findings are accurate. While they do not charge him with data fabrication, they do question the results. An article from *The New York Times* reports this controversy:

> Dr. Laura Manuelidis, a neuropathologist at the Yale University Medical School, one of Dr. Prusiner's most vocal critics, said the prion strain that turned up in the experiment looked like a mouse prion frequently used in Dr. Prusiner's laboratory. She said that meant that something else might have caused the infection. "Basically I think the data look like contamination," Dr. Manuelidis said, possibly stemming from "inadequately washed instruments." Dr. Prusiner, who won the 1997 Nobel Prize in Physiology or Medicine for his prion research, said some of his critics would never be satisfied. "They'll say we need to do 10 more years of experiments," he said. "It's just silly." (Blakeslee 2004)

Uncertainty arises due to disagreements within the scientific community, such as the dispute described above. These definitional struggles illustrate the profound problems with policy-making and are at the heart of current work on risk in social theory. "It is argued that the neglect of questions of agency which are central to definitional struggles has led to some theorists presenting risks as inevitable concomitants of technological and cultural developments leaving them in the grip of political quietism" (Miller 1999). For example, government agencies involved with protecting farm animals and consumers purportedly use sound scientific reasoning to support decisions. Without scientific agreement on disease structure, however, it is difficult to obtain financial support for BSE test development, which creates a policy lag, as well.

This uncertainty was the focus of a study by Beck et al. (2007), who compared the handling of the BSE crisis in the UK and Germany. Kewell and Beck (2008) further described the construction of risk in regard to the vCJD-BSE link. They analyzed how the UK handled the uncertainty and communicated risk in the public health context over time.

Other public health researchers have raised concerns about the widespread impact should BSE become a threat to our food supply. To verify the link between BSE and the brain-wasting disease vCJD, researchers studied the disease among inbred mice (Brown et al. 2003). They found what they believe to be incontrovertible evidence that "vCJD arose from BSE." A study by Belay and Schonberger (2005) described the threat of secondary transfer of BSE, e.g., the danger of vCJD transmission to patients receiving infected blood through transfusions. Although species-to-species transfer of the prion-based disease has not been verified or agreed upon, the researchers found evidence that secondary transference of the disease was a possibility, human to human.

These three issues – the economics of beef, consumer protection, and scientific uncertainty surrounding BSE – were measured in the US's BSE policy decisions in absence of a domestic case discovery. Maintaining this uncertainty stance was perhaps an effective strategy to guide conservative domestic policy actions in response to BSE crises abroad. This all changed when the first case of a BSE-infected US cow was discovered.

Timeline of Relevant US BSE Policy Events

UK-Prompted Policy

The mad cow scare in the United Kingdom began in 1985 when a cow that died was observed exhibiting symptoms of "novel progressive spongiform encephalopathy in cattle" (Wells et al. 1987). BSE as a disease was officially recognized in the UK in 1986; as a result, several new policies were enacted. Most important, in 1987, officials identified meat and bone meal (MBM) as the root cause of BSE, which led to the 1988 law banning MBM. Officials also created a new slaughter policy to prevent BSE from entering the food supply and implemented import bans for countries with high risk for BSE.

In 1988, in response to the UK outbreak, a USDA inter-agency work group was established to study BSE. As a result, several firewalls were put in place in the US to prevent the spread of BSE to cattle and humans. In 1989, the United States banned all imports of live cattle and other ruminants (plus some ruminant products) that came from countries identified as having high BSE risk (Veneman 2004). This act was enforced by the USDA's APHIS. The US began its own BSE testing program in 1990; 40 cattle brains were tested. In 1997, the US extended the import ban to include all of Europe for live cattle, cattle feed, and beef products. While the UK banned MBM in feed shortly after the BSE outbreak, the US did not enact any feed bans until nearly 10 years later; in 1997, the FDA banned animal feed containing mammalian proteins (Becker 2005). With the feed ban came FDA requirements for registration and monitoring of feed mills, renderers, pet manufacturers, and related businesses. During 2002, the US's BSE testing program expanded and nearly 20,000 cattle brains were tested.

In 2003, the US restated its import restrictions and allowed what were considered lower-risk products from formerly BSE high-risk countries, if they had enacted adequate BSE controls since 1989. Canada was one of the first countries for which US lifted its ban.

During the UK BSE crisis, The *Office of International des Epzootees* (OIE or the World Organization for Animal Health) created international standards for animal surveillance, which the US adopted in 1990. According to the US Secretary of Agriculture, Ann Veneman, who served from 2001 to 2005, the US had met or exceeded these guidelines since 1992. In 1998, a BSE risk analysis was commissioned by the USDA to assess the possibility of BSE becoming an epidemic in the US. The Harvard Center for Risk Analysis determined it was very unlikely that BSE would be a threat in the United States. Although considered a credible source, the director of the Center was accused by consumer advocacy organizations of having a research bias or conflict of interest, due to government funding (see below).

In spite of this positive review, in 2002 and 2003, the USDA tested 20,000 cows each year, which was an increase of previous testing rates. At this time, only cows that showed "suspicious neurological symptoms," such as "downer" cows that were non-ambulatory after transfer to the slaughterhouse, or those that appeared disoriented, were separated for testing (Becker 2005). Some of these tested cows died on farms. Although a larger number of cows were tested than was deemed necessary by previous government-sponsored studies, less than 1% of the 35 million cows that were slaughtered each year were tested for BSE.

Canada's BSE Case and Trade Relations

No other US policies were developed until May 2003, when the first BSE case in North America was discovered. A sick cow in Alberta, Canada, was confirmed to have BSE. Immediately following this announcement, the US banned all Canadian beef imports and live cattle. In response to the crisis, the Canadian government launched a mass media appeal to its citizens, reassuring the public that no parts from the infected cow entered the food system and Canadian beef was safe to eat. This strategy apparently worked, as domestic beef sales increased in the months following the crisis. An article in the *Los Angeles Times* reported, "Canada is the only country hit by BSE where domestic beef consumption has risen. According to Cindy McCreath of the cattlemen's association, 'People quickly grasped that this was an economic crisis, not a food-safety crisis' " (Simon 2004).

The US ban on Canadian beef was highly criticized by Canada, since cattle frequently were traded between the two countries and they roamed freely on both sides of the border. Some argued it was hard to distinguish their origin – they are simply North American cattle. Provincial and territorial governments argued unsuccessfully for trade sanctions against the US (and Japan, as well) or limits on US beef in response to its ban on Canadian beef (CBC News 2006a).

Within 3 months of the ban on Canadian beef and live cattle, the US began issuing import permits for certain beef products (e.g., Canadian origin boneless beef from cattle less than 30 months old), and the list of acceptable imports grew monthly. R-CALF (Ranchers-Cattlemen Action Legal Fund) and the American Meat Institute (representing meat processors and packers) filed lawsuits and temporarily blocked the USDA's efforts to resume trade with Canada, but the US eventually was able to support its rule that live cattle under 30 months old and from "minimal risk regions" in Canada, could be traded (APHIS).

Complicating trade relations, in the midst of the Canadian beef crisis, a US cow tested positive for BSE. The US relaxed its ban on Canadian beef. In April of 2004, the US lifted its import restrictions on most Canadian beef products from animals younger than 30 months, but retained its ban on the import of live cows as well as meat from older animals. The Canadian agriculture minister lobbied Washington for several months. Following a December 2004 visit to Canada where he dined on Alberta beef, President George W. Bush announced the US would begin allowing the trade of young live cattle from Canada in March of 2005. Five days before this trade agreement was to be enacted, the action was temporarily blocked by a US District Court, in response to a lawsuit filed by The Ranchers-Cattlemen Action Legal Fund, United Stockgrowers of America (R-CALF USA). This trade organization argued that opening the borders to live cattle trade would cause "immediate and irreparable damage" to the US beef industry (CBC News 2006b). On July 14, 2005, the US Ninth Circuit Court of Appeals overturned this temporary injunction; 4 days later the first truckload of live cattle crossed over the border from Canada to the US, ending a ban that lasted 26 months (R-CALF tried again, unsuccessfully, to appeal the decision).

Although temporary injunctions were sought by these two organizations, beef and cattle trade with Canada resumed to about 90% of previous levels by July of 2005. Within the time frame of the Canadian BSE discovery and US announcement to transition from "enhanced" to "ongoing" BSE surveillance practices, six more Canadian cows tested positive for BSE and three US cows tested positive (APHIS).

A significant result of the BSE conflict was the development of a "harmonized approach to BSE risk mitigation" between Canada, the US, and Mexico (APHIS). Each country believed that cooperation was necessary to assess and address BSE risk in all of North America.

BSE Case in the United States

On December 23, 2003, the claim that Canadian and US cattle were co-mingled was confirmed: The US government announced that a cow in Washington State tested positive for BSE, which led to a widespread international ban on US beef

(Wald & Lichtblau 2003). Over fifty countries banned US live cattle or beef products following the discovery. This finding was just 7 months following the discovery of BSE in a Canadian cow. The news was certainly devastating to the beef industry. One month earlier, *The New York Times* headline was notably different: "Cattle rushed to market as the price of beef soared." (Kilborn 2003). The article claimed, "Cattle ranchers, too, have never seen it so good..."

In response to this serious problem, and dissociate from the Canadian situation, the government employed two key arguments. These arguments were designed to address the beef crisis and reassure both domestic and international publics that US beef was safe. First, the government emphasized the fact that the infected dairy cow was born prior to the US's 1997 feed ban. Second, the government scapegoated the issue by placing the blame on Canada. On January 6, 2004, DNA tests confirmed the cow was Canadian born. Neither of these announcements, however, led countries to reconsider their bans on US beef. In June of 2005, a second cow tested positive for BSE. It was the first BSE case of a US-born cow.

At the time of this first US BSE discovery, suspect cows were tested, but they were still slaughtered and sent to processors and meat packers before test outcomes were known. This particular cow became part of hundreds of pounds of ground beef that became mixed with other beef and widely distributed to grocers in early December. Although a massive beef recall was issued, officials concluded that most of the meat was probably purchased and eaten, before the confirmatory test came back positive for BSE:

> Also unknown is what happened to all 10,410 pounds of meat that was recalled from the slaughterhouse that processed the Holstein and mixed its meat with that of 19 other animals. Officials theorize people may have eaten some of that meat. The meat supply is safe, Dr. DeHaven said. The Agriculture Department had said the parts of the Holstein that could have had infectious material like the brain and spinal cord were removed before processing (Associated Press 2004).

New Policies Resulting from International BSE Trade Disputes

US Domestic BSE Policy Action

In response to the BSE finding and the resulting trade bans on US beef, several policy changes were enacted. Agricultural Secretary Ann Veneman summarized these actions when called to testify before the US House of Representatives Committee on Agriculture and Committee on Government Reform in the summer of 2004. She listed the following evidence to support the US's claims of a swift and effective response to the public health crisis:

> – An immediate ban on non-ambulatory disabled (downer) cattle from going into the food chain;– A "test and hold" policy, which mandates that meat from cattle tested for BSE cannot enter into the food chain until test results come back negative;– A requirement to remove specified risk materials (SRMs), which can carry the infectivity, from the food supply in order to protect public health;– Enhanced requirements on the use of advanced meat recovery systems. Product produced using advanced meat recovery cannot contain

spinal cord or dorsal root ganglia;– A ban on the use of mechanically separated beef from the human food supply;– And a ban on air-injection stunning.– In addition, we announced the expedited implementation of a national verifiable animal identification system. Our goals are to achieve uniformity, consistency and efficiency across the national ID system. (Veneman 2004)

These actions were explained in a series of articles in *The New York Times*. On December 31, 2003, the Food Safety and Inspection Service (FSIS) banned specified risk materials (SRMs) from cows aged 30 months and older and ordered SRMs removed and disposed of separately at slaughterhouses, to ensure they did not enter the human food chain. To enforce this rule, new procedures for verifying age were implemented, so inspectors could determine age if records were not available. The US also banned the use of advanced meat recovery technology (AMR methods) and air injections to stun cattle (those 30 months and older), as these methods spread infected SRMs on the remains of the cow to be processed for human consumption. Also, a "test and hold" rule was put in place in March of 2004, 3 months after meat from a BSE-infected cow entered the food system, to prevent potentially BSE-contaminated beef from being processed and eaten by consumers. This meant that non-ambulatory ("downer") cows were not released until receiving a confirmatory negative BSE test (O'Neill 2005).

BSE testing practices

In addition to the bans and new slaughter procedures to prevent BSE-infected cows from reaching the food supply, in early 2004, new BSE testing practices were established. The USDA announced a 12–18-month program, during which a high percentage of downer cows were tested to determine the rates of BSE. At the end of the testing period (November 2005), only one cow of 535,000 tested positive for BSE. As a control, the USDA also tested 21,000 healthy cows and none was infected with BSE (Becker 2005, p. 2). Although this represented a major change, if only temporary, other governments perceived the US's actions weak and inadequate to ensure consumer safety. For example, in Western Europe, most countries test every cow over 2 years old as well as all sick cows, and Japan tests all cows that are intended for human consumption (McNeil 2003).

Also in 2004, the USDA considered adding new restrictions to the feed rules to include bans of "poultry litter, plate waste, and bovine blood from cattle feed" (Becker 2005, p. 3), but this was not officially enacted. The USDA adopted the scientific view that only cows over 30 months old were at risk for BSE. Thus, the USDA banned SRMs such as spine and brain matter from cows older than 30 months in animal and pet foods, rather than the more restrictive ban previously considered (Becker 2005).

BSE testing methods also were changed. In early 2004, the US adopted the rapid screening test that was widely used in Europe; inconclusive results were retested using the immunohistochemistry (IHC) test that was much slower but more accurate than the rapid screening method. In June 2005, when the second US BSE cow was identified, testing procedures were scrutinized once again, because it had taken 7 months for the BSE case to be confirmed. The US agreed to use both the Western blot test and the IHC following an inconclusive rapid test. If the Western blot was positive, officials would know much sooner. If also inconclusive, the more sensitive IHC would be the back up. It was believed the Western blot would capture more BSE cases earlier.

In summary, the US added a few new permanent restrictions or monitoring policies after the BSE case was announced. The measures listed below are what remained following the initial government actions that Veneman reported above.

1. Downer cows were removed from the food supply until a negative BSE result was found.
2. A brief program was initiated where a high percentage of downer cows and a small percentage of healthy cows were tested to compare samples. The program lasted about 18 months.
3. SRMs were ordered removed from all cows over 30 months old that were slaughtered.
4. AMRs and air injection stunning methods were banned for at-risk older cows to prevent SRMs from mixing with beef destined for human consumption.
5. The FDA banned the inclusion of SRMs (mostly brain and spinal matter) from pet food and farm animal feed and ordered registration and monitoring of these industries.
6. Rapid screening tests became the standard for testing at-risk cattle and the slower IHC was used as a back-up.
7. Western blot tests were required as a second BSE test, along with the IHC, to speed up the confirmation process.

Cooperation with International Advisory Panels

In an attempt to harmonize trade relations, the US worked with international advisory panels and consulted with established international organizations. For example, during the UK BSE crisis, The *Office of International des Epzootees* (OIE) recommended surveillance standards to prevent BSE. The US adopted these in the early 1990s and in some ways exceeded the standards (United States Department of State 2004; Veneman 2004).

Following the 2003 BSE discovery in the US, the USDA also formed its own advisory panel on BSE, by creating a subcommittee of the US Secretary of Agriculture's Foreign Animal and Poultry Disease Advisory Committee. Members were scientists from the US, New Zealand, Switzerland, and the United Kingdom. Although the US was praised for its open process to investigate BSE with an international panel, the panel did say that the US's safeguards are not strong enough to keep BSE from spreading. Some of the panel's recommendations, such a national identification system, were not immediately accepted by the USDA then, but are today in the process of implementation. The panel did recommend that the US work with other countries to adopt international standards.

US Reaction to BSE Announcement

The US policy actions described above were influenced by many entities that had a vested interest in US beef; these were both domestic and international sources. Some groups lobbied for stronger US BSE policy, specialized privileges for testing, and others lobbied for less restrictive policies. Additionally, the mainstream media provided a steady diet of BSE-related stories illustrating the different viewpoints, but they also served to provide alternative perspectives of the crisis.

Meat and Agriculture Industry

Government policies in reaction to the US BSE incident were not as restrictive as expected by some in agriculture and ag-related businesses. They were concerned that US policy would respond to consumers' fears rather than employ sound scientific reasoning. Most organizations associated with agri-business argued that more restrictive policies would be a financial burden on the industry; they believed the risk reduction afforded by stricter policies would not be worth the money (National Cattlemen's Beef Association 2008c). In the end, most leaders in the cattle and beef industry supported USDA and FDA actions, as they exceeded OIE standards. The Harvard University risk study, prepared in the late 1990s, also was cited frequently in news stories and government reports in support of the US's BSE-related policy decisions. Considered a credible and independent external source, Harvard held more esteem with some constituents than did the international organization, the OIE. Both the ag-industry and government emphasized science-based facts as the foundation for BSE policy. This science-based message continues to be applied in trade negotiations with other nations. For example, the Office of the United States Trade Representative released its 2008 National Trade Estimate Report and a press release announcing its contents. They described efforts to reopen export markets to US beef: "We have made considerable progress on achieving the goal of eliminating non-science-based regulatory barriers to exports of beef and beef products from the United States" (Office of the United States Trade Representative 2008). While it is appropriate to use scientific data in policy development, the government was selective about which reports it used as evidence for policy making. For example, the credibility of the Harvard report was later questioned (see discussion below).

Still, there were three alternative viewpoints revealed in the news coverage following the BSE announcement. First, some parties in the agriculture industry argued that any new restrictions were costly and unnecessary. For example, many thought it illogical to ban all downer cows from reaching the meat supply. Meat processors and cattle ranchers argued that cows could get disoriented when transported and may be trampled in the rush to leave the trucks, which disables some of the otherwise healthy animals (i.e., downer cows with broken legs should be distinguished from the neurologically impaired).

A second view was held by those who believed testing should not be limited only to sick cows, suggesting that downer cows were the wrong target for the USDA altogether. This viewpoint was supported by Dave Louthan, an employee who worked at the slaughterhouse when the BSE-infected cow was found. Considered a whistle-blower, Louthan claimed that this particular cow was not a "downer." If true, it would call to question the US policy of testing only downer cows and cows that die on farms:

> Contrary to reports from the federal Department of Agriculture, he (Louthan) asserts that the cow he killed was not too sick to walk. And it was caught not by routine surveillance, he says, but by a "fluke": he killed it outdoors because he feared it would trample other cows lying prostrate in its trailer, and the plant's testing program called for sampling cows killed outside only."Mad cows aren't downers," he said. "They're up and they're crazy." The Agriculture Department disputes his account. Dr. Kenneth Petersen, a food safety official, faxed copies of the Dec. 9 inspector's report saying the cow was "sternal," or down on its chest. Mr. Louthan said he believed the government changed the report on Dec. 23, during the panic at Vern's when a positive test was found. The "smoking gun," he said, is that it is the only one on the page marked "unable to get temp" while other cows' temperatures were recorded. It is easy, he said, to get a rectal temperature from a downed cow but hard from a jumpy one (McNeil 2004).

Louthan's claims were strongly disputed. Inspectors denied changing the report; instead, they argued the rectal temperature was impossible to conduct due to the heavy and immobile animal's positioning near a wall. If Louthan's testimony had been supported, the US's BSE policy response might have been different. The test samples would not be confined to non-ambulatory cows.

A third view was held among a minority of US cattle ranchers and meat processors who believed that BSE testing should be more rigorous. They did not disagree with scientific reports necessarily, but instead wanted to comply with Japanese standards, which would allow beef exports to resume. Ranchers also knew that 100% testing would set them apart from other US beef, which would give them a consumer niche. For example, Creekstone Farms in Kansas wanted to test every cow in his herd that was slaughtered, but the US government denied its request, asserting that it would send an inaccurate and confusing message to consumers that some beef is safer than others.

In the US, no company or individual can make BSE testing decisions on its own. "It (Creekstone Farms) has all the equipment it needs, but it does not have the kit of chemical reagents needed to run the tests. In the United States, the USDA controls the sale of those kits, and the agency ruled last week that only labs in the US government's testing program can buy them" (Kaufman 2004). The USDA invoked the international panel's recommendation that only cows older than 30 months are at risk, and concluded that BSE testing would be limited to a small percentage of these at-risk animals.

Consumers and Retailers

The pre-Christmas discovery of a BSE-infected cow shook up the beef industry, if not consumer confidence. While international markets were closed, Americans continued to eat beef. This was aided by three efforts. First, the National Cattlemen's Beef Association launched an effective crisis campaign that had been developed based on consumer research and case study analysis following European BSE incidents. Second, the low-carbohydrate/high-protein Atkins diet was at the peak of its popularity after gaining widespread acceptance for its effectiveness. Finally, at the time of the BSE announcement, the public was distracted by the war in Iraq, which was launched in March 2003 and highly debated; mad cow disease did not stay on the front page for long.

Of these, the National Cattlemen's Beef Association's crisis plan often is credited with allaying consumer fears and preventing panic (Hendrix & Hayes 2006). In 1996, shortly after BSE became linked with human illness, the association initiated extensive consumer research to determine people's knowledge about BSE and attitudes toward beef safety in general. Close to thirty tracking surveys were conducted to examine consumer opinion, so when the BSE news hit in 2003, the trade association was ready. It already had identified audiences for a public relations campaign designed to protect assets of member publics. Thus, the organization targeted the mass media, relevant policymakers, beef export markets, consumers, and beef marketers. The association learned from its research that messages must be science-focused to be credible and bolster consumer confidence.

Some of the programming tactics the organization was able to execute in a very timely manner following the BSE announcement was an e-conference with media within 15 minutes of the breaking news. Additionally, the organization's BSE crisis response section was activated on its website and it became a source of credible information for media and consumers, as well. A Super Bowl radio promotion was also launched to get people interested in beef. The association actively monitored the Internet to identify inaccurate information so it could ensure accurate information was widely available to counterbalance it. With relevant government publics, the organization essentially trained officials to stick to science-based information in their communications with the public, as well. All state chapters of the organization were on the same page and held accountable for the campaign's key messages based on science. Throughout the crisis campaign, the association watched for other competing yet related topics that threatened to distract from its main theme. An example is the country-of-origin product labeling.

The National Cattlemen's Beef Association's campaign was highly successful. The year 2004 became the best year in decades for beef consumption; sales rose 7.74% in spite of the BSE (National Cattlemen's Beef Association 2008d). Surprisingly, survey tracking research prior to the BSE crisis indicated that 88% of US citizens were confident in US beef, whereas following the announcement, consumer confidence in beef was even higher, at 93% (National Cattlemen's Beef Association 2008d; Slagle 2004). Further, no surveys since then have shown consumer confidence drop below pre-BSE levels.

This finding, that consumer confidence in beef was high, is especially interesting, because consumers' attitudes toward food safety in general have become more negative in recent years. For example, a survey of 7,976 Ohioans by Tucker et al. (2006) found that people perceive contamination of drinking water and pesticide residue in food to offer significantly greater risk than mad cow disease and genetically modified organisms. In fact, these two items received the lowest risk evaluations of seven items. Although published in 2006, the data were collected in 2004 and 2005 when the mad cow crisis was at its peak.

Certainly the beef trade association was successful in its BSE response plan, but the organization was bolstered by increased consumer demand for high protein diets. The Atkins diet is one of the popular low-carbohydrate/high-protein diets associated with quick initial weight loss. With the Atkins diet, people lose weight by eating high levels of protein and reducing carbohydrate intake dramatically. A search on LexisNexis of US newspapers and wire services articles between the dates of December 23, 2003 (when the BSE case was announced) and July 31, 2006, yielded 1,362 articles that mentioned the Atkins diet either as a main topic or as a subtopic in regard to consumers' reaction to the mad cow disease incident.

For example, an article in Nebraska's *Omaha World-Herald* described how Nebraska suffered dramatically from the international ban on US beef but was buffered by the fad diet. "Creighton University economist Ernie Goss said Nebraska was fortunate that domestic beef demand remained strong, buffering the impact of the export loss. 'Luckily, the Atkins diet picked up some of the slack,' Goss said. Now that the national diet fad is waning, the timing of Japan's expected resumption of trade could be good, he said."

Many of the news articles were about the restaurants and fast-food places that had launched "Atkins-Friendly" menu items. For example, at Hardee's, consumers could purchase a "monster-size" burger wrapped in a lettuce leaf rather than a bun. Scores of chain restaurants, including fast food, had launched new menu items and expensive advertising campaigns to promote these Atkins Diet-friendly menus. Rather than respond to the BSE announcement by lifting these items from menus, restaurants continued to promote them. These businesses invested a lot of money in these products and made no effort to encourage public scrutiny of beef.

Finally, the media present a wide array of world and local affairs, and it is difficult to know how much interest any given topic will raise. Certainly the war in Iraq was a worthy distraction from health and other news. The Iraqi war was mentioned in 809 headlines in *The New York Times* during the first 6 months following the announcement of the BSE-infected cow in December 2003, and appeared on the front page of the front section of the newspaper five times during the first week following the BSE announcement. Comparatively, the beef crisis was headlined in 73 articles during the same 6-month period and appeared on the front page three times the first week. While obviously the two topics have very different news values, both represent international conflict. It could be argued, however, that the beef issue held greater relevance to the average American at the time, due to its broad consumer safety implications and potential for personal impact. Still, interest in the topic waned shortly after the announcement, as the media coverage diminished.

Another plausible explanation for why the BSE crisis had little impact on domestic beef sales is consumer skepticism. The public is becoming more and more wary of media stories that warn of food-related dangers (Tucker et al. 2006). To consumers, the mass media are not a credible source of health news. One day they may read a seemingly compelling health study report regarding the benefits of red wine and within weeks may read an equally convincing story about the hazards of red wine. Consumers become desensitized to these stories of new risks that just a moment ago were considered beneficial.

Critics have long blamed the media for overstating public health issues and invoking fear unnecessarily. In his book, *Are we Scaring Ourselves to Death? How Pessimism, Paranoia, and a Misguided Media are Leading us toward Disaster*, author H. Aaron Cohl (1997) describes the "chicken little" phenomenon; reporters take insignificant risks and blow them completely out of proportion. He argues that media employ these tactics because most journalists don't understand science and risk, sensational news sells newspapers and attracts viewers, and space and time are too limited to provide the necessary details to achieve public understanding.

It's even more surprising that a public panic did not emerge in the US following the BSE announcement, given some of the news story angles. An article published in *The Oregonian* within 3 weeks of the BSE discovery ominously described the cow byproducts we can find in anything from mayonnaise to medicine and from sutures to soccer balls (Woodward 2004). In *One Cow, Hundreds of Uses*, Woodward says one can avoid the T-bone but warns "there is no escaping the humble cow."

Consumer Activist Groups

While the agriculture industry was well-prepared to respond to the BSE incident and restaurants and other affected retailers were relatively silent about the beef scare, several consumer advocacy groups used the BSE incident to leverage attention for their key issues. They sought to raise awareness of general food safety issues, the benefits of a vegetarian diet, the problems of obesity, and the appalling conditions of factory farms where cattle, chickens, and pigs are fattened up en masse. Mass slaughter policy has been also criticized from the standpoint of animal welfare (Murphy-Lawless 2004).

In the US, the most aggressive campaign was launched by People for the Ethical Treatment of Animals (PETA). PETA used the BSE incident as a gateway to educate the public about the poor conditions surrounding beef production. The first US cow diagnosed with BSE was described as a "downer" cow, which allowed PETA to describe slaughterhouse conditions and call for a beef boycott. Vivid news accounts described how cattle stunning devices spread banned tissue to cattle carcasses, raising the risk that BSE-contaminated beef reached the food supply. Within 2 weeks of the December 2003 BSE announcement, PETA had rented the domain name "www.beef.com," where confused consumers would find messages about boycotting beef rather than recipes for chateaubriand. Instead, PETA urged consumers visiting the site to become vegetarians. Over two dozen stories in US newspapers and wire services mentioned PETA in regard to the BSE incident, within a 2-month time frame following the BSE announcement (LexisNexis Academic).

PETA also tried to leverage BSE worries to raise concerns about eating poultry. An article in the *Denver Post* on February 10, 2004, described a PETA pro-vegetarian marketing campaign that featured a gun-toting chicken on a billboard that warned consumers not to eat chicken (caption: "If the cow doesn't get you, I might"):

> The animal-rights organization is urging Americans to stop eating chicken. The group claims that poultry given feed containing cattle byproducts, such as cow's blood, could carry a variant of bovine spongiform encephalopathy, also known as mad cow disease. "The best thing anyone who doesn't want to support horrific cruelty to animals or put their family in harm's way can do with any meat in their refrigerator is to throw it in the trash can," PETA spokesman Bruce Friedrich said. Scientists at two universities and the Denver-based National Cattlemen's Beef Association rejected PETA's claims. No published research shows that poultry is susceptible to BSE, they said. (Tatum 2004)

Another well known yet less colorful organization supporting healthy lifestyles is the Center for Science in the Public Interest. A LexisNexis search revealed over 60 news stories published in US newspapers that mentioned this organization in regard to general food safety issues and regulations in the 2 months following the BSE incident, but the beef crisis is what got the organization's foot in the door. Some articles described BSE as yet the tip of the iceberg of a plethora of food safety issues that the federal government is failing to address (Lindquist 2004). The majority of initial articles following the BSE incident feature the organization as an expert source to respond to government reports and actions.

The Consumers Union[1] is a nonprofit organization most known for publishing the magazine *Consumer Reports*, one of the US's most trusted sources of product ratings and reviews. Ralph Nader, an independent presidential candidate during the past two national elections, founded the organization and publication. Consumer safety and value are key measures for assessing products and services, anything from shampoo to insurance companies to automobiles. The Consumers Union also has a powerful lobbying arm advocating stronger public safety and health policies.

Within 3 weeks of the BSE announcement, members of the Consumers Union, along with a coalition of other consumer safety and health groups called the "Safe Food Coalition," met with Agriculture Secretary Ann Veneman to urge stronger testing and cattle identification practices. A letter from the coalition dated January 15, 2004, requested the meeting and listed the coalition members, which included American Public Health Association, Center for Science in the Public Interest, Consumer Federation of America, Government Accountability Project, National Consumers League, Public Citizen, STOP (Safe Tables Our Priority), as well as the Consumers Union. The coalition requested a meeting with Veneman to discuss new surveillance, testing, and slaughtering procedures.

The group praised the US for its immediate ban on downer cows from the meat supply. It also urged expanding testing to include asymptomatic cows, citing the example of Europe, where thousands of cows that tested positive for BSE displayed no outward signs of the disease. Specifically, the nonprofit Safe Food Coalition urged Veneman to take the following regulatory steps: approve rapid tests for BSE;

[1] See Consumers Union website: http://www.consumersunion.org/

create a national system where all cattle are tagged and tracked "from birth to slaughter"; enforce existing regulations to prevent any spread of tissue from the nervous system to parts that will be processed for meat; ban spinal matter and neck bones from processing through AMR; disclose all locations that received the BSE-contaminated meat and eliminate various states' confidentiality rules that prevent this; and urge the Bush administration and Congress to give full authority to the USDA to recall beef without other checks and balances currently in place; offer public forums to keep all informed about the government's actions to monitor cattle and prevent BSE from entering the food supply.

Several of these steps eventually were taken by the USDA (as listed earlier in this case study), but these consumer groups really had little impact on BSE policy development. For example, the Consumers Union's demand for testing samples of healthy animals as a standard procedure was not met. The US tested a small sample of healthy cattle but just temporarily (for 18 months). Disclosure and confidentiality regulations were not overturned and a national identification system to track cattle from birth to slaughter was never pushed. SRMs were ordered removed, but only from cows over 30 months old and AMR and air injection stunning methods were banned only for older at-risk cows, rather than all cows as requested by the coalition. Instead, the agriculture industry successfully used what they offered as "science-based" messages to shape policy development. They argued that earlier cattle feed regulations dictate what cows remained at risk, in order to build rational arguments to shape new regulations.

Other major nonprofits not part of this coalition that also had a voice in the BSE policy discussion included the National Association of Consumer Advocates, National Consumer Law Center, and US Public Interest Research Group. These are just a few of the many organizations that serve as watchdogs for the public in food safety issues.

Front Groups for Industry

While grassroots consumer groups and other public advocacy nonprofits clearly communicated their views in the BSE policy debate, other forces combined their efforts to support ag-related business. The most famous advocate for the beverage and food industry is the Center for Consumer Freedom (CCF). The CCF is an astroturf group, a fake grassroots organization that pretends to be a consumer group, when in reality it is funded by industry with a financial interest in policy outcomes related to their respective businesses. The CCF is funded by the food and beverage industry. Founded by Richard Berman in 1995 with a $600,000 grant from tobacco giant Philip Morris, the CCF began spreading its message of individual responsibility, using the key message of "personal freedoms" to combat government or other consumer advocates from suggesting that certain foods, products, or beverages could be harmful. According to its website, "The Center for Consumer Freedom is a nonprofit coalition of restaurants, food companies, and consumers working together to promote personal responsibility and protect consumer choices. The growing cabal of 'food cops,' health care enforcers, militant activists, meddling bureaucrats, and violent radicals who think they know 'what's best for you' are pushing against our basic freedoms. We're here to push back."[2]

The CCF is so powerful it was able to pressure the Centers for Disease Control and Prevention (CDC) to restate its statistics on the link between obesity and mortality. The CCF claimed the CDC over-calculated deaths from obesity, and CDC agreed and restated the figures. The organization is well financed, too. Many of its contributors (e.g., Coca-Cola, Wendy's, Tyson, Outback Steakhouse, and Cargill) are also clients of Berman's lobbying arm of his Washington DC firm that bears his name (Klinkenborg 2005).

After the BSE incident in December 2003, the CCF dominated the media with headlines such as "Mad cow rant short on science," and "Mad cow: Scaremongers not beef the real problem." Although several pejorative exposes of Berman have been written, he and the CCF continue to make headlines in support of the food and beverage industry, including the beef industry. Because of his industry funding, Berman is able to reach consumers quickly and powerfully with professionally produced messages using mainstream advertising. These are tools not readily available to real grassroots organizations that struggle with funding. With his team of professionals, Berman effectively persuaded consumers that BSE presented no risk to them and that beef was safe. Downgrading the risk reassured consumers.

Some legitimate consumer advocacy organizations have attempted to "out" front groups such as the CCF, but without major funding they lack the ability to purchase space and time in the media to communicate their messages. Consequently, much of their health campaign efforts are relegated to organization websites and press releases to media. Recall that the Harvard risk study was cited frequently by the US government as a credible, scientific foundation for BSE policy formation. The "SourceWatch" website of the Center for Media and Democracy, however, discredits the Harvard report as an industry buy-out: "Corporate-funded think tanks and food industry front groups are rushing into the debate over mad cow disease. This Wednesday, January 28 (2004), the George C. Marshall Institute is hosting a presentation by Harvard Center for Risk Analysis director George Gray, hyping him as 'the nation's foremost mad cow expert,' although his expertise is in getting money from government agencies and Fortune 500 companies to produce risk studies that favor their positions" (Stauber 2004).

The Center for Media and Democracy opposes individuals and organizations that accept research funding from institutions that have a financial or otherwise vested interest in research findings. For example, SourceWatch posts a copy of a letter from Dr. Gray to the US Office of Management and Budget, in which he argues against a standing policy that would limit research funding by those institutions that may be considered to have a conflict of interest (SourceWatch 2009). In his letter of December 2003, Gray says, "I prefer the notion of disclosure of potential conflicts

---

[2] See the Center for Consumer Freedom website: http://www.consumerfreedom.com/

of interest, including work as an expert witness and institutional funding, to strict rules of disqualification in the required agency guidelines (Section 4(b)). Complete and widespread disclosure will allow interested parties to make judgments about the appropriateness of reviewers. Although I recognize that it will sometime be necessary and appropriate, disqualification has the potential to raise questions of agency bias in the choice of experts" (Gray 2003).

The Center for Media and Democracy is not the only voice questioning scientific sources. In *Trust Us, We're Experts: How Industry Manipulates Science and Gambles with Your Future* (Rampton & Stauber 2002), the authors investigate expert sources and their funding or backing to illustrate what the authors believe reflect widespread practices of public deception by the government and corporations. They call for increased citizens' involvement in the democratic process. While some of the authors' claims have been challenged, many of the incidents described accurately portray corporate influences on public policy.

Books and Popular Media

A final influence on the public debate on BSE policy comes from the numerous books that were published predating the US beef crisis and afterwards. Several books, including *The Mad Cow Crisis: Health and the Public Good* (Ratzan 1998), *Sacred Cow, Mad Cow: A History of Food Fears* (Ferrieres 2005), *Mad Cow U.S.A.: Could the Nightmare Happen Here?* (Rampton & Stauber 1997), *How the Cows Turned Mad* (Schwartz 2003), and the recently published comprehensive UK case study, *The Politics of BSE* (Packer 2006), illustrate the importance of this topic in public health and international debate. Some of these books, such as Ratzan's, blame the media in part for sensationalizing health topics and creating irrational fears about mad cow disease. Other books, such as Rampton and Stauber's, accuse the US of covering up critical evidence pointing to an imminent mad cow crisis. Packer's book represents a scholarly view of the politicization of the BSE crisis. With these contradictory voices, it is hard to decide what voices the public finds credible. The presence of multiple contradicting voices may push mass audiences to trust sources that have name recognition and prestige.

Managing Uncertainty in the Case of the US BSE Crisis

In this risk management case, agri-business successfully communicated its messages to government and to the general public. Instead of adopting the precautionary principle in regard to public safety, the government strengthened testing measures only incrementally. Officials ignored pressures from boycotting countries or consumer advocacy groups that argued for even stricter BSE testing policies. Other countries have responded differently to domestic BSE crises.

For example, Japan adopted a more precautionary approach following the discovery of several BSE cases. The government ordered that every single cow destined for human consumption be tested for BSE, in spite of the absence of scientific support for this zero-risk measure.

The US beef industry successfully used science-based arguments to influence publics, including US policy analysts. They strategically invoked "uncertainty" surrounding testing effectiveness (e.g., it is uncertain that increasing testing will yield more BSE cases, given current practices) and uncertainty regarding human risk (e.g., science has not shown direct causal links) to shape messages. Conversely, emphasizing high levels of certainty regarding the low risk of BSE-human transmission, given current slaughtering and beef production practices, led to public acceptance of government actions. While government policy actions ultimately were supported and informed by the beef industry, these actions also reflected a need to reassure publics that "something was being done." Consumer advocacy groups and international groups were calling for stricter controls to prevent BSE, but the US was able to reassure publics by announcing the many changes being made and the "science-based reasoning."

It is also true that societal trends such as the Atkins diet and public skepticism about media health reports contributed to the public's response to the BSE crisis – there was little public outcry for stronger testing, other than by marginal consumer advocacy groups. Front groups supporting the beef industry, such as the Center for Consumer Freedom, also shared a role in shaping policy.

Public health case studies provide a mechanism for understanding issue outcomes and effects on organizations and their many publics. They provide a structure for identifying underlying motivations that guide important decisions that may have significant effects on public health. By examining policy development in the context of risk management, we can better predict government actions in regard to public health.

References

Associated Press. (2004, February 10). U.S., voicing confidence, ends search for more mad cow cases. *The New York Times*, p. A20.

Beck, M., Kewell, B., & Asenova, D. (2007). *BSE crisis and food safety regulations: a comparison of the UK and Germany* – Working Paper 38. York, UK: University of York, The York Management School. Retrieved, March 12, 2009, from http://eprints.whiterose.ac.uk/3477/1/beckm42007.pdf

Becker, G. S. (December 8, 2005). *BSE ("mad cow disease"): a brief overview (CRS Report for Congress, RS22345)*. Washington, DC: Congressional Research Service.

Belay, E. D. & Schonberger, L. B. (2005). The public health impact of prion diseases. *Annual Review of Public Health, 26*, 191–212.

Blakeslee, S. (2003, June 3). Mad cows, sane cats: Making sense of the 'species barrier.' The New York Times, p. F3.

Blakeslee, S. (2004, July 30). Study lends support to mad cow theory. *The New York Times*, p. A13.

Bradbury, J. (1996, October 26). Latest results strengthen link between BSE and vCJD. *The Lancet, 348*(9035), 1157.

Brown, D. A., Bruce, M. E., & Fraser, J. R. (2003, June). Comparison of the neuropathological characteristics of bovine spongiform encephalopathy (BSE) and variant Creutzfeldt-Jakob disease (vCJD) in mice. *Neuropathology and Applied Neurobiology, 29*(3), 262–272.

CattleFax. (2008, April). *CattleFax Trends*. Centennial, CO: CattleFax.

CBC News. (2006a, August 24). Mad cow in Canada: The science and the story. *CBC News*. Retrieved, March 12, 2009, from http://www.cbc.ca/news/background/madcow/

CBC News. (2006b, October 23). Timeline of BSE in Canada and the U. S. *CBC News*. Retrieved, March 12, 2009, from http://www.cbc.ca/news/background/madcow/timeline.html

Coghlan, A. (2006, April 15). Are prions the real cause of BSE and vCJD? *New Scientist, 189*(2547), 18.

Cohl, H. A. (1997). *Are we scaring ourselves to death? How pessimism, paranoia, and a misguided media are leading us toward disaster*. New York: St. Martin's Press.

Ferrieres, M. (2005). *Sacred cow, mad cow: a history of food fears*. New York: Columbia University Press.

Ghani, A. C., Donnelly, C. A., Ferguson, N. M., & Anderson, R. M. (2002, January). The transmission dynamics of BSE and vCJD. *Comptes Rendus Biologies, 325*(1), 37.

Gray, G. (2003). *Email from George Gray to Mabel E. Echols: peer review comments*. Retrieved, March 12, 2009, from http://www.whitehouse.gov/omb/inforeg/2003iq/132.pdf

Hendrix, J. A., & Hayes, D. C. (2006). *Public relations cases (7th ed.)*. Belmont, CA: Wadsworth.

Hill, A. F., Desbruslais, M., Joiner, S., Sidle, K. C. L., Gowland, I., Collinge, J., et al. (1997, October 2). The same prion strain causes vCJD and BSE. *Nature, 389*(6650), 448–450.

Hoppe, R. A., Korb, P., O'Donoghue, E. J., & Banker, D. E. (2007). *Structure and finances of U. S. farms: family farm report, 2007 edition*. Retrieved, March 10, 2009, from http://www.ers.usda.gov/Publications/EIB24/

Kaufman, M. (2004, April 16). Company's mad cow tests blocked: USDA fears other firms' meat would appear unsafe. *The Washington Post*, p. A01.

Kewell, B., & Beck, M. (2008, April). The shifting sands of uncertainty: risk construction and BSE/vCJD. *Health, Risk & Society*, 10(2), 133–148.

Kilborn, P. T. (November 15, 2003). Cattle rushed to market as the price of beef soars. *The New York Times*, sect. 1, p. 22.

Klinkenborg, V. (2005, July 24). The story behind a New York billboard and the interests it serves. *The New York Times*. Retrieved, March 12, 2009, from http://www.nytimes.com/2005/07/24/opinion/24sun3.html?_r=1

Lindquist, D. (2004, February 7). Safeguarding our food: preventive measures scrutinized as outbreaks of illness skyrocket. *The San Diego Union-Tribune*, p. A1.

McNeil, D. J. (2003, December 26). Mad cow disease in the United States: the overview; Mad cow case leads Government to consider greater meat testing. *The New York Times*, p. A1.

McNeil, D. J. (2004, February 3). Man who killed the mad cow has questions of his own. *The New York Times*, p. F2.

Miller, D. (1999, November). Risk, science and policy: definitional struggles, information management, the media and BSE. *Social Science & Medicine, 49*(9), 1239–1255.

Morgan, M. G. & Henrion, M. (1990). *Uncertainty: a guide to dealing with uncertainty in quantitative risk and policy analysis*. Cambridge, UK: Cambridge University Press.

Murphy-Lawless, J. (2004). The impact of BSE and FMD and democratic process. *Journal of Agricultural and Environmental Ethics, 17*, 385–403.

National Cattlemen's Beef Association. (2008a). *Leading markets for U.S. beef and veal*. Retrieved March 12, 2009, from http://www.beefusa.org/uDocs/leadingmarketsforusbeefandveal737.pdf

National Cattlemen's Beef Association. (2008b). *Cattle and beef products, exports versus imports*. Retrieved, March 10, 2009, from http://www.beefusa.org/uDocs/cattleandbeefproductsexportsversusimports.pdf

National Cattlemen's Beef Association. (2008c). *Cattle numbers and beef production*. Retrieved, March 10, 2009, from http://www.beefusa.org/uDocs/cattlenumbersandbeefproduction725.pdf

National Cattlemen's Beef Association. (2008d). *U.S. cash receipts from farming*. Retrieved March 12, 2009, from http://www.beefusa.org/uDocs/uscashreciepts778.pdf

O'Neill, K (2005). U.S. beef industry faces new policies and testing for mad cow disease. *California Agriculture*, 59(4), 203–211.

Office of the United States Trade Representative. (2008, March 28). *Bush administration submits annual trade report to Congress*. Retrieved, March 12, 2009, from http://www.ustr.gov/Document_Library/Press_Releases/2008/March/Bush_Administration_Submits_Annual_Trade_Report_to_Congress.html

Packer, R. (2006). *The politics of BSE*. Hampshire, UK: Palgrave MacMillan.

Rampton, S., & Stauber, J. (1997). *Mad cow U.S.A.: could the nightmare happen here*? Monroe, ME: Common Courage.

Rampton, S., & Stauber, J. (2002). *Trust us, we're experts: how industry manipulates science and gambles with your future*. New York: Tarcher/Penguin.

Ratzan, S. C. (1998). *The mad cow crisis: health and the public good*. New York: New York University Press.

Scheier, L. M. (2005). The safety of beef in the United States. *Journal of the American Dietetic Association, 105*(3), 339–340.

Schwartz, M. (2003). *How the cows turned mad*. Berkeley, CA: University of California Press.

Simon, B. (2004, August 16). Beef industry in Canada remains 'upside down.' *Los Angeles Times*.

Slagle, M. (2004). *Consumer confidence in beef strong in the face of single case of BSE*. Centennial, CO: Cattlemen's Beef Board. Retrieved, March 12, 2009, from http://www.beefboard.org/news/Release_2004_01_06_a.asp

Sobal, J. & Maurer, D. (1995). Food, eating, and nutrition as social problems. In D. Maurer & J. Sobal (Eds.), *Eating agendas: food and nutrition as social problems* (pp. 3–7). New York: Aldine De Gruyter.

SourceWatch. (2009). *George M. Gray*. Retrieved, March 12, 2009, from http://www.sourcewatch.org/index.php?title=George_M._Gray

Stauber, J. (2004). *Agribusiness pumping money into front groups to sway mad cow disease*. Madison, WI: Organic Consumers Association. Retrieved, March 14, 2009, from http://www.organicconsumers.org/madcow/stauber012704.cfm

Tatum, C. (2004, February 10). Animal rights group's ads claim poultry presents mad cow risk. *The Denver Post*.

Tucker, M., Whaley, S. & Sharp, J (2006). Consumer perceptions of food-related risks. *International Journal of Food Science & Technology, 41*(2), 135–146.

United States Department of State. (2004, July 4). *U. S. beef safer than year ago, Agriculture's Veneman says: banning risk materials heightened meat's safety, secretary notes*. Bureau of International Information Programs, U. S. Department of State. America.gov – Telling America's Story. Retrieved, March 12, 2009, from http://www.america.gov/st/washfile-english/2004/July/20040714160024AKllennoCcM0.4283563.html

Veneman, A. M, (July 14, 2004). *Report on behalf of the United States Department of Agriculture to the U.S. House of Representatives*. Washington, DC: United States Department of Agriculture.

Wald, M. L., & Lichtblau, E. (2003, December 24). U.S. is examining a mad cow case, first in country. *The New York Times*, p. A1.

Wells, G. A., Scott, A. C., Johnson, C. T., Gunning, R. F., Hancock, R.D., Jeffrey, M., et al. (1987, October 31). A novel progressive spongiform encephalopathy in cattle. *The Veterinary Record, 121*(18), 419–420.

Woodward, S. (2004, January 9). *One cow, hundreds of uses*. The Oregonian. Retrieved, March 14, 2009, from http://www.oregonlive.com/news/oregonian/index.ssf?/base/front_page/1073135194312870.xml

Part IV
Conclusions

Chapter 11
Conclusions: Policies, Politics, and Communications of Health Risks: In Search of Safety and Public Reassurance

Hajime Sato

Emergence of Health Hazards, and Policies Against Them

Examination of asbestos- and BSE-related policies in several countries has revealed several commonalities, concerning the emergence of health hazards, and the regulation of these products (and their manufacturing processes) for health protection. In the early stages of the crises, when health hazards were not evident or just emerging, the industries had close relationships with the government agencies that were regulating them. At that time, there was much debate about the causes of the problems, and even if there were any health hazards at all. When it became evident that there were indeed problems (caused by industrial products) and many cases started to be reported, the experts and the government agencies increasingly paid attention to these cases as a medical problem. The industries and their regulatory agencies tended to have reservations about stricter regulations, as there was still scientific uncertainty about the causality and scale of the hazards. They argued that it is not appropriate to create a significant economic burden based on uncertain health hazards that were not evidently their responsibility. Sometimes the mass media highlighted the issues, but the general public remained largely out of the policy loop. The general population was regarded as lacking scientific and technical expertise, and therefore incapable of behaving rationally. Only when the issues became socially visible, especially through media reports of policy failures, were policies put in place to reassure the public.

As the mass media paid more attention to the issues, the public, especially the real and potential victims of asbestos or BSE, became mobilized. People sometimes resorted to political outcries, demonstrations, and litigation. This was the stage during which the issues were manifestly politicized and rather drastic policies were adopted. Government policies, past and present, were critically reviewed by newly established ad-hoc committees in several countries. Consequently, the protection of public health replaced economic development and the protection of industry as the

H. Sato
Department of Public Health, Graduate School of Medicine, The University of Tokyo, Japan
e-mails: hsato-tky@umin.net; hsato@post.harvard.edu

H. Sato (ed.), *Management of Health Risks from Environment and Food*,
Alliance for Global Sustainability Bookseries 16,
DOI 10.1007/978-90-481-3028-3_11,

top priority. Regulatory and risk-assessment agencies and departments that were independent of the industries they were to assess or regulate were (re)established.

Given the scientific uncertainty, and with the ever-changing status of the scientific evidence, no standard operating procedures were developed to define what roles the government, industry, public organizations, the mass media, workers, and consumers should assume in managing the health hazards. The absence or failure of risk-communication efforts was repeatedly pointed out.

Science, Policy, and Politics in the Management of Health Risks

Science and Society

Scientific knowledge unfolds over time. For health risks to be scientifically and effectively handled, these risks must be predicted quantitatively, and the predictions must be based on repeated testing (Belton 2001; Knight 1957). In the case of asbestos, the evidence on its health hazards accumulated over decades. It came from clinical case reports, epidemiological observations of workers, the discovery of pathogenesis and the rarer reports of other health problems spread among the general population. It took more than a century to get to this point, and scientific research still has a way to go in clarifying the uncertain and undiscovered aspects of the asbestos problems. In the case of BSE, scientific advancements were reported every few years. They included case reports of diseased animals, hypotheses about how the disease is transmitted, and suspected cases of human transmission that were later confirmed.

In both cases, generation of knowledge about the health hazards and the publication of research findings and their application to further research and policy development all depended on the socio-political environment, especially insofar as government policies were concerned. In the UK and France, as our study suggests, research on the health hazards of asbestos, as well as the utilization of this research, was hampered in one way or another by the actions of quasi-public (and sometimes quasi-scientific) steering committees or the inertia of policymakers. Both of these groups were under the influence of the asbestos industry and its workers. Even in other countries, government decisions on research funding, as well as official consultations with research institutions, could not escape political considerations. In the UK, academics did not fully exercise leadership in promoting research.

When BSE cases began to emerge, the governments of the countries addressed in our case study launched efforts to accumulate scientific data on it, both as an animal disease and (at first, to different degrees) a potential human disease. In the UK and Japan, committees of government experts in one way or another took the lead in research. In France, insufficient government commitment to research was retrospectively criticized as the cause of delays in policy. In the US, the government's consultation with a university research center marks its commitment to scientific research.

It should be noted that science is not an unbiased observer of external reality; rather, its output is dependent upon various social and cultural networks and interactions.

In other words, science is socially constructed (Gergen 1999, pp. 46–61; Hacking 1999, pp. 63–99), and it interacts extensively with the social order (Jasanoff 2004). Scientific knowledge is not a transparent mirror of reality (Jasanoff 2004; Newton 1997). It both embeds and is embedded in social practices, identities, norms, conventions, discourses, instruments, and institutions (Jasanoff 2006). Knowledge generation is incorporated in the establishment of government institutions and governance more broadly; conversely, government practices influence the generation and application of knowledge (Etzkowitz 2002). In this vein, we expect knowledge and society to unfold in tandem, a point that applies to the judgment and management of risks as well.

As has been pointed out, predicting risk can be reflexive (Adams 1995; Kerr & Cunningham-Burley 2000). The identification of a risk and any attempt to quantify its potential adverse events will result in an increased awareness of the risk and attempts to avoid it. These measures in turn will tend to decrease the number of adverse outcomes (Burns et al. 1998). If this is true, we would expect the industries concerned about possible economic losses resulting from the regulation of their products to be reluctant to promote scientific research that might play a role in the introduction of said regulations. Accordingly, risk management, and the production and management of the scientific knowledge informing it, are inevitably sensitive to the social context in which risks are managed.

Our case studies support the propositions of researchers on “Mode 2” science (Gibbons 2000; Gibbons et al. 1994; Nowotny et al. 2001, 2003) that knowledge is produced, transferred, and utilized in the context of application, for the (sometimes cosmetic/symbolic) management of health risks. Research priorities were sometimes steered by the conglomerate of government, industry, scientists, and other experts, as well as various other groups in society. In the case of asbestos-related research, this industry-academia link is especially evident in the UK and France. In the UK, BSE research was gradually fortified by government support, but this was not the case in France. In both cases, many scientific researchers, even if their links to industry and government were not obvious, were driven in their research by the emergence of health issues, and they took great pains to apply their findings to the policy arena. Good-quality scientific research is valuable so long as it helps policy making and the development of regulations. In this sense, scientific knowledge generated and used for these purposes is action- and issue-driven, socially mandated, and robust.

Thus, the scientific knowledge generated and utilized for the management of the asbestos and BSE hazards had many of the characteristics “regulatory science.” The initiation of research was basically issue-driven. The aim of the research was largely to inform regulatory decisions and resolve political policy conflicts in a limited amount of time; in other words, the research and the utilization of its outcomes were sometimes under political influence (Ashford 1984). The development and use of science as the basis for public decisions are inevitably dependent on the socio-political context (Jasanoff 1995). The influence of this context on science, both the research itself and the utilization of its outcomes, is not always political or the result of intentional manipulation of the facts by political interests; it can also be caused by the relative, and sometimes implicit, dominance of different disciplines in their social context.

Science and Policy

In addition to the production and dissemination of scientific knowledge, the interpretation and the use of scientific data are determined by the socio-political environment. This is especially so when the interpretation of scientific findings has significant distributional consequences for different groups – health risks for some and economic dislocation for others. How the risks were perceived, handled, accepted, or tolerated, either explicitly or implicitly (mostly implicitly), affected the fate of the products, the management of their risks, and the social response to these risks (Cranor 1993; Foreman 1994; Harremoes et al. 2002; Ilchman & Uphoff 1969; Sato 1999a,b; Taylor 1985). Consumers are likely to demand protection from what they perceive as a danger, despite the assurances of scientists.

Practical economic calculations are commonly regarded as especially suited to the needs of bureaucracies exposed to constant public scrutiny. Economic theories, accounting rules, and official statistical data suffice to make some parts of a quantitative cost–benefit analysis relatively straightforward. However, hedonic calculations or the estimation of utilties, which inevitably depends on the assumption and projections of values, can greatly differ across social groups. Risk cannot be quantified in a straightforward way, because it combines both prediction and a concern about outcomes: different people will judge the same risk in different ways, according to their personal circumstances and the context within which the risk is presented (Ashworth 1997; Frewer et al. 1996, 1997; Lofstedt & Frewer 1998). In these cases, the appearance of objectivity can always be exploited. Therefore, it is often quite difficult, or even impossible, to know when policy concerns are influencing scientific judgments (Graham et al. 1988, pp. 189–192; Porter 1994).

This phenomenon is illustrated by our cases. In the French asbestos case, the industry and the unions argued against asbestos regulations on the grounds of insufficient scientific evidence. In the initial stages of the BSE crisis in the UK, the agriculture department and the industry opposed stricter health policies because of the economic burden they might impose on the government and a loss of profits for the industry. Another possible example is the complete ban on asbestos in the 2000s. This ban was introduced in the last stage of the asbestos crisis in the absence of any consideration of the relative costs and benefits of the policy alternatives, especially the marginal ones resulting from an additional ban on products exempted from the previous regulations. Rather, both the workers and the public were reassured that the ban was risk-free.

It is possible that science can ameliorate policy conflicts by reducing uncertainty. However, it is also possible that increases in scientific knowledge can sharpen, or at least make explicit, the underlying political disagreements. By more clearly identifying winners and losers in the regulation of chemicals, science can arouse conflicting interest groups and make it more difficult to resolve policy disagreements (Graham et al. 1988, p. 198; Morgan & Henrion 1990, 141–171). When risk assessment, policy making, and the implementation of regulations are left to the same agencies (or to the same individuals), the distinction between the generation and interpretation of scientific findings and the application of these findings to policy making becomes blurred (Hinchliffe 2001).

Policy and Politics

Policies are not made in a political vacuum. It has been observed that the regulation of products becomes difficult after they are widely marketed. At least in the short term, all products have their own benefits in the market as regards to quality (both perceived and real), quantity, and price, even though in retrospect these benefits might be seen as myopic or illusory. Moreover, the business practices of each industry have wide-ranging ramifications. In the case of asbestos, many corporations and individuals are involved in the mining, importation, transport, manufacturing, marketing, sales, and abatement of asbestos products and byproducts. For beef products, there are farms, slaughterhouses, the rendering industry, meat processors, retailers, consumers, and so on.

Eliminating the tangible benefits people already enjoy – familiar products and traditional jobs, with their identifiable and self-aware constituencies – is more difficult politically than stopping something new that has yet to surround itself with a self-protective belt of interests. The management of health risks operates largely on a trial-and-error basis (Holling 1979; Douglous & Wildavsky 1982; Wildavsky 2000). The burden of proof in justifying any regulation falls on the government, not the producers. Though it is sometimes considered desirable for producers to prove beforehand that their product will do no harm (Huber 1983), that does not always happen. Asbestos had been widely used before its adverse health effects were confirmed. Meat-rendering technology and the use of MBM had become popular worldwide before their possible association with BSE, and later human vCJD, were suggested (Department for Environment, Food and Rural Affairs, DEFRA, 2005).

Thus, scientific research does not always create knowledge and, even when it does, this knowledge does not necessarily reduce policy conflicts. When it comes to regulating certain products or processes, differences in ideology and interests are both common and intense. This is especially true if the product or process already has market value and confers economic privileges. We can expect these factors to sustain policy conflicts even in the presence of clear-cut technical results – assuming the stakes are high enough.

In addition, the relationships between risk-assessment bodies, regulatory agencies, and industry can impede the swift implementation of health-oriented product regulations. Regulatory agencies are frequently assigned the role of guardian of products and industries. They are considered to have two missions: the utilization of product benefits and the management of their potential negative side effects. Coordination of these missions is also sometimes a task left to those agencies (Pearce 2005). These sometimes competing policy objectives can be reconciled in many ways. The reconciliation may occur explicitly within a group of people or an agency clearly assigned the task. Otherwise, it happens implicitly, either through a legislative body, litigation, or the market. In the implicit case, groups, institutions, or agencies with competing interests are allowed to freely fight one another in society and in the market, using their own resources.

When regulatory agencies are expected to reconcile among themselves, decisions are reached either unanimously or by compromise, but the process is not adversarial. Policies are adopted only if they do not threaten the development, and thus the existence, of the industry (Jordan 1972). The introduction of new and more stringent health regulations is postponed until the industry develops and markets alternate products. Therefore, modifications in both products and the policies formulated to regulate them tend to be minor. Effective and efficient regulation of the industry for the purpose of preventing hazards can be spoiled if the development of policies and their implementation are wanting.

We also observed in our study a lack of commitment to public health, as well as delays in the introduction of strict regulations and loose enforcement of these regulations by bureaucrats. Asbestos regulation in the UK and France is a good example. With regard to mitigating the BSE hazard, it has been suggested that the mission of the MAFF was to support the industry. This led to the MAFF's initial optimism that human vCJD is not linked to bovine BSE, resulting in a delay in the UK government's response to the BSE problem.

This phenomenon has been depicted in the political science literature as regulatory capture or bureaucratic clientelism. A regulatory agency, which is expected to act in the public interest, acts instead to support the commercial or other special interests that it is supposed to regulate (Laffont & Tirole 1991; Levine & Forrence 1990; Stigler 1971). In the production of knowledge, power is exercised from the explicit assertion of interests to the woven fabric of social existence where the power is only visible through its consequences (Foucault et al. 2001). To understand these regulatory situations, it may be necessary to understand how the hearts, minds, and emotions of the people making the policy subtly manifest in their favorable discretionary decisions, or in their tardiness or reluctance to implement certain regulations. It has been argued for a long time that the involvement of third parties that are independent of the industry to be regulated and have clear jurisdiction and sufficient resources is a prerequisite for the effective enforcement of regulations (Bernstein 1955).

Uncertainty as a (Political) Playground

Both in the case of asbestos and the case of BSE, scientific uncertainty was always a key component of the environment in which the policies were made. Certainty denotes the ability to predict accurately the consequences of actions, whereas uncertainty implies that the probability of a certain kind of event or class of events occurring is not known, although the probability is known to be greater than zero (the event itself is known). Ignorance involves knowing neither the class nor the probability of an event. The controversies about possible health risks were long-running because they were scientifically unresolved and irresolvable within the required time frame for making policy decisions. Furthermore, once the preferred framing of the issues in terms of the perception of risk, potential causalities, potential impact, and remedies is decided on, significant influence can be exerted over the subsequent course of events, including government action (Bierman 1989; Tversky 1972; Tversky & Kahneman 1982).

The degree of certainty required for action depends on the competing values and interests of different actors who claim different degrees of certainty about the desirability of the action. Those who favor more stringent regulation base their arguments on the high likelihood of a health hazard, whereas those who do not appeal to the uncertainty. Thus, the battles are fought on the interpretation of scientific evidence, sometimes using scientific language (Majone 1992). Certainty and uncertainty are also related to the margin of error allowed for policy decisions. These clamorous debates are characterized not by irrationality, but by multiple forms of rationality; the contending parties argue logically, but from different premises (Thompson et al. 1990).

The role and the importance of the (un-)certainty in policy making depend on the consequences of the policy chosen, and vice versa (von Schomberg 1993). Therefore, certainty and its importance in policy making are considered important political resources to be maneuvered, or strategically managed, along with the perception of the policy's consequences, both in decision circles and among the general public (Heazle 2006). Fear, reassurance, demands (and even needs) can be crafted, stirred or soothed, depending on the ideological or substantive interests of social groups.

This maneuvering of uncertainty was remarkable in the case of asbestos regulation. The argument that more research is required to verify, quantify, and understand the hazardous effects of different kinds and different quantities of asbestos was repeatedly presented by those who opposed more stringent regulation of asbestos products. These players included the industry, unions, and quasi-public committees of experts in many countries, especially the UK and France. In addition to the long latency periods for the manifestation of asbestos-related diseases, these political efforts highlighted the fact that there were still scientific questions that remained to be answered. Those who engaged in these efforts also claimed that more deliberate steps by the government and non-governmental health-oriented groups were also needed. This appeal for certainty extended to a concern about the safety of asbestos substitutes, leading to a further delay in the development of regulatory policies.

Uncertainty was also an important issue in the case of BSE throughout the various stages of the crisis. The UK government (MAFF) was criticized for choosing to err on the side of the industry rather than public safety. The government was trying to placate the public concerns about BSE, and uncertainty regarding the human transmission of BSE was also an issue at that time. The French government behaved in a similar way when it sluggishly adopted the ban on the use of MBM, motivated at least in part by a fear of hurting the industry. Thus, the risk of MBM appeared on the agenda.

The import ban was the easy part, at least politically. The economic losses mainly affected exporters and, therefore, resistance from the domestic industry was limited.

In many cases, therefore, there was not much of a call in the exporting countries for rigid scientific assessments of the risks or the effects of policy implementation. Rather, the domestic producers, especially before the domestic BSE cases started to be reported, were trying to promote their products, appealing to consumers that their products were safe. The scientific basis for the relative safety of beef products was rather unclear. The prolonged ban by the French government on the importation of British beef and the ban on the importation of US beef by the Japanese government were sometimes justified by uncertainty about how effective the policies adopted by the exporting countries would be. This applied to both the policy measures themselves (e.g., herd culling, testing of all cattle, and a cattle tracking system) and the rigor of their implementation (e.g., enforcement and monitoring).

In an adversarial process of rule making, knowledge claims are sometimes deconstructed, exposing areas of weakness or uncertainty. These weaknesses provide political decision makers with a justification to assert their right to interpret scientific data, especially in controversial areas. Thus, the uncertainty within science itself again became a topic for social negotiation, debate, and decision making (Meyer et al. 2005). This partial transfer of cognitive authority to the legal and political arena may be seen as a way of assuring that the interpretation of indeterminate facts reflects the public values embodied in the legislation and the norms of the scientific community. The combination of scientific data, expert interpretation of these data, and policy content is like a chain of linked arguments and beliefs. The process might be termed co-evolution or the mutual validation of policy and science (Davenport & Leitch 2005; Nowotny et al. 2001). When science and policy are co-constructed through processes that occur in tandem, it becomes difficult to explain one by appealing to the other.

The handling of the uncertainty brought about by inconclusive scientific evidence has thus become an important aspect of policy management (Adams 1995). To the extent that scientists suppress uncertainty, they are in effect arrogating to themselves a political decision for which they are not accountable. This is because, in an uncertain situation, which data options are chosen to build one's case create different winners and losers, and hence the process is inherently political (Shapiro 2005). In other words, an artificial consensus among scientists is undemocratic in its implications. Likewise, to the extent that policymakers and regulatory agencies suppress uncertainty, they are leaving its management to the markets and other actors.

Political Processes in the Pursuit of Health and Safety

The cases presented in this book illustrate two different patterns of policy making: incremental policy making, which is dominated by bureaucrats and administrative agencies, and policy making leading to major changes, which follows from political initiatives. In addition to the political mobilization of organized interest groups, especially victims and workers' unions, litigation sometimes becomes a factor that facilitates the introduction of health-protecting regulations, highlighting the pertinent issues, and inviting political intervention. In the background, the socio-economic environment and the international nature of the issues work either for or against the introduction and implementation of stricter policies.

Incremental Policy Making

In social life, it is not possible to completely avoid risks and errors in policy making. Except for a very few special cases in which nearly complete information is available and the possible ways to err are limited, errors are unavoidable in the management of complex technologies. In such cases, maximizing the expected value of a choice could be the best alternative. To do so, a decision maker must search exhaustively for alternate choices, identify the consequences of each alternative, and predict the likelihood that it will succeed. Complete information, ample time and money, sufficient control over the environment, powerful causal models, and a consistently effective and powerful method for ordering preferences are required (Morone & Woodhouse 1986, pp. 125–137). However, most decision makers are denied these conditions.

When one is confronted with a complex problem, one alternative to the analytic model of policy making, which requires complete information and resources, is a classic trial-and-error strategy. The steps one takes to apply this strategy are the following: (1) establish a policy, (2) observe the effects, (3) correct for undesired effects, (4) observe the effects of the correction, and (5) correct again. Many elaborate variations of the trial-and-error strategy have been proffered (Morone & Woodhouse 1986, pp. 132–138). Examples include general problem solving (Simon 1947, 1955), disjointed incrementalism and partisan mutual adjustment (Lindblom 1959), decision making in formal organizations (March & Simon 1958), the garbage can model (Cohen et al. 1972; March & Olsen 1979), the cybernetic paradigm (Steinbruner 1974), the mixed scanning perspective (Etzioni 1976), and the disjointed-incremental and rational model (Dror 1971).

The trial-and-error strategy is a widely used and effective means to discover both the benefits and the drawbacks of new products, technologies, and regulatory policies (Wildavsky 2000, Chap. 2). The criteria suggested for applying the strategy to political and social decision making – namely, errors that are small, recognizable, and reversible – are similar to those that usually apply to the alternate method of incrementalism, formerly called piecemeal social engineering (Popper 1971). For incrementalism, the errors should (1) be small in magnitude, (2) limited in number, and (3) follow a consistent pattern of relationships. As incrementalism cannot anticipate sharp discontinuities, society must try out innovations on a small scale, gradually introducing and assessing various products and technologies, and the policies to regulate them. This approach enables incrementalism to better sample the unknown and assess the relevant benefits and risks (Wildavsky 1988, p. 221).

Given the need to acknowledge the practical limitations imposed by bounded rationality, insufficient resources, and satisficing behavior, bureaucrats in regulatory agencies are the people usually assigned the task of formulating safety regulations. It is expected that their decisions will be expert-driven, based on scientific and technical assessments, and approved by political overseers. In accordance with the principle of incrementalism, policy modifications are usually minor so as to avert major disruptions of the policy already in place. The recent tendency to shift the burden of proof to the proponents of policy changes, including the introduction of new policies, products and technology, could help bolster this tendency. This is the politics-as-usual that is frequently observed in safety regulation.

Many countries have followed the above approach to deal with asbestos and BSE. In the case of asbestos, its use was widespread before health concerns arose and preventive measures were systematically introduced. As the recognition of the health hazards gradually increased, both domestically and internationally, the use of asbestos was partially regulated; for example, asbestos spraying and the use of crocidolite and amosite were banned. However, the effectiveness of these regulations, and their implications for the products still allowed, could not, and still cannot, be quickly and easily be determined. Besides the ever-changing scientific evidence on asbestos risks – evidence that is necessary to assess the nature of the risks – the benefits and risks of asbestos, which in all likelihood are not static over time, cannot be easily balanced and quantified. Favoring regulation was a loose, informal coalition of environmentalists, public health advocates, labor union leaders, and some activist scientists. Their opposition was led by the industry institute(s), funded by companies with a commercial interest in the production and use of asbestos and asbestos products. As a result, decisions, if they were made at all, were largely implicit, at least in the objective, quantified, and scientific sense.

In the case of BSE, it is now thought that the introduction of rendering technology and the use of cattle MBM in the 1970s caused the BSE epidemic and the entrance of abnormal prions into the human food chain. When this farming technology was introduced, no serious concerns were raised about possible adverse health effects on either cattle or humans, nor were any measures taken to detect them. When it was suggested that MBM posed a risk for cattle and other animals, and that cattle BSE might be transmissible to humans, policymakers were faced with the question of how much in the way of preventive measures could be justified. However, it was not sufficiently clear, at any stage of the BSE crisis, how to handle the problem, either substantively or procedurally.

Learning from experience too often has been a purely reactive strategy – regulators wait for errors to occur and then make corrections. Therefore, to build a well-designed catastrophe aversion system, regulators must anticipate the need to employ this strategy before the errors occur and then structure the regulatory system so that the errors receive immediate attention when they do occur (Morone & Woodhouse 1986, p. 133). Furthermore, to protect against the most catastrophic consequences, it is desirable to employ the zero risk option, which means the prohibition of any product or technology that poses any risk whatsoever of an uncontainable catastrophe. Another option is to restrict the use of such products or technologies to levels that are presumed to be safe or safer. Yet another strategy is catastrophe mitigation, in which one assumes that errors will occur and takes steps to prevent those that might result in undesirable outcomes (Myers & Raffensperger 2006).

However, there is no way to fully anticipate the changes in and the reaction of society, the flowering of the industry, the popularity of the products, and the social, economic, and political implications of their effects on health. These are the results of a political process. They affect the acceptability of the risks, and therefore, the policy choices (Durodie 2000). Even if one is determined to err on the side of caution, the question of how much regulation can be justified, and how much is acceptable to the different actors in society, cannot be consistently answered.

Policy Change by Crisis Politics

A crisis is a disruption that affects the social system as a whole and threatens its basic assumptions, its subjective sense of self, and its existential core (Pauchant & Mitroff 1992, p. 11). A crisis is also an event that surprises people, limits the time they have to develop a response, and threatens their high-priority goals (Habermas 1975; Hermann 1963; Perrow 1984). Similar to the general concept of crisis, a political crisis can be defined as a situation in which the large-scale political landscape changes, the legitimacy of political parties and/or the government is threatened, and their status and ability to function becomes endangered. A crisis is usually driven by a focus on particular events. These focusing events highlight certain adverse conditions, increase public concern, trigger political mobilization, define the issues as serious, and get them a high priority on the political agenda (Kingdon 1995; Sapolsky 1990).

To restore its legitimacy and status in society, the government must change its policies, usually drastically, to conform to public concerns and expectations. Policy decisions are made by legislators, transferred from the bureaucrats in regulatory agencies, and if any, from the subgovernment of regulatory policy (Grindle 1996). Greater importance is attached to the political implications of policy choices than to established precedents, scientific or technocratic assessments, and socio-economic effects, which had been given higher priority in the past. The resultant policy shifts are in many cases major, rather than small and incremental. This is called crisis politics, or the politics in crisis situations. It sometimes indicates the failure of incremental policy making (Sato 2002).

In the case of asbestos, our case study revealed that on several occasions policy changes came about as the result of crisis politics. This mode of policy making was invoked by political mobilization, which took place when the health hazards of asbestos became socially visible as occupational hazards (UK, France, Japan), and the potential risks of asbestos were recognized as applying to school children and hospitalized patients (France and Japan) and the general public (Japan). When policy making entered this mode, the decisions were not always informed by scientific considerations, but rather by the anticipated practical and political effects of the policies. The heightened public concern and the political mobilization highlighted the gap between public expectations and existing policies. As public confidence in the government and its policies tanked, the public became outraged.

This response demanded that measures be taken that could not be justified scientifically.

Likewise in the case of BSE, legislators sometimes took over policy making from the bureaucrats when the issue became a visible public concern and the politics entered crisis mode. Examples include the policies adopted by the national government in the UK after it initially failed to contain BSE, and the ban by the local governments in the UK on the use of beef products for school lunches. Import bans on UK beef products by many of the European countries, and the ban on US beef by the Japanese government, are other possible examples. In these cases, rigorous scientific assessments were lacking at the time the policies were introduced, but emergency measures and precautions were adopted nonetheless.

During the processes of issue building and agenda setting in this crisis mode, pressure groups sometimes sought to attract media attention to their campaigns for safety measures, while experts complained of the media's scare-mongering (Miller & Reilly 1995; Peters 1995; Singer & Endreny 1993). To garner or maintain public confidence, both the industry and the government employed risk-communication and public-relations experts (Adams 1992; Friedman et al. 1986; Hansen 1994; Sandman 1988). The mass media paid increasing attention to the scientific uncertainty and were instrumental in generating public concern about particular threats (Schanne & Meier 1992). However, media reporting does not always advance communication to a level that improves the public's understanding of the issues and policies, as well as their understanding by the policymakers themselves (Goodell 1987; Campbell & Sato 2009).

Litigation

Litigation can be another facilitator of policy change. Examination of the lawsuits and the court deliberations provides an opportunity to review the scientific evidence, the relevant legal frameworks and jurisdiction, and the state of administrative adjudication. Thus, such an examination can suggest remedies to problems that cannot be solved by other means. As court battles are fought with the resources sometimes different from those required for other political struggles, litigation is an attractive political tactic for groups with limited resources (Bouwen & Mccown 2007; Olson 1990; Rosenberg 2008; Upham 1989).

Unintentionally inflicted injuries are generally considered to be negligence, and proof of fault is important (White 1980). On the other hand, under strict liability, or liability regardless of fault, you pay if you do harm, whether you were negligent or not. It is noteworthy that the recent movement toward strict liability has tended to keep new products and services off the market. If a product has been marketed for a long time, the question often becomes whether it continued to be marketed after the harm becomes evident. Judgments in tort cases are thus made only after an accident occurs, or after some unfavorable outcome results from using the product or technologies and implementing the associated policies. In other words, even though regulation is preventive and standards are thus established before the fact, personal-injury (or tort) law is reactive, which means it provides redress after the fact. In this sense, tort law is a classic example of a resilient response (Douglous and Wildavsky 1982, pp. 186–198; Wildavsky 1988, p. 170).

Litigation played a significant role in the asbestos case, especially in France. Lawsuits and their subsequent adjudication highlighted the asbestos-related health hazards to workers, criticized the government for its negligence, and demanded more appropriate regulation and prompt compensation of victims by the government. In the US, the Supreme Court did not take up the issue of whether or not the government introduced appropriate regulations to prevent asbestos-related health hazards, but its decisions ordered the asbestos companies to indemnify their employees for these hazards. In Japan, Lawsuits were filed after the government acknowledged its negligence and that its past policies might actually have increased the asbestos-related health hazards. By exercising chilling (deterrence) effects, the fact that victims could sue the government (and, of course, the companies) also put pressure on the asbestos-handling companies (Zeegers et al. 2005). They knew that the risk of being sued itself would increase if they continued to operate without disclosing information on the health hazards emanating from their factories and products.

The Socio-Economic Environment

It has been suggested that in a given society at any given time, safety is a function of the availability of resources such as education, wealth, energy, and communication tools (Stigler 1988; Wildavsky 1988). In other words, especially at the national level, safety measures are dictated by national priorities, specifically the cost/benefit ratio, which is arrived at, either explicitly or implicitly, by a struggle between competing forces. The value of the products, and therefore the socio-political influence of those associated with these products, depends on the popularity of the products, the breadth of the industry and how it is organized, the economic stakes for the industry, the social and regulatory environment, and the valuation of the adverse outcomes arising from the industry and its products.

The problem is that the validity of the above thesis can only be judged retrospectively and by cross-national comparisons; thus, it is hard to establish. Government policy might clearly state which goals are most desirable, but the government usually does not (manifestly) calculate whether the negative side effects of these policies are acceptable. However, it is frequently noted that economically developing countries tend to attach less importance to the protection of health and the environment than do more developed ones (Calder 1995; Johnson 1983). The more a country advances economically and socially, the more attention it pays to public safety and the protection of public health (Motta & Thisse 1994; von Homeyer et al. 2001).

An examination of past asbestos-related policies suggests that these socio-economic factors affected the relative importance attached to the negative effects (costs) of asbestos use. The increased demand for asbestos and its products during war times, as well as rapid industrialization, pushed the health and environmental issues arising from them to the back burner in the UK, France, and Japan, at least for a while. More consideration was given to these adverse outcomes later on, when the development of the state had reached a certain level and society became politically structured in a way that reflected the concerns of the victims, mostly workers. Our study documents that in the 1970s, French workers chose the security of employment over the protection of their health from possible environmental hazards.

Although there was a 10 to 30 year lag in policy development across countries, a global trend can be discerned in the handling of occupational and environmental hazards. A spate of labor disputes emerged in the 1950s and 1960s, as public concern about the hazards of environmental pollution rose. In the 1960s and 1970s, the governments claimed they were taking more action. In the 1970s to 1990s, litigation in the form of class action suits became popular in some countries. Finally, in the 1990s and 2000s, private corporations were expected to assume more responsibility for health and the environment, including the liability for the defects of their products (Davis & Frederick 2000; Wells 1994).

To help develop its economy and protect the health and safety of workers, Korea introduced laws and regulations similar to those previously adopted in other countries. The government intented to learn from the experiences of these countries so that comparable disasters could be avoided in Korea. However, the enforcement of these measures in Korea was quite loose and their effectiveness questionable. Malaysia, which also adopted laws to protect workers, continued to lack official commitment, as evidenced by the absence of statistics on asbestos and the lack of a registry for asbestos-related health hazards (Kazan-Allen 2003; Rampal & Feitshans 2000).

Changes in the Ideological Matrix

An environmental factor that can both affect and reflect policy is the ideological matrix, a set of doctrines and values widely accepted in decision-making circles and among the public (Hsu 2005; Sabatier & Jenkins-Smith 1993; Sato and Narita, 2003). The acknowledgement of workers' rights, corporate responsibility for employee protection, product liability, and the role of the government in such efforts can all influence the balancing of competing policy objectives concerning the safety and health of workers and the protection of the environment (Smith 1991). Recently, as concern about the accumulation of hazardous materials in the environment increased, the PRTR (Pollutant Release and Transfer Register) principle was gradually embodied in laws that further increased the awareness and control of hazardous substances in the environment.

Risk aversion is sometimes justified on the grounds that reducing the probability of irreversible harm or a disaster overwhelms all other policy considerations. This principle has been named the "no safe dose" argument (Morris 2000). It follows from the precautionary principle, which states that once a risk has been identified, a lack of scientific proof of the causality (cause and effect of a potentially hazardous effect) shall not be used as an excuse to avoid taking action (Freestone & Hey 1996). This principle in turn is based on German domestic law, in which the term *Vorsorgeprinzip* denotes the avoidance of damage to the natural environment, even when there is no scientific certainty that a threat exists (European Commission 1998). In 1990, this principle was adopted by the UN Economic Conference for Europe (ECE) and the Second World Climate Conference, as well as in the Ministerial Declaration of the UN Conference on Environment and Development in Rio de Janeiro in 1992 (the so-called Earth Summit). Principle 15 of this declaration states that "where there are threats of serious or irreversible damage, lack of full scientific certainty shall not be used as a reason for postponing cost-effective measures to prevent environmental degradation (Rio Declaration 1992)" (European Commission 2000).

An early example of applying the precautionary principle might be the 1985 EC decision to ban hormones used for promoting growth in animals. At this time, however, the EC did not argue that its ban on beef hormones was precautionary. In 1999, discussion started on the first international rule regarding the precautionary principle and human health as applied to food (both seeds and commodities). The rule became part of the Cartagena Protocol to the Convention on Biological Diversity, which was adopted in 2001. The protocol states that a government may act to avoid or minimize the potential adverse effects of these hormones, even if there is not enough scientific information about their potential adverse effects to create scientific certainty (Echols 1998).

The recent popularity of the precautionary principle can be traced to the increasing recognition that there is always uncertainty about the validity of scientific claims and that public interest in risk aversion is widespread. However, these factors caused intense unease and debate about the extent to which precautionary measures should be included in science-based trade agreements. Precaution is a pivotal issue in the debate about food safety regulation, and application of the principle can be arbitrary. The EC cited precaution (the precautionary principle) as the basis for its initial regional bans on the importation of British beef, which it adopted at the time the existence of BSE became public knowledge. France also cited the precautionary principle in 2000 when it refused to lift a ban on imports of British beef, ignoring the EC's control system (*European Report*, January 6, 2000).

However, we could find no evidence that the precautionary principle was a powerful impetus for either the introduction of more stringent regulation on asbestos or the adoption of measures to prevent BSE. Morris, unable to find any documentation that the European Court of Justice appealed to the precautionary principle to uphold the EC's ban on beef exports from the UK (Case C157/96 and C-180/96, 5 May 1998), concluded that there is no clear legal precedent regarding how the principle might be used. In other words, the precautionary principle was not employed to justify the safety measures under consideration (Morris 2000, p. 6). In 1999, the European Council of Ministers adopted a resolution stating that in the future the EC should show a greater willingness to be guided by the precautionary principle in preparing legislative proposals and in its other consumer-related activities (European Commission 2000, p. 8).This resolution is further evidence that up to that time the precautionary principle had not really been a driving force in policy making.

In retrospect, the use of the precautionary principle was instrumental, in that it was employed to support policies that were preferred for other reasons, although it was not a driving force. Rather, the initial impetus for the EU to use the precautionary principle came from crises such as BSE (Durodie 2000, Chap. 8).

International Contexts

Government decisions are determined not only by local politics; they are also bound by their international context. International organizations such as the WHO, IARC, FAO, OIE, WTO, and EU took up a series of issues, compiled and published pertinent scientific information, and set guidelines and rules regarding health, safety, the environment, and trade. The activities of these organizations have provided numerous standards – scientific, traditional, economic, and political – and in so doing have provided the environment in which national governments make their policies. As a result, the regulations justifiably imposed by each country gradually were formalized, a process that facilitated international trade and market liberalization.

The results of the Uruguay Multilateral Trade Negotiations in 1994 include the Agreement on the Application of Sanitary and Phytosanitary Measures (SPS Agreement). However, a sanitation measure may not be maintained indefinitely in the absence of sufficient scientific evidence (Echols 2001). Such a measure does allow the member states to consider scientific and other factors that are not susceptible to quantitative analysis by the empirical methods normally used in scientific laboratories. If there is a relevant international standard, members must base their regulatory measures on this standard, unless the member state determines that a different level of protection is more appropriate for its citizens (Kazan-Allen 2004). A critical decision for regulators is whether to impose a sanitation measure when there is uncertainty or disagreement about whether a danger exists, and when the scientific evidence about the existence or extent of the risk is inconclusive or there is no such evidence at all. It is also quite important whether the decision is accepted as legitimate by the other member states (Naubauer 2005). There are innumerable historical examples of food safety measures being criticized as a disguised form of protectionism (Johnson 1983; Sato 2000; Vogel 1992).

The WHO and the IARC have published technical bulletins and expert reports on the health hazards of asbestos, and the WHO and the OIE did so on BSE and vCJD. When national policies involve international trade, the WTO and the international courts sometimes play a prominent role in resolving the disputes. An example is a dispute on the banning of asbestos imports that was brought to the WHO. In Europe, the committees of experts and the ministerial councils of the EU set standards for regulating BSE. An embargo on British beef products was discussed by EU panels. The US and Japan established a bilateral panel on

BSE-related trade negotiations. When trade disputes are brought before these international forums, national governments are provided with an opportunity to reassess policy and search for a new rationality.

The possibility of transferring technology, and the associated risks, exists whenever resources are available (Taylor 1999). The international harmonization of risk regulation is arguably desirable, as is the orchestration of risk-communication efforts.

The case of asbestos in Korea is an example of the international transfer of hazards, and the industries that create them, from economically developed countries to less developed ones that tend to be more lenient in regulating health hazards. As hazardous and polluting industries have become increasingly regulated in industrialized nations, they migrated to the developing countries where cheap and uninformed labor is abundant (Castleman 1979). The health hazards these countries incur can be considered a necessary cost of industrialization (LaDou 1992). Our case study documents that in the 1970s, when environmental regulation was tightened in Japan and other economically developed countries, asbestos-handing companies moved their factories to Korea, where the regulation was loose. Another possible example is the case of the French battleship, for which a maintenance regimen including the spraying of the walls with asbestos was planned in India.

Choice of Policy Alternatives

Regulatory policy shapes the policy that determines the distribution of costs and benefits of regulation over time, as well as the process of policy making, which affects the power relationships among the actors that might be involved in decision making (Montgomery 1974). The choice of instruments for enforcing policy compliance determines how a policy is to be implemented to attain its goals, that is, which procedures are employed and how they are employed. The policy instruments include (1) direct restrictions and bans concerning the composition and the use a product, where and how it can be used, etc., (2) economic procedures such as price caps, taxation, and insurance premiums, and (3) education and information in the form of warnings, guidelines, recommendations, educational programs, official reports, etc. (Walsh & Gordon 1986). Compliance inspections and sanctions against violators are also key to policy effectiveness. Depending on how a policy distributes costs and benefits among the different actors in society, different patterns and degrees of political mobilization are expected (Noll 1985; Schattschneider 1975).

Hazard Detection, Prevention, Monitoring, and Compensation

Measures to prevent and mitigate the health hazards from asbestos exposure, the improvement of working conditions, and the health checkup programs for the early detection of asbestos hazards were introduced early on. Environmental regulations were tightened over time, and the health checkup programs became more substantial. Also installed fairly early on were compensation programs for the victims, which included payments for their medical costs. At first, however, the effectiveness of these programs was quite limited. The regulatory standards were loose and enforcement was not rigorous. Furthermore, the health checkups and compensation applied only to current workers; ex-workers who might develop health problems after a long latency period were not covered. Although the situation in all these areas improved over time, the shortcomings continued in most countries until recently.

With these loose regulations, the asbestos industry could avoid paying the full cost of preventing health hazards, and that of compensating for the losses its products caused. This was true for the several decades during which the adverse consequences of asbestos use were not recognized. Also, if a company went out of business before these consequences were exposed, they never had to pay anything. This inherently myopic policy allowed the true costs of the products, namely the costs resulting from the hazards other than the costs of asbestos manufacture, maintenance, and disposal, to largely be put off to the future.

Later, in the 1960s and 1970s, when the court ordered the asbestos companies in the US and France to indemnify victims for their financial losses resulting from asbestos exposure, the situation changed in these countries, but only slightly. Under a pure comparative negligence standard, anyone who contributes in any way to an accident can be sued. If found liable, the person must pay a percentage of the costs based on how much they contributed to the accident (Miceli 1997). Especially in the US, some asbestos corporations went bankrupt because of these court-ordered payments. Also in dispute was the negligence of these corporations in failing to institute necessary health protection measures when the hazards became known. In Japan, where many of the asbestos-handling corporations are no longer present, their remnants and the main users of their products established a fund, to which the government also contributed.

In any case, direct regulation of asbestos-containing products and asbestos spraying was limited for quite a long time. The use of asbestos in construction materials such as pipes and slate boards was very popular almost up to the moment asbestos was banned. Thus, the market for asbestos products was largely maintained under the policy of "controlled use." Consequently, total consumption decreased very slowly over the ensuing few decades. Only after the size of the market reached a low point was a total ban of asbestos use put on the policy agenda.

Another issue that threatens the (techncial) rationality of safety and regulatory policies is the desire for zero risk. Prospect theory suggests that outcomes that are merely probable are given less weight than certain other outcomes. Therefore, the policy of unconditional protection (pseudo-certainty) appears more attractive than conditional protection (protection only against specified risks). Pseudo-certainty highlights the contrast between the reduction and the elimination of risk (Fischoff et al. 1984; Miljkovic 2005; Slovic 2000; Slovic et al. 1988, 1982). Therefore, a total ban could be psychologically more appealing and reassuring to the public than a limited ban or calculating the relative effectiveness and efficiency of a policy.

For both asbestos and BSE, the governments in our case study often proposed a variety of "total" bans on the manufacture, use, and trade of the products, especially when the issues appeared on the political agenda and the government faced public outrage. Bans on the use of asbestos, of whatever type and for whatever purpose, have been in effect in France and Japan since the 1990s. A ban on the use of ruminants' organs and the mandate of herd culling by the UK, France, and Japan, as well as import bans by France and the US on British beef and by Japan on US beef products, were introduced without waiting for verifiable scientific evidence to justify these measures and without considering their relative costs and benefits. They were justified mostly as emergency measures and/or as necessary for consumer safety and reassurance.

Administrative Reforms and Independent Agencies

Apart from the phenomenon called regulatory capture or bureaucratic clientelism, a bureaucracy commonly defines an issue in a way that fits the available organizational resources and strategies (Reich 1991, pp. 224–231). Therefore, it has been argued that improving the efficacy of regulatory agencies requires providing them with sufficient jurisdiction and resources, as well as keeping them independent of the industries to be regulated and immune from any political pressure these industries might apply (Bernstein 1955).

In many countries, one of the most remarkable recent changes in public policy making is the official appraisal of past policies. This popular approach has been adopted in response to public criticism of government policies and in accordance with the government's desire to demonstrate that it is accountable to the public. Such appraisals have been conducted by the governments of the France and Japan in relation to asbestos and by the UK, France, and Japan in relation to BSE. They were motivated by heightened public concern about these issues and public outcries about perceived government incompetence in managing them.

The appraisals of asbestos policies more or less acknowledged the lack of full government commitment to the management of the asbestos hazard, especially the monitoring of workers' health and the compensation of victims. The appraisals of BSE policies pointed out the factors to which policy failures, if any, could be attributed. These factors include, above all, the structure of the regulatory authorities, which assumed the dual role of industry support and consumer protection. For both asbestos and BSE, the handling of scientific uncertainty was also questioned, although not necessarily criticized. Also, the lack of involvement in decision making by the general public and the victims was considered inappropriate.

The BSE crisis, and the policy appraisals that followed, acted as the catalyst for a thorough reorganization of the official agencies (Millstone & van Zwanenberg 2002). In the UK, France, and Japan, new agencies were established to deal with food safety, independent of the agencies that promoted the industry. The agriculture ministries were divided into separate sections to address these sometimes competing policy objectives. Coordination mechanisms between the agriculture and health ministries was also built. During the course of these administrative reforms, the opinion that the regulatory (or risk-management) bodies should be separate from the risk-assessment bodies gained momentum in many countries, the rationale being that risk assessment should be based on science, whereas risk management, or regulatory policy making, is more a matter of values and politics (FAO [Food and Agricultural Organization] & WHO [World Health Organization] 2002).

In the UK and France, semi-governmental committees, consisting of scientists, government officials, and representatives from the asbestos industry and the unions, were considered to be inappropriate organs. Meeting mostly behind closed doors, they were considered to be susceptible to a bias in favor of protecting the asbestos industry. Following the BSE turmoil, the MAFF was dismantled in the UK, and its responsibilities were largely transferred to the new Department of the Environment, Food and Rural Affairs (DEFRA). In addition, a new Food Standards Agency (FSA) was established to provide the government with expert advice. Similar administrative reforms were introduced in France and Japan.

The purpose of this type of reorganization was to make policy through the competition and coordination of different departments and sections, each representing clear values and assigned a coherent policy objective. In so-called bureaucratic policy making, these different values become clear in the process of arguing policy alternatives (Allison 1999). Political leadership is often expected and desirable if the value and policy conflicts cannot be resolved through bureaucratic competition. However, these reforms and reorganizations have yet to be assessed (Campbell 1992; Muramatsu & Krauss 1983).

Safety, Reassurance, and Risk Communication

Communication and Policy Accountability

Risk communication, defined as the exchange of information about risks, has become an important subject in many countries since the early 1990s. It includes not just scientific and technical dimensions, but also the communication of information about legal frameworks, public policies, and the behavior and concerns of corporations, the general public, and a range of groups and individuals in the society (National Research Council 1989). Its increase in importance following the alignment of government policies and performance with the expectations and behavior of the public became a critical issue for two reasons. First, if the public does not understand the goals of a policy and how it is being implemented, the implementation cannot be effective. Second, when either the objectives or the effectiveness of a policy does not meet public expectations, the public might not see the policy as justified. As a result, the government's actions are criticized and the government itself discredited (Kasperson 1999; Sidall 1980).

In the context of risk management, hazards (as are predicted technically), causality (as is explained scientifically), adversarial events (as are revealed as such), government actions, corporate behavior, and the risk perception and behavior of individuals all interact with one another (Pidgeon 1999). Decisions on hazard management depend on the importance and probability of the events, both as predicted scientifically by experts and as perceived by the public. They also depend on the political importance of the problem, as reflected by the achievements and reputations of the policymakers (Holzmann & Joergensen 2001; Plough & Krimsky 1987). The wisdom of these decisions, and of course the consequences of policy influences the government's legitimacy. To different degrees, therefore, all governments try to assure the technical effectiveness of their policies as well as the practical effectiveness of the enforcement of these policies. Every government also tries to reassure members of the public about the seriousness and effectiveness of its efforts.

Inevitably, many of these factors are not predetermined and fixed but change over time. The best we can hope for is to intelligently muddle through (Lindblom 1959). Uncertainty about both facts and values implies that determining the accountability of a hazard must be an iterative process. One reason is that as time goes by, we learn more about how a hazard behaves and how much we like or dislike its consequences. In other words, it takes experience and acknowledgement of the experimental nature of life to tell us what the facts are and what we really want (Fischoff et al. 1984, 1978; Schwing & Albers 1980; Slovic 2000). A critical issue is determining the society's goals, which are (to be) set by the interactions among the government and various groups and individuals. The formal analyses must be open so as to avoid suspicion and rejection of whatever conclusions are finally reached. Thus, the goals might include developing an informed citizenry and preserving democratic institutions.

Successful communication with the public about risks is a hard task to accomplish. The history of policy making in the UK regarding BSE suggests that before the link between BSE and vCJD was announced, many government officials held the view that the public was unable to understand scientific uncertainty, which is an inherent part of science and the assessment and management of risk. The government was also concerned about the public panic that might result from the disclosure of the uncertainty about the link between BSE and vCJD. Because of its possible effects on exports and the political implications of discovering the disease, the dissemination of BSE-related information was restricted in one way or another (HM government 2001). In practice, the government's messages to the public stressed that beef was safe, while avoiding the suggestion of possible but empirically unverifiable risks (Phillips 2000, vol. 10). For example, in November 1989, the UK government, in response to Germany's ban on imports of beef from Britain, made an announcement stating that the ban was unreasonable and unjustified and reiterating that British beef was entirely safe (*BBC*, November 1, 1989). In 1990, the Agriculture Minister, John Gummer, was pictured eating a hamburger and feeding one to his daughter.

At the same time, the UK government initiated measures to prevent BSE or stem its spread. For example, in 1988, the Agriculture Minister, John MacGregor, banned the sale of milk from cattle with BSE as a precautionary measure, even though there was no evidence at the time that the disease could be transmitted to humans. The British news media frequently reported these activities to the public (Eldridge 1999).

Despite these efforts to reassure the public, distrust of the government increased. When the linkage between BSE and vCJD was finally announced, the public responded by boycotting beef and beef products. The government's communication policy was based on "the deficit model," which assumes that if the public understood science, they would respond to technical risks in the same way the experts do (Hilgartner 1990).

The deficit model implies that policy should be guided by scientific facts, and continuous efforts should be made to educate the public about science. It further implies that the public is generally incapable of handling uncertainty, and that if the deficit is such that the public cannot be brought to understand it through increased communication, the information, especially threatening one, should be suppressed to prevent the public from reacting irrationally and inappropriately. Thus, it is justified to craft the communications so they reassure the public and induce it to accept the prescribed policies (Irwin & Michael 2003).

However, the UK government exacerbated the BSE communication problem by sending mixed or contradictory messages through its statements and actions. It tried to reassure the public that beef was safe to eat, but at the same time it bowed to pressure from the EU to stop beef exports (Reilly 1999). Our analysis of the BSE case demonstrates that the UK government failed to discuss the uncertainty surrounding the safety of beef and beef products with the public. This failure resulted in increased public distrust in the government's risk-management practices (Frewer 2003, pp. 63–64; Frewer et al. 2001; Miles & Frewer 2001). The rampant public criticism of the ostensible policy failures can thus be attributed to a failure of communication. This disruption of the trust in government institutions aggravated public concern about the health risks of beef and caused the public to question the effectiveness of government policies to address the problem (Cvetkovich & Lofstedt 1999; Peters et al. 1997; Siegrist & Cvetkovich 2000).

Over the years, researchers have pointed out differences in how experts and the lay public conceptualize risk (Barke & Jenkins-Smith 1993; Mertz et al. 1998). The public tends to judge risks psychologically rather than through a logical estimation of the probability of harm: in particular, perceived control (the extent to which the assumption of a risk is perceived to be voluntary), dread, and the perceived potential for catastrophe were found to be important psychological determinants of people's responses to risk (Slovic 1993). Members of the public who have experienced direct benefits from technology are likely to have qualitatively different opinions about the relative advantages and drawbacks of technology than do those who have not experienced such benefits (Sparks & Shepherd 1994). However, those findings and advice did not significantly help to improve government (communications) policy.

Shaping the Public's Perception of Risk

As discussed earlier, risk communication is ideally a two-way process. In other words, the communications should reshape the perceptions of both the public and the policymakers about the risk and the issues underlying it. Interpretations of scientific evidence and the choices of policy are inevitably value-laden, never value-free. Furthermore, values are not fixed; they are flexible, changeable, manageable, and influenced by many factors. They can be conscious or unconscious, and they can be influenced, intentionally or unintentionally, by interactions among different actors in society (Fiorino 1989; Jasanoff 1990).

Thus, any actor can act consciously and strategically with the intent to shape the risk landscape and the perception of which policy options are best (Latour 1987). However, the consequences of these actions are not always promised, and sometimes unintended. For example, when people are scared about a risk or hazard there are two conceivable policy alternatives: one promises and provides technical solutions and reassures the public, whereas the other persuades the public to accept the risk as unavoidable or tolerable. On the other hand, if people are at ease and are unaware of, minimize, or accept the risk as defined technically, the best choice is to do nothing that would irritate and arouse them. Specifically, one should not inform the public about the extent of the risk as inferred from scientific research, not take aggressive action to induce the public to behave in the desired way, and not send out frightening communications that might exaggerate the risk (Kasperson 1999; Pidgeon 1999; Renn et al. 1992).

Regarding BSE, the governments' approaches to risk communication combined action and inaction (Dora 2006). The actions included announcements that beef and/or beef products were safe, notification of risk, guidelines, education and information about appropriate behavior, and regulatory actions such as BSE testing, SRM removal, and herd culling. The enactment of laws and regulations, even when they were poorly implemented, could convey messages to the public about the risks, the safety of beef, and the governments' management of the problem. Also considered to be message-laden, either explicitly or implicitly, were opinion surveys about safety concerns and public confidence in government policy, hearings and conferences to which industry representatives, experts, and the public were invited, and reports from the monitoring of marketing activities and the consumption of beef products.

The official programs for BSE testing and the disclosure of the results certainly gave people the impression that their government was implementing the necessary measures and that government actions were open and credible. Furthermore, they illustrated the role of science in determining government policy. It should be noted here that regulation of testing, not only the implementation of testing, is an important "testing" policy. In the US, official regulation of BSE testing by the Department of Agriculture is in place. BSE test kits cannot be used on cattle under a certain age. The government introduced this policy to send a clear message that its policies were based on scientific judgments of the reliability and effectiveness of the tests used. Moreover, during the trade disputes between the US and Japan, the US government consistently repeated this argument, insisting that policy should be based strictly on science. Even when some beef producers planned to test cattle under the official age limit to distinguish their products in the marketplace, the USDA disapproved, lest the enforcement of government's policy be seen as inconsistent. Such a message could make the public skeptical about both the policy and the government.

In Japan, on the other hand, the government's first step was to introduce a program that required testing all cattle for BSE. The intention was to contain public concerns and to assure greater safety, even though the precise ages of the cattle were unknown. It was only after the trade dispute that discussions visibly started about whether age limits could be set for BSE testing. Only after meticulous reports by panels of experts and a series of public conference measures did the government adopt a policy that exempted from BSE testing cattle under a designated age that were intended to provide beef products for export. Later, the Japanese government ceased its subsidiary payments to local governments for BSE testing of cattle under the age limit. However, many of these local governments continued to fund BSE testing regardless of cattle ages as a way to reassure the consumers.

The relevant policies in these countries were derived not solely from government initiatives, but also from lobbying and communication efforts by non-government entities. As described in our case study, the US cattle and beef industries have long strived to make sure that government policy adheres to scientific evidence and reasoning. Although the local governments responded to the public's concern about safety, they did not completely abandon their concern for local producers. BSE testing was expected to make the products of these producers appealing in the market. It is not always easy to determine which actors, the governments or others, dominate in policy discussions and in shaping the public's perception of risk.

In the case of asbestos, on the other hand, the US government sometimes behaved differently than the case of BSE, deviating from its commitment to science-based policy making. At the time the public became concerned about asbestos in schools, there appeared to be insufficient scientific evidence to warrant the immediate abatement of asbestos from school buildings. Adverse health effects stemming from asbestos in those buildings had not been verified, or even firmly indicated. Nonetheless, as happened in Japan, the government decided to support the abatement in response to the increasing public concern. In retrospect, it appears that in the 1970s and 1980s both governments were reacting to the public's risk perception rather than proactively reshaping it. Accordingly, the government's actions did not deny, or even symbolize, the risk of asbestos hazards in such places.

In the countries under study, statements about the risks were included in laws, regulations, and guidelines, in the policies and programs based on them, and in the official risk assessments. These statements were addressed primarily to the government officials who made and enforced policy, the employers responsible for worker protection, and, secondarily, past and present employees. However, the communications about potential health hazards have never been intentionally directed to end-users, including the people who live near asbestos factories and the general public who use asbestos-containing products. This restriction of government communications steepened the abrupt rise in public concern following the disclosure of an increase in asbestos-related diseases in those living near the factories.

The general public remained more or less unconcerned about the asbestos-containing materials used by providers, such as house builders and car dealers, who supply them with goods and services. Those balancers who see a cost/benefit ratio close to one might argue that many consumers continue to use possibly hazardous products because of their benefits, including low prices. Consumers therefore do not appear to regard health protection as an absolute value. According to this balancing philosophy, regulators should devise standards that approximate the outcomes of a perfectly functioning consumer market (Graham et al. 1988, p. 106; Smith 1976; Viscuti 1979). The public, however, remained largely ignorant of possible asbestos risks or confused by different, and sometimes contradictory, arguments and policies.

Partnership and Cooperation in Risk Management

As our case study has disclosed, every step of policy making and regulation regarding health hazards, including problem definition, research promotion, knowledge transmission, and the development, implementation, and appraisal of policy, cannot escape the influence of politics and the social context. From the perspective of "objectivist or Mode 1 science," insufficient, delayed regulations can be attributed to the myopia or incompetence of scientists and policymakers. From the perspective of "constructivist or Mode 2 science," these defects can be seen as the result of political negotiations reflecting the combined interests and preferences of the society at large. In this case, every decision, so long as it emerges from the application of due process in the polity, can be regarded as the product of rational choice, namely, the implicit acceptance of risks after considering the tradeoff between benefits and hazards (Fuller 2001).

In many cases, the effective management of an environmental hazard depends on reconciling the objectivist and constructivist perspectives by integrating the technical information about the hazard with the interests and values of the affected parties toward a common solution (Cvetkovich & Earle 1992). Post-normal science implies that various sorts of uncertainty and value commitment enter into any decision on risks. Therefore, in policy making, scientific considerations should be complemented by other considerations (Bohnenblust & Slovic 1998; March & Ravets 1999). Management of health risks is not the sole prerogative of the government; it also requires the participation of civil society, and interactions and negotiations between interest groups, networks, and sectors.

Risk communication consists of exchanges of information among individuals, groups and institutions. Some of these messages concern the nature of risk, others express concerns, opinions, or reactions to the initial messages, and yet others address the legal and institutional arrangements for risk management (National Research Council 1989). Successful communications efforts are expected to stimulate appropriate risk-management planning and actions, good decision making regarding the acceptability of risk, and if necessary, anxiety reduction in exposed persons. Communication might also promote insight into risk management in the society (Vlek 1995).

Much advice has been offered to improve the provision of information about risks, increase the involvement of the public by making the information personally relevant, and enhance the ability of the various actors to process the information (Browde et al. 2004). Technical and cultural rationality must be combined with a broad orientation to risk among the actors (Hamilton 2003). When scientists attempt to use science to address public policy issues, they are usually not well prepared to consider the complexities (Hammond et al. 1983; Pool 1990).

At the minimum, it is imperative to inform the public about the hazard and then consult with the public about possible solutions (Baird & Thomas 1985). Concealment of key information, and above all, the intentional delivery of purposefully selected and skillfully crafted information, might produce short-term success, but they could later erode public confidence and hamper the cooperative efforts needed to manage the health risks successfully. Moreover, such an approach raises ethical questions (Johnson 1999). After the BSE crisis was over, government agencies in the UK officially recognized their failures in risk communication and acknowledged the importance of consultation, consistency, and transparency (McCrea 2005). The Japanese government followed suit.

Public involvement is one way to help identify and articulate the values implied by the choices that can be made about managing and communicating risks (Roux et al. 2006; Renn et al. 1991). The problem is not necessarily limited to the involvement of ordinary citizens in risk assessment, policy making, and risk management; private organizations, such as commercial interests and non-profits, must also be invited to participate. Our study revealed that in the UK, the US, and Japan, private companies, such as hamburger chains, food shops, and restaurants, appeared to sense the public concern about BSE more quickly than the government, and they promptly acted to alleviate this concern. Their public-relations activities were quite successful in reassuring the public and maintaining sales of beef products (Hendrix & Hayes 2007, pp. 368–394). Although the conformance of these activities to scientific standards remains to be assessed, they seem to have been effective in communicating with the public.

The Future of Health-Risk Management

Applying scientific knowledge can help enhance the rationality of health policy, which is especially desirable when actions inconsistent with such knowledge could produce adverse effects. However, when policy changes follow from the convergence of the three largely independent streams of problem, policy, and politics, we cannot assume that a rational, linear process will automatically incorporate science in policy making (Kingdon 1995; Sato & Frantz 2005). As discussed earlier, a variety of dynamics influence the utilization of scientific research in policy making. For policies to reflect scientific input, these models suggest that interactions between researchers and policymakers are critical. Research is more likely to be utilized in a significant way when effective networks and mechanisms are established at the interface between researchers and policymakers (Frantz & Sato 2005; Sabatier & Jenkins-Smith 1993). However, even that may not be sufficient.

To open the window of opportunity, individual events in the three streams should be made readily understandable and visible, either with the hope that the streams will converge serendipitously or so that policy entrepreneurs can aid that convergence. The problems should be moved to the foreground, the political influences should be elucidated, and feasible and attractive policy alternatives should be envisioned. Experts should be mobilized to take authoritative roles in linking science and policy, although it should always be kept in mind that experts too are susceptible to bias in one direction or the other. Political empowerment of the people who suffer from the negative policy outcomes, and therefore might be most attentive to them, could also help secure the desired objectives (Sato 2002; Sato & Frantz 2005).

Policy is always formed by both scientific and nonscientific considerations, as this study illustrates. The possibilities for the instrumental use and/or avoidance of science should be made explicit to the public. Mobilization of different social views and values, as part of a comprehensive analysis of the direct and indirect impact of a policy decision, could lead to a more explicit and formal social process that could eventually secure and propel the muddling through (Davern & Eitzen 1995). The public's oversight of this process, along with its mobilization and an appeal to its wisdom, might facilitate, check, and complement the input of the experts. The analysis of policy and politics is an indispensable part of the process.

Policies intended to prevent or overcome health hazards are usually made by trial and error (Morone & Woodhouse 1986). Because correcting an error requires first that the error and its consequences be recognized, there is an inevitable time gap between the commission of the error and feedback about its consequences. This time gap is filled by the emergence of information about the error, the focus of appropriate attention on it, and the adoption and implementation of effective corrective measures. The losses created by the hazardous events are thus systematic, and those who enjoy the benefits of the system should be expected to fully compensate its victims. Further delays in feedback, caused by the factors examined in this study, result in the accumulation of undesired consequences during the delay. It is a maxim that the greater the danger, the greater the duty to exercise care; if extraordinary damage could result, the care had better include taking all possible precautions (Steinbruner 1974; Wildavsky 1988).

Many factors – scientific, economic, social, and cultural – were found to determine the course of events surrounding the management and communication of health risks. Science, and the generation, dissemination, and utilization of its findings, are deeply embedded in the socio-political context, and therefore they have not always been given a privileged position in policymaking (Breakwell & Barnett 2001; Kasperson 1999). Likewise, scientists, policymakers, and other elite groups are not immune to criticism and are not always accepted as the chief guides in a society.

There are many actors that have played a role in the health crises addressed in this book. Collectively they form a culture that includes the existence, behaviors, and exchanges of regulatory organizations, industries, workers, consumers, legislators, scientists, and the general public, both material and symbolic. The basic assumptions that the members of this culture make about themselves, their organizations, their environments, and the nature of humankind and life in general, color their views of the risks and their management. Consequently, these assumptions significantly affect the perception of the issues and of the policies developed to address them (Schein 1985). The bounded rationality inherent in both organizations and individuals can result in policy myopia and thereby distort policy choices and the actions that follow from them (Pauchant & Mitroff 1992).

Until recently, government officials dealing with health and the environmental have not sought a total ban on asbestos-containing products. But policymakers, sometimes in cooperation with experts, did draft regulations that were reasonable in light of the need to balance the usefulness of asbestos products and their adverse social impact on society. The policymakers and experts then carefully watched the public response and adjusted their actions accordingly. These actions consisted of a cautious implementation of technology and communications intended to increase the public's understanding of the problem (Sato et al. 2006). However, such efforts have largely followed from an elitist and technocratic perspective. The traditional view is that decisions regarding technical issues are best left in the hands of experts and regulatory agencies. This view has been justified by appeals to the ostensible deficiencies in the knowledge and reasoning abilities of ordinaries citizens (Earle & Cvetkovich 1995).

However, agencies and experts have gradually become less confident about their ability to know what is desirable for and expected by the public, while at the same time the public has become aware that agencies are not always unbiased and omnipotent and that experts do not necessarily represent the public interest (Pratchett 1999; Veatch 1991). As a result, the active interactions among the state, corporate and non-corporate private interests, consumers and other social interests, conceptualized as a model of contested regulation, are sometimes considered more desirable for food regulation (Smith et al. 2004).

In a democratic society, social consensus is considered important for two reasons. First is the practical reason that policies cannot be implemented effectively and efficiently without public consent. The second reason is that, from the perspective of democratic ideology, policies must be based on the judgment of a rational, informed and willing public, or at least on their voluntary delegation of autonomy to expert authority (King et al. 1998). The efforts to arrive at a legitimate and effective policy are also attempts to form a consensus (Delvin 1968). The importance of the public's trust in government, especially its confidence in the government's legitimacy and competence, has been shown repeatedly in our case studies.

Public participation, involvement, and consultation in health-care decisions has come to be considered desirable, and sometimes even necessary, by both policymakers and the public (Richardson et al. 1992; Sato et al. 2005). A variety of public-participation measures have become part of technology-related policy making. These measures range from conventional opinion polls and public hearings to deliberative opinion polls, citizen juries, scenario workshops, and consensus conferences (Abelson et al. 2002; National Science Board 2004; Sato & Akabayashi 2005). Such measures are intended to stimulate discussion among members of an informed public and elicit their judgments about policies, not just to aggregate their preferences (Smith 1993). This dialectic and discursive activity would be expected to cultivate in society the sense of common good toward acceptable policies while at the same time encouraging a tolerance of differences in individual preferences (Jennings 1991; Sato et al. 2006).

The management of risks is thus inevitably a process of muddling thorough, just as it frequently is in politics. Partnership, interaction, and cooperation among different actors in society should form the basis for the development of an effective risk-management strategy. The search for "rational" risk management, like the quest for social consensus, is two-way, multi-focused, and co-evolutionary. This reasoning implies that the hierarchical mode of governance, with government agencies are in command, should be transformed into the coordinative mode (Mayntz 1993). Governance should foster interaction between the state and society as a way to coordinate the activities of multiple and fragmented social actors and thereby render policy making more effective. It must be recognized that not even a government is a monolithic, unified entity; its decisions are made and its actions undertaken by a multiplicity of authoritative agencies embodying different objectives and preferences (Dean 1999).

References

Anonymous (2000). Beef: commission takes action and France counteraction. *European Report*, January 6. Section 2463.

Abelson, J., Forest, P. G., Eyles, J., Smith, P., Martin, E., & Gauvin, F. P. (2002). Obtaining public input for health systems decision making: past experiences and future prospects. *Canadian Public Administration, 45*(1), 70–97.

Adams, W. (1992). The role of media relations in risk communication. *Public Relations Quarterly, Winter*, 28–32.

Adams, J. (1995). *Risk.* London: UCL.

Allison, G. T. (1999). *Essence of decision: explaining the Cuban missile crisis, second edition.* New York, NY: Longman.

Ashford, N. A. (1984). Advisory committees in OSHA and EPA: their use in regulatory decision-making. *Science, Technology, and Human, 9*(1), 72–82.

Ashworth, J. (1997). *Science, policy and risk: science in society.* London: Royal Society.

Baird, I. S., & Thomas, H. (1985). Toward a contingency model of strategic risk taking. *The Academy of Management Review, 10*(2), 230–243.

Barke, R. P., & Jenkins-Smith, H. C. (1993). Politics and scientific expertise: scientists, risk perception, and nuclear waste policy. *Risk Analysis, 13*, 425–439.

Belton, P. S. (2001). Chance, risk, uncertainty and food. *Trends in Food Science and Technology, 12*(1), 31–35.

Bernstein, M. (1955). *Regulating business by independent commission.* Princeton, NJ: Princeton University Press.

Bierman, H. (1989). The allais paradox: a framing perspective. *Behavioral Science, 34*(1), 46–52.

Bohnenblust, H., & Slovic, P. (1998). Integrating technical analysis and public values in risk-based decision making. *Reliability Engineering and System Safety, 59*, 151–159.

Bouwen, P., & Mccown, M. (2007). Lobbying versus litigation: political and legal strategies of interest representation in the European Union. *Journal of European Public Policy, 14*(3), 422–443.

Breakwell, G. M., & Barnett, J. (2001). *The impact of social amplification of risk on risk communication.* London: Health and Safety Executive.

Browde, S., Szabo, A., & Persensky, J. (2004). *The technical basis for the NRC's guidelines for external risk communication* (NUREG/CR-6840). Washington, DC: US Nuclear Regulatory Commission, Office of Nuclear Regulatory Research.
Burns, W. J., Slovic, P., Kasperson, R. E., Kasperson, J. X., Renn, O., & Emani, S. (1998). Incorporating structural models into risk research on the social amplification of risk. In R. Lofstedt & L. Frewer (Eds), *Risk and modern society*. London: Routledge.
Calder, K. E. (1995). *Strategic capitalism: private business and public purpose in Japanese industrial finance*. Princeton, NJ: Princeton University Press.
Campbell, J. C. (1992). *How policies change: the Japanese government and the aging society*. Princeton, NJ: Princeton University Press.
Campbell, R. G., & Sato, H. (2009). Examination of a global prohibition regime: a comparative study of Japanese and US newspapers on the issue of tobacco regulation. *The International Communication Gazette, 71*(3), 161–179.
Castleman, B. I. (1979). The export of hazardous factories to developing nations. *International Journal of Health Services, 9*(4), 569–606.
Cohen, M. D., March, J. G., & Olsen, J. P. (1972). A garbage can model of organizational choice. *Administrative Science Quarterly, 17*(1), 1–25.
Cranor, C. F. (1993). *Regulating toxic substances: a philosophy of science and the law*. New York and Oxford: Oxford University Press.
Cvetkovich, G., & Earle, T. C. (1992). Environmental hazards and the public. *Journal of Social Issues, 48*(4), 1–20.
Cvetkovich, G., & Lofstedt, R. E. (1999). *Social trust and the management of risk*. London: Earthscan Publications Ltd.
Davenport, S., & Leitch, S. (2005). Agoras, ancient and modern, and a framework for science-society debate. *Science and Public Policy, 32*(2), 137–153.
Davern, M.E., & Eitzen, D.S. (1995). Economic sociology: An examination of intellectul exchange. *Amercian Journal of Economics and Sociology, 54*(1), 79–88.
Davis, K.,& Frederick, W. C. (2000). *Business and society: corporate strategy, public policy, ethics*. New York, NY: Mcgraw-Hill.
Dean, M. (1999). *Governmentality: power and rule in modern society.* London: Sage.
Delvin, P. (1968). *The enforcement of morals.* London: Oxford University Press.
Department for Environment, Food and Rural Affairs (DEFRA). (2005). *Transmissible spongiform encephalopathies (TSE) in Great Britain 2005: a Progress Report*. London: Author. Retrieved, March 03, 2009, from http://www.defra.gov.uk/animalh/bse/pdf/tse-gb_progressreport12-05.pdf
Dora, C. (2006). *Health, hazards and public debate: lessons for risk communication from the BSE/CJD Saga* (A EURO Publication). Geneva: World Health Organization.
Douglous, M., & Wildavsky, A. (1982). *Risk and culture: an essay on the selection of technological and environmental dangers*. Berkeley, CA: University of California Press.
Dror, Y. (1971). *Ventures in policy sciences.* New York: American Elsevier.
Durodie, B. (2000). Plastic panics: European risk regulation in the aftermath of BSE. In J. Morris (Ed), *Rethinking risk and the precautionary principle* (pp. 140–166). Oxford, UK: Butterworth-Heinemann.
Earle, T. C., & Cvetkovich, G. T. (1995). *Social Trust.* Westport, CT: Praeger.
Echols, M. A. (1998). Food safety regulation in the European Union and the United States: different cultures, different laws. *Columbia Journal of European Law, 4*, 525–542.
Echols, M. A. (2001). *Food safety and the WTO: the interplay of culture, science and technology*. London; Kluwer Law Internationa.
Eldridge, J. (1999). Risk, society and the media: now you see it, now you don't. In G. Philo (Ed), *Message received* (pp. 106–127). London: Addison Wesley Longman
Etzioni, A. (1976). *Social problems.* Englewood Cliffs, NJ: Prentice-Hall.
Etzkowitz, H. (2002). Incubation of incubators: innovation as a triple helix of university-industry-government networks. *Science and Public Policy, 29*(2),115–128.
European Commission. (1998). *Measures concerning meat and meat products-hormones*. WT/DS58/AB/R and WT/DS48/AB/R. 16 Jan 1998, adopted 13 Feb 1998.

European Commission. (2000). *Communication from the Commission on the Precautionary Principle: COM (2000) 1*. Brussels: Commission of the European Communities.
Fiorino, D. J. (1989). Technical and democratic values in risk analysis. *Risk Analysis, 9*, 293–299.
Fischoff, B., Lichtenstein, S., Slovic, P., Derby, S. L., & Keeney, R. (1984). *Acceptable risk*. Cambridge, UK: Cambridge University Press.
Fishohoff, B., Slovic, P., Lichtenstein, S., Read, S., & Combs, B. (1978). How safe is safe enough? A psychometric study of attitudes towards technological risks and benefits. *Policy Sciences, 9*(2), 127–152.
Food and Agricultural Organization & World Health Organization. (2002). Food safety and quality in Europe: emerging issues and unresolved problems, France. Conference paper, Pan-European Conference on Food Safety and Quality. Budapest, Hungary 25–28, 2002. http: www.fao.org/docrep/meeting/004/y2741e.htm.
Foreman, C. H. (1994). *Plagues, products, and politics: emergent public health hazards and national policymaking*. Washington DC: The Brookings Institution.
Foucault, M., Hurley, R., Faubion, J. D., & Rabinow, P. (2001). *Power: Essential works of Foucault, 1954–1984, volume III*. New York, NY: New Press.
Frantz, J. E., & Sato, H. (2005). The fertile soil of policy learning: Hansen's disease policy in US and Japan. *Policy Sciences 38*(2/3), 159–176.
Freestone, D., & Hey, E. (1996). The precautionary principle and international law: the challenge of implementation. The Hague: Kluwer Law International.
Frewer, L. (2003). Science, society and public confidence in food risk management. In P. S. Belson & T. Belson (Eds), Food, science and society: exploring the gap between expert advice and individual behavior. Berlin, Heiderberg: Springer.
Frewer, L. J., Howard, C., Hedderley, D., & Shepherd, R. (1996). What determines trust in information about food-related risks? Underlying psychological constructs. *Risk Analysis, 16*, 473–486.
Frewer, L. J., Howard, C., & Shepherd, R. (1997). Public concerns about general and specific applications of genetic engineering: risk, benefit and ethics. *Science, Technology and Human Values, 22*, 98–124.
Frewer, L. J., Hunt, S., Miles, S., Brennan, M., Kuznesof, S., Ness, M., & Ritson, C. (2001). *Communicating risk uncertainty with the public: final report project February 2001*. Newcastle: University of Newcastle.
Friedman, S. M., Dunwoody, S., & Rogers, C. L. (1986). *Scientists and journalists: reporting science as news*. New York, NY: The Free Press.
Fuller, S. (2001). *Knowledge management foundations*. Oxford, UK: Butterworth-Heinemann.
Gergen, K. J. (1999). *An invitation to social construction*. London: Sage.
Gibbons, M. (2000). Mode 2 society and the emergence of context-sensitive science. *Science and Public Policy, 27*(3): 159–163.
Gibbons, M., Limoges, C., Nowotny, H., Schwatzman, S., Scott, P., & Trow, M. (1994). *The new production of knowledge: the dynamics of science and research contemporary societies*. London: Sage.
Goodell, R. (1987). The role of mass media in scientific controversy. In T. Engelhardt, & A. Caplan (Eds), *Scientific controversies: case studies in the resolution and closure of disputes in science and technology*. Cambridge: Cambridge University Press.
Graham, J. D., Green, L. C., & Roberts, M. J. (1988). *In search of safety: chemicals and cancer risk*. Cambridge, MA: Harvard University Press.
Grindle, M. S. (1996). *Challenging the State: crisis and innovation in Latin America and Africa*. Cambridge: Cambridge University Press.
Habermas, J. (1975). *Legitimation crisis*. Boston, MA: Beacon Press.
Hacking, I. (1999). *The social construction of what?* Cambridge, MA: Harvard University Press.
Hamilton, J. D. (2003). Exploring technical and cultural appeals in strategic risk communication: the Fernald Radium case. *Risk Analysis, 23*(2), 291–302.
Hammond, K. R., Mumpower, J., Dennis, R. L., Fitch, S., & Crumpacker, W. (1983). Fundamental obstacles to the use of scientific information in public policy making. *Technological Forecasting and Social Change, 24*, 287–297.

Hansen, A. (1994). Journalistic practices and science reporting in the British press. *Public Understanding of Science, 3*, 111–134.
Harremoes, P., Gee, D., MacGarvin, M., Stirling, A., Keys, J., Synne, B., & Guiedes Vaz, S. (2002). *The precautionary principle in the 20th century: late lessons from early warnings*. London: Earthscan.
Heazle, M. (2006). *Scientific uncertainty and the politics of whaling*. Seattle, WA: University of Washington Press.
Hendrix, J. A., & Hayes, D. C. (2007). *Public relations cases* (7th ed.). Belmont, CA: Thomson Wadworth.
Hermann, C. F. (1963). Some consequences of crisis which limit the viability of organizations. *Administrative Science Quarterly, 8*, 61–82.
Hilgartner, S. (1990). The dominant view of popularisation: conceptual, problems, political uses. *Social Studies of Science, 20*, 519–539.
Hinchliffe, S. (2001). Indeterminancy in decision: science, policy and politics in the BSE (Bovine Spongiform Encephalopathy) crisis. *Transaction Institute of British Geographers NS, 26*, 182–204.
HM Government. (2001). *The interim response to the report of the BSE Inquiry by HM Government in consultation with the devolved Administrations*. London: The Stationary Office.
Holling, C. S. (1979). Resilience and stability of ecological systems. *Annual Review of Econology and Systematics, 4,* 1–23.
Holzmann, R., & Joergensen, S. (2001). Social risk management: a new conceptual framework for social protection, and beyond. *International Tax and Public Finance, 8*(4), 529–556.
Hsu, S.H. (2005). Terminating Taiwan's fourth nuclear power plant under the Chen Shui-bian administration. *Review of Policy Research, 22*(2), 171–186.
Huber, P. (1983). Exocists vs. gatekeepers in risk regulation. *Regulation, November/December*, 23–32.
Ilchman, W. F., & Uphoff, N. T. (1969). *The political economy of change*. Berkeley, CA: University of California Press.
Irwin, A. & Michael, M. (2003). *Science, social theory and public knowledge*. Maidenhead, UK: Open University Press.
Jasanoff, S. (1990). *The fifth branch: science advisers as policy makers*. Cambridge MA: Harvard University Press.
Jasanoff, S. (1995). Procedural choices in regulatory science. *Technology in Society, 17*(3), 279–293.
Jasanoff, S. (2004). *States of knowledge: the co-production of science and social order.* London: Routledge.
Jasanoff, S. (2006). Just evidence: the limits of science in the legal process. *Journal of Law, Medicine and Ethics, 34*(2), 328–341.
Jennings, B. (1991). Possibilities of consensus: toward democratic moral discourse. *Journal of Medicine and Philosophy,16*(4), 447–463.
Johnson, C. A. (1983). *MITI and the Japanese miracle: the growth of industrial policy, 1925–1975*. Stanford, CA: Stanford University Press.
Johnson, B. B. (1999). Ethical issues in risk communication: continuing the discussion. *Risk Analysis, 19*(3), 335–348.
Jordan, W. A. (1972). Producer protection, prior market structure and the effects of government regulation. *Journal of Law and Economics, 15*(1), 151–176.
Kasperson, R. E. (1999). *Communicating risks to the public: technology, risk and society*. The Netherlands: Kluwer.
Kazan-Allen, L. (2003). Asbestos issues in Australia and Southeast Asia. International Ban Asbestos Network. Retrieved, January 10, 2009, from http://www.btinternet.com/~ibas/lka_aus_rep_jan03.htm
Kazan-Allen, L. (2004). Asbestos dispatches. *International Journal of Occupational and Environmental Health, 10*(2), 111–120.
Kerr, A., & Cunningham-Burley, S. (2000). On ambivalence and risk: reflexive modernity and the new human genetics. *Sociology, 34*, 283–304.
King, C. S., Felty, K. M., & Susel, B. O. (1998). The question of participation: toward authentic public participation in public administration. *Public Administration Review, 58*(4), 317–326.

Kingdon, J. W. (1995). Agendas, alternatives, and public policies: second edition. Boston: Little Brown.
Knight, F. (1957). *Risk, Uncertainty and Profit: Reprint series no. 16*. London: London School of Economics and Political Science.
LaDou, J. (1992). The export of hazardous industries to newly industrialized countries. *J Occup Med Environ Health, 5*(3): 223–226.
Laffont, J. J., & Tirole, J. (1991). The politics of government decision making: a theory of regulatory capture. *Quarterly Journal of Economics, 106*(4), 1089–1127.
Latour, B. (1987). *Science in action: how to follow scientists and engineers through society*. Cambridge, MA: Harvard University Press.
Levine, M. E., & Forrence, J. L. (1990). Regulatory capture, public interest, and the public agenda: toward a synthesis. *Journal of Law Economics & Organization, 6*, 167–198.
Lindblom, C. E. (1959). Science of muddling through. *Public Administration Review, 19*, 79–88.
Lofstedt, R., & Frewer, L. (1998). *Risk and modern society*. London: Earthscan.
Majone, G. (1992). *Evidence, argument and persuasion in the policy process*. New Haven, NJ: Yale University Press.
March, J. G., & Olsen, J. P. (1979). *Ambiguity and choice in organizations, 2nd edition*. Bergen: Universitetsforlaget, 1979.
March, B. D., & Ravets, J. R. (1999). Risk management and governance: a post-normal approach. *Futures, 31*, 743–757.
March, J. G., & Simon, H. A. (1958). *Organizations*. New York, NY: Wiley.
Mayntz, R. (1993). Governing failures and the problems of governability. In J. Kooiman (Ed.), *Modern governance: new government-society interactions*. London: Sage.
McCrea, D. (2005). Risk communication related to animal products derived from biotechnology. *Revue Scientifique et Technique (International Office of Epiootics), 24*(1): 141–148.
Mertz, C. K., Slovic, P., & Purchase, I. F. H. (1998). Judgements of chemical risks: comparisons among senior managers, toxicologists, and the public. *Risk Analysis, 18*, 391–404.
Meyer, G., Folker, A., Joergensen, R., Krauss, M., Sandoe, P., & Tveit, G. (2005). The factualization of uncertainty: risk, politics, and genetically modified crops – a case of rape. *Agriculture and Human Values, 22*(2), 235–242.
Miceli, T. J. (1997). *Economics of the law: Torts, contracts, property and litigation*. Oxford, UK: Oxford University Press.
Miles, S., & Frewer, L. J. (2001). Investigating specific concerns about different food hazards – higher and lower order attributes. *Food Quality and Preference, 12*, 47–61.
Miljkovic, D. (2005). Rational choice and irrational individuals or simply an irrational theory: a critical review of the hypothesis of perfect rationality. *Journal of Socio-Economics, 34*(5), 621–634.
Miller, D., & Reilly, J. (1995). Making an issue of food safety: the media, the pressure groups and the public sphere. In D. Maurer, & J. Sobal (Eds), *Eating agendas: food and nutrition as social problems*. New York, NY: Aldine de Gruyter.
Millstone, E., & van Zwanenberg, P. (2002). The evolution of food safety policy-making institutions in the UK, EU and Codex Alimentarius. *Social Policy and Administration, 36*(6), 593–609.
Montgomery, J. D. (1974). *Technology and civic life: making and implementing development decisions*. Cambridge, MA: MIT.
Morgan, M. G., & Henrion, M. (1990). *Uncertainty: a guide to dealing with uncertainty in quantitative risk and policy analysis*. Cambridge, UK: Cambridge University Press.
Morone, J. G., & Woodhouse, E. J. (1986). *Averting catastrophe: strategies for regulating risky technologies*. Berkeley, CA: University of California Press.
Morris, J. (2000). *Rethinking risk and the precautionary principle*. Oxford, UK: Butterworth-Heinemann.
Motta, M, & Thisse, J-F. (1994). Does environmental dumping lead to delocation? *European Economic Review, 38*(3–4), 563–576.
Muramatsu, M., & Krauss, E. S. (1983). Bureaucratic politics and politicians in policy making: The case of Japan. *American Political Science Review, 78*, 126–146.
Myers, N. J., & Raffensperger, C. (2006). *Precautionary tools for reshaping environmental policy*. Cambridge, MA: MIT.

National Research Council. (1989). *Improving risk communication.* Washington, DC: National Academy.

National Science Board. (2004). Science and technology: public attitudes and understanding. In Subcommittee on Science and Engineering Indicators, National Science Board (Ed), *Science and engineering indicators 2004: Chapter 7* (pp. 1–37). Arlington, VA: National Science Foundation.

Naubauer, D. E. (2005). Globalization and emerging governance modalities. *Environmental Health and Preventive Medicine, 10*, 286–294.

Newton, R. G. (1997). *The truth of science*. Cambridge, MA: Harvard University Press.

Noll, R. G. (1985). *Regulatory policy and the social sciences.* Berkeley, CA: University of California Press.

Nowotny, H., Scott, P., & Gibbons, M. (2001). *Re-thinking science: knowledge and the public in an age of uncertainty.* Cambridge, UK: Polity.

Nowotny, H., Scott, P., & Gibbons, M. (2003). Mode 2 revisited: the new production of knowledge. *Minerva, 41*, 179–194.

Olson, S. M. (1990). Interest-group litigation in federal district court: beyond the political disadvantage theory. *The Journal of Politics, 52*(3), 854–882.

Pauchant, T. C., & Mitroff, I. I. (1992). *Transforming the crisis-prone organizations: preventing individual, organizational, and environmental tragedies.* San Francisco, CA: Jossey-Bass.

Pearce, F. (2005). Responsible corporations and regulatory agencies. *Political Quarterly, 61*(4), 415–430.

Perrow, C. (1984). *Normal accidents: living with high-risk technologies.* New York, NY: Basic Books.

Peters, H. P. (1995). The interaction of journalists and scientific experts: co-operation and conflict between two professional cultures. *Media, Culture and Society, 17*, 31–48.

Peters, R. G., Covello, V. T., & McCallum, D. B. (1997). The determinants of trust and credibility in environmental risk communication: an empirical study. *Risk Analysis, 17*(1), 43–54.

Phillips, N. (2000). *The BSE inquiry: return to an order of the Honourable the House of Commons dated October 2000 for the report, evidence and supporting papers of the inquiry into the emergence and identification of Bovine Spongiform Encephalopathy (BSE) and variant Creutzfeldt-Jakob Disease (vCJD) and the action taken in response to it up to 20 March 1996.* Norfolk, UK: The Stationary Office.

Pidgeon, N. (1999). Risk communication and the social amplification of risk: theory, evidence and policy implications. *Risk, Decision and Policy 4*(1): 1–15.

Plough, A. & Krimsky, S. (1987). The emergence of risk communication studies: social and political context. *Science, Technology, & Human Values, 12*(3/4), 4–10.

Pool, R. (1990). Struggling to do science for society. *Science, 248*, 672–673.

Popper, K. (1971). *The open society and its enemies.* Princeton, NJ: Princeton University Press.

Porter, T. M. (1994). Objectivity as standardization: the rhetoric of impersonality in measurement, statistics, and cost–benefit analysis. In A. Megill (Ed), *Rethinking objectivity* (pp. 197–237). Durham, NC: Duke University Press.

Pratchett, L. (1999). New fashions in public participation: towards greater democracy? *Parliamentary Affairs, 52*(4), 617–633.

Rampal, K. G., & Feitshans, I. L. (2000). Legal requirements for medical surveillance of asbestos workers in Malaysia, the USA and under international law. SSRN: DOI: 10.2139/ssrn.241093.

Reich, M. R. (1991). *Toxic politics: responding to chemical disasters.* Ithaca, NY: Cornell University Press.

Reilly, J. (1999). Just another food scare: public understanding and the BSE crisis. In G. Philo (Ed), *Message received* (pp. 128–146). London: Addison Wesley Longman.

Renn, O., Webler, T., Johnson, B. B. (1991). Public participation in hazard management: the use of citizen panels in the US. *Risk: Issues in Health and Safety, 2*, 197–226.

Renn, O., Burns, W. J., Kasperson, J. X., & Slovic, P. (1992). The social amplification of risk: the theoretical foundations and empirical applications. *Journal of Social Issues, 48*(4): 137–160.

Richardson, A., Charny, M., & Hammer-Lloyd, S. (1992). Public opinion and purchasing. *BMJ, 304*, 680–682.

Rosenberg, G. N. (2008). *The hollow hope: can courts bring about social change?* Chicago: University of Chicago Press.
Roux, D. J., Rogers, K. H., Biggs, H. C., Ashton, P. J., & Sergeant, A. (2006). Bridging the science-management divide: moving from unidirectional knowledge transfer to knowledge interfacing and sharing. *Ecology and Society, 11*(1), 4 (online).
Sabatier, P. A., & Jenkins-Smith, H. C. (1993). *Policy change and learning: an advocacy coalition approach*. Boulder, CO: Westview.
Sandman, P. (1988). Telling reporters about risk. *Civil Engineering, 58*, 36–38.
Sapolsky, H. M. (1990). The politics of risk. In E. J. Burger (Ed), *Risk* (pp. 83–96). Ann Arbor: The University of Michigan Press.
Sato, H. (1999a). The advocacy coalition framework and the policy process analysis: the case of smoking control in Japan. *Policy Studies Journal 27*(1), 28–44.
Sato, H. (1999b). Policy and politics of smoking control in Japan. *Social Science and Medicine, 49*(5), 581–600.
Sato, H. (2000). Conflict of trade versus health: problems and prospects. *Addiction, 95*(12), 1870–1872.
Sato, H. (2002). Abolition of leprosy isolation policy in Japan: policy termination through leadership. *Policy Studies Journal, 30*(1): 29–46.
Sato, H., & Akabayashi, A. (2005). Bioethical policymaking for advanced medical technologies: institutional characteristics and citizen participation in eight OECD countries. *Review of Policy Research, 22*(4), 571–587.
Sato, H., & Frantz, J. E. (2005). Termination of the leprosy isolation policy in the US and Japan: science, policy changes, and the garbage can model. *BMC International Health and Human Rights, 5*(3), 1–16.
Sato, H., & Narita, M. (2003). Politics of leprosy segregation in Japan: the emergence, transformation and abolition of the patient segregation policy. *Social Science and Medicine, 56*(12): 2529–2539.
Sato, H., Akabayashi, A., & Kai, I. (2005). Public appraisal of government efforts and participation intent in medico-ethical policymaking in Japan: a large scale national survey concerning brain death and organ transplant. *BMC Medical Ethics, 6*(1), 1–12.
Sato, H., Akabayashi, A., & Kai, I. (2006). Public, experts, and acceptance of advanced medical technologies: the case of organ transplant and gene therapy in Japan. *Health Care Analysis, 14*(4), 203–214.
Schanne, M., & Meier, W. (1992). Media coverage of risk. Results from content analysis. In J. Durant (Ed), Biotechnology in public: a review of recent research. London: Science Museum.
Schattschneider, E. E. (1975). The semisovereign people: a realist's view of democracy in America. New York, NY: Harcourt Brace College.
Schein, E. H. (1985). *Organizational culture and leadership: a dynamic view.* San Francisco: Jossey-Bass.
Schwing, R. C., & Albers, W. A. (1980). *Societal risk assessment*. New York, NY: Springer.
Shapiro, M. (2005). Deliberative, independent technocracy v. democratic politics: will the globe echo the EU? *Law and Contemporary Problems, 68*, 341–356.
Sidall, E. (1980). *Risk, fear and public safety*. Ontario, Canada: Atomic Energy of Canada.
Siegrist, M, & Cvetkovich, G. (2000). Perception of hazards: the role of social trust and knowledge. *Risk Analysis 20*(5), 713–719.
Simon, H. A. (1947). *Administrative behavior.* New York: Macmillan.
Simon, H. A. (1955). A behavioral model of rational choice. *Quarterly Journal of Economics, 99*, 99–118.
Slovic, P. (1993). Perceived risk, trust and democarcy. *Risk Analysis, 13*, 675–682.
Slovic, P. (2000). *The perception of risk*. London: Earthscan.
Slovic, P., Fischoff, B., & Lichtenstein, S. (1982). Facts versus fears: understanding perceived risk. In D. Kahneman, P. Slovic, & A. Tversky (Eds.), *Judgment under uncertainty: heuristics and biases*. Cambridge, UK: Cambridge University Press.
Slovic, P., Fischoff , B., & Lichtenstein, S. (1988). Response mode, framing, and information-processing effects in risk assessment. In D.E. Bell, H. Raiffa, & A. Tversky (Eds), *Decision*

making: descriptive, normative, and prescriptive interactions. Cambridge, UK: Cambridge University Press.

Smith, R. S. (1976). *The occupational safety and health act*. Washington DC: American Enterprise Institute.

Smith, J. A. (1991). *The idea brokers: think tanks and the rise of the new policy elites*. New York, NY: The Free Press.

Smith, L. G. (1993). *Impact assessment and sustainable resource management*. Harlow, UK: Longman.

Smith, E., Marsden, T., Flynn, A., & Percival, A. (2004). Regulating food risks: rebuilding confidence in Europe's food? *Environment and Planning C: Government and Policy, 22*, 543–567.

Sparks, P., & Shepherd, R. (1994). Public perceptions of the potential hazards associated with food production and food consumption: an empirical study. *Risk Analysis, 14*, 799–806.

Steinbruner, J. D. (1974). *The cybernetic theory of decision*. Princeton, NJ: Princeton University Press.

Stigler, G. (1971). The theory of economic regulation. *Bell Journal of Economics, and Management Science, 2*, 3–21.

Stigler, G. J. (1988). *Chicago studies in political economy*. Chicago: University of Chicago Press.

Taylor, P. (1985). *The smoke ring: tobacco, money, and multinational politics*. London: Little Brown UK paperbacks.

Taylor, A. L. (1999). Globalization and biotechnology: UNESCO and an international strategy to advance human rights and public health. *American Journal of Law and Medicine, 25*, 479–541.

Thompson, M., Ellis, R., & Wildavsky, A. (1990). *Cultural theory*. Boulder, CO: Westview.

Tversky, A. (1972). Elimination by aspects: a theory of choice. *Psychological Review, 79*, 281–299.

Tversky, A., & Kahneman, D. (1982). The framing of decisions and the psychology of choice. *Science, 211*, 1453–1458.

Upham, F. K. (1989). *Law and social change in postwar Japan*. Cambridge, MA: Harvard University Press.

Veatch, R. M. (1991). Consensus of expertise: the role of consensus of experts in formulating public policy and estimating facts. *The Journal of Medicine and Philosophy, 16*(4), 427–446.

Viscuti, W. K. (1979). *Employment hazards: an investigation of market performance*. Cambridge, MA: Harvard University Press.

Vlek, C. A. J. (1995). Understanding, accepting and controlling risks: a multistage framework for risk communication. *European Review of Applied Psychology, 45*(1): 49–54.

Vogel, D. (1992). Consumer protection and protectionalism in Japan. *Journal of Japanese Studies, 18*, 119–154.

von Homeyer, I., Klaphake, A., & Sohn, H-D. (2001). EU accession: negotiating"environmental dumping"? *Intereconomics, 36*(2), 87–97.

von Schomberg, R. (1993). *Science, politics, and morality: scientific uncertainty and decision making*. Norwell, MA: Kluwer Academic.

Walsh, D. C., & Gordon, N. P. (1986). Legal approaches to smoking deterrence. *Annual Review of Public Health, 7*, 127–149.

Wells, C. (1994). Corporate liability and consumer protection: Tesco v Nattrass revisited. *The Modern Law Review, 57*(5), 817–823.

White, G. E. (1980). *Tort law in America: an intellectual history*. New York: Oxford University Press.

Wildavsky, A. (1988). *Searching for safety*. New Brunswick and Oxford: Transaction.

Wildavsky, A. (2000). Trial and error versus trial without error. In J. Morris (Ed), *Rethinking risk and the precautionary principle* (pp. 22–45). Oxford, UK: Butterworth-Heinemann.

Zeegers, N., Witteveen, W. J., & Klink, B. V. (2005). *Social and symbolic effects of legislation under the rule of law*. Lewiston, NY: Edwin Mellen.

Singer, E., & Endreny, P. M. (1993). *Reporting on risk: how the mass media portray accidents, diseases, disasters, and other hazards*. New York, NY: Russell Sage Foundation.

List of the Project Members

The AGS Research Project "Strategic Management and Communications of Health Risks, 2005–2009"

Principal Investigator

Hajime Sato is Assistant Professor in Department of Public Health, Graduate School of Medicine, the University of Tokyo. He holds M. D. (medicine) and Ph. D. (social medicine) from the University of Tokyo, and M. P. H. (public health policy) and Dr. P. H. (population and international health/ health policy), both from Harvard University. Besides a variety of topics in medicine, epidemiology, law, human rights, and ethics, his research encompasses policy, politics, management, and political economy in the areas of health, environment, and safety, especially from an international comparative perspective.

Project Researchers

Andrew Webster is Professor and Head in Department of Sociology, University of York, and Director of Science and Technology Studies Unit (SATSU) in the Department. He holds B. Sc. in Social Sciences from Polytechnic of the South Bank, and Ph. D. in Sociology of Science, from University of York. His research interests encompass sociology of science and technology, science policy studies, innovative health technologies (especially genetics, stem cells and informatics), sociology of innovation, pan-European innovation analysis, intellectual property and the commercialization of research, technology foresight and social science - the dynamics of knowledge flow, acquisition and use.

Photo (*from left to right*): Andrew Webster, Hajime Sato, and Conor Douglas. Conor Douglas was a Research Fellow of the SATSU, and is now Research Technician at W. Maurice Young Centre for Applied Ethics, The University of British Columbia.

Pierre-Benoit Joly is Directeur de Recherches at the French National Institute for Agronomic Research (l'institut scientifique de recherche agronomique, INRA), and Director of its Research Unit on Social and Political Changes related to Life Sciences and Life Forms (Transformations Sociales et Politiques liées au Vivant, TSV). He holds Ingénieur en Agriculture from ESAP (Toulouse), DEA d'Economie de la Production from Université Toulouse I, and Doctorat Nouveau Régime en Sciences Economique from University of Toulouse. Areas of his research include socio-economics of innovation, sociology of risk, and public policy studies, especially concerning biotechnologies, biodiversity, seed industry, patents, and organizations of research and development.

INRA, Lury-sur-Seine, Paris

Bernard Reber is Chargé de recherche at the CNRS, Centre de Recherche Sens, Ethique et Société (CERCES), and the UMR 8137 of the CNRS -Université Paris V. He completed 1er cycle de théologie et de philosophie, Université de Fribourg (Switzerland), holds Licence de philosophie from Institut supérieur de Théologie et de Sciences Religieuses du Centre Sèvres (Paris), Maîtrise de sciences économiques et sociales from Institut d'Études Économiques et Sociales (Institut catholique de Paris) and Maîtrise de théologie from Institut Catholique de Sciences Théologiques Systématiques et Bibliques (Paris), and Doctorat en recherches politiques (philosophie morale et politique) from Centre de Recherches Politiques Raymond Aron, École des Hautes Études en Sciences Sociales (Paris). He is interested in moral philosophy, philosophy of science, and public policies concerning health, risk, safety, and science.

L'Universite' Paris Descartes, Paris

Rose G. Cambpell is Associate Professor of Journalism, Eugene S. Pulliam School of Journalism, Butler University. She holds B. S. in journalism/advertising from University of Kansas, M. S. in mass communication, and Ph. D. in health communication in the mass media, from Purdue University. Her research covers mass- and interpersonal- communications and related topics, including those areas such as media studies, consensus building, public relations, and health and risk communications.

The Universdity of Tokyo, Tokyo

Domyung Paek is Professor in Department of Environmental Health, School of Public Health, Seoul National University. He holds M. D. from College of Medicine, Seoul National University, M. S. in occupational medicine from London School of Hygiene and Tropical Medicine, and D. Sc. in Environmental Health Sciences (Occupational Health) from Harvard University. His research covers a wide range of environmental and occupational health issues, from clinical and biological studies to epidemiology and policy in the area.

The Seoul National University, Seoul

Index

MIX
Papier aus verantwortungsvollen Quellen
Paper from responsible sources
FSC® C105338

If you have any concerns about our products,
you can contact us on
ProductSafety@springernature.com

In case Publisher is established outside the EU,
the EU authorized representative is:
Springer Nature Customer Service Center GmbH
Europaplatz 3, 69115 Heidelberg, Germany

Printed by Libri Plureos GmbH
in Hamburg, Germany